Snowflake Recipes

A Problem-Solution Approach to Implementing Modern Data Pipelines

Dillon Dayton
John Eipe

Apress®

Snowflake Recipes: A Problem-Solution Approach to Implementing Modern Data Pipelines

Dillon Dayton
Box Elder, SD, USA

John Eipe
Austin, TX, USA

ISBN-13 (pbk): 979-8-8688-0937-8
ISBN-13 (electronic): 979-8-8688-0938-5
https://doi.org/10.1007/979-8-8688-0938-5

Managing Director, Apress Media LLC: Welmoed Spahr
Acquisitions Editor: Celestin Suresh-John
Development Editor: Jim Markham
Coordinating Editor: Gryffin Winkler
Copy Editor: Kim Burton

Cover image designed by Freepik (www.freepik.com)

Distributed to the book trade worldwide by Springer Science+Business Media LLC, 1 New York Plaza, Suite 4600, New York, NY 10004. Phone 1-800-SPRINGER, fax (201) 348-4505, e-mail orders-ny@springer-sbm.com, or visit www.springeronline.com. Apress Media, LLC is a California LLC and the sole member (owner) is Springer Science + Business Media Finance Inc (SSBM Finance Inc). SSBM Finance Inc is a **Delaware** corporation.

For information on translations, please e-mail booktranslations@springernature.com; for reprint, paperback, or audio rights, please e-mail bookpermissions@springernature.com.

Apress titles may be purchased in bulk for academic, corporate, or promotional use. eBook versions and licenses are also available for most titles. For more information, reference our Print and eBook Bulk Sales web page at http://www.apress.com/bulk-sales.

Any source code or other supplementary material referenced by the author in this book can be found here: https://www.apress.com/gp/services/source-code.

If disposing of this product, please recycle the paper

Table of Contents

About the Authors

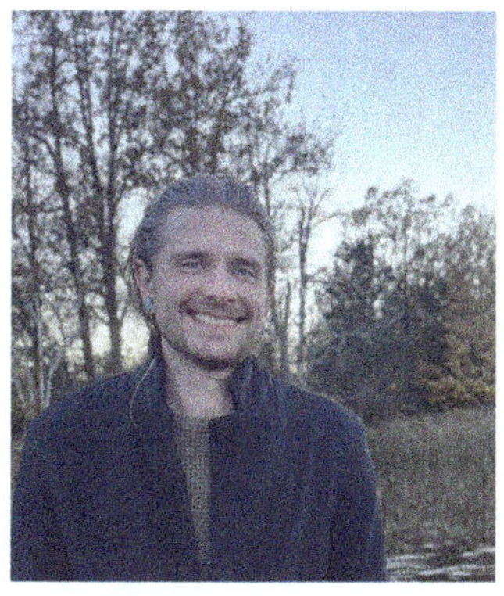

Dillon Dayton is a data expert with a proven track record in architecting robust and scalable data solutions. He is passionate about turning data into a strategic asset and specializes in data engineering, data products, and data governance. Dillon's deep understanding of the Snowflake Data Cloud and Snowflake SnowPro Core certification, coupled with his extensive experience in professional services, empowers him to help organizations harness the full potential of their data. When not immersed in the world of data, Dillon enjoys exploring the great outdoors through fishing and traveling with his wife.

John Eipe is a senior solutions specialist for CDW and has over 12 years of experience in various roles, from enterprise application development to data engineering. He worked primarily with customers from the ecommerce and insurance domain. John is SnowPro® Advanced Architect certified and has been working extensively on Snowflake in recent years. Apart from work, he enjoys cooking and time with his kids.

About the Technical Reviewer

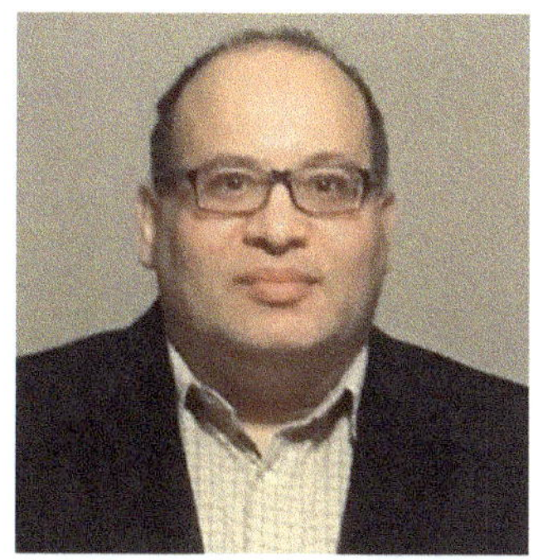

Nadir Doctor is a database and data warehousing architect and a DBA who has worked in various industries with multiple OLTP and OLAP technologies. He has also worked on primary data platforms, including Snowflake, Databricks, CockroachDB, DataStax, Cassandra, ScyllaDB, Redis, MS SQL Server, Oracle, Db2 cloud, AWS, Azure, and GCP. His major focus is health-check scripting for security, high availability, performance optimization, cost reduction, and operational excellence. He has presented at several technical conference events, is active in user group participation, and can be reached on LinkedIn.

Preface

Welcome to this guide on maximizing the potential of Snowflake, a data cloud platform that has been shown to offer a return on investment of 616% over three years, according to a Total Economic Impact Study by Forrester. Snowflake is a highly flexible and powerful tool, but it is important to ensure that it is used to its full potential to maximize it.

The authors of this book have a background in professional services, with expertise in data and digital, and have worked on a wide range of projects across various industries, companies, and technologies. They understand the challenges of finding the right solution to complex problems and have created this book as a resource to help teams quickly bridge the gap between technology and business objectives so they can deliver value more efficiently.

This book is not intended to replace Snowflake's detailed documentation, which is regularly updated. Instead, it offers a collection of easily digestible recipes that can be used to inspire solutions to specific problems. From setting up the Python Connector to leveraging Python for data science and implementing an internal chargeback model to supporting data as a product, this book covers a wide range of topics for users of all levels.

As Snowflake grows in popularity and attracts a diverse range of users with different levels of expertise, we hope that this book will be a valuable resource for those looking to make the most of the platform's capabilities.

Welcome to Snowflake

Snowflake is a cloud-native data platform. If you are reading this, we assume you know what Snowflake is, and you are here to learn more and how to practically implement Snowflake into real-world business problems and use cases. In a saturated and competitive landscape, businesses often have issues finding the right tool for the business objective. Our goal is to highlight the core concepts of Snowflake and key differentiators against industry competitors, such as Azure Synapse and Google BigQuery. These topics are leveraged or discussed in later chapters while we show how to use them and why the features exist.

Snowflake differentiates itself from industry competitors by offering a fully managed, cloud-native data warehouse that is highly scalable and can handle structured, semi-structured, and unstructured data. It also has built-in support for multi-cloud environments, allowing customers to easily access and analyze data across different cloud platforms. Additionally, Snowflake has a unique architecture that separates storage and computing, enabling customers to scale up or down their compute resources independently of their storage, which can help to optimize costs. Finally, Snowflake provides an SQL-based query interface that is compatible with most data analytics tools and has a wide range of security and data management features.

Software as a Service (SaaS)

Software as a Service (SaaS) is a software delivery model in which a software application is hosted by a third-party provider and made available to customers over the Internet. In this model, the customer does not have to install or run the software on their own computers or servers but instead accesses it through a web browser or API.

SaaS providers like Snowflake typically handle all aspects of the software's maintenance, including updates, security, and backups, allowing customers to focus on using the software to achieve their business goals. SaaS is often paid on a subscription basis, with customers paying a monthly or annual fee for access to the software.

Database Architecture

Snowflake's database architecture is based on a unique, multi-cluster, shared data architecture that separates storage and compute resources. This allows for independently scaling compute and storage resources, which can help to optimize costs, concurrency, and performance.

Snowflake Layers

Snowflake's data warehouse is composed of several layers, each with its specific role and function.

Data storage layer: This layer stores the data in a hybrid compressed-columnar format optimized for data warehousing and analytics workloads. It also includes built-in support for semi-structured data, such as JSON and Avro, which allows Snowflake to handle a wide range of data types and workloads alongside analytics. These include AI/ML, transactional data using hybrid tables, applications, and containerization.

Compute layer: This layer handles the processing and querying of the data. It comprises several individual compute clusters, which can be scaled up or down as needed. These clusters are connected to the shared data storage layer and perform query execution.

Cloud services: The management pane for Snowflake that unifies and connects the different services. Cloud services manage many critical services, such as authentication, metadata management, query planning, and infrastructure management.

Overall, Snowflake's architecture is designed to be highly scalable, flexible, and secure, enabling customers to easily access and analyze large amounts of data, both structured and semi-structured, across different cloud platforms.

Snowflake Data Sharing

Snowflake data sharing is a feature that enables customers to share their data with other Snowflake customers or external organizations in a secure and controlled manner. This feature allows customers to share their data without copying or moving it while maintaining full control over access and security.

When a share is accessed, the data remains in the original location, and the shared user can query and analyze the data using their Snowflake account or a reader account for non-Snowflake customers. Access to the data is controlled by using Snowflake's built-in security features, such as role-based access control, and data is encrypted both in transit and at rest.

In summary, Snowflake data sharing is a powerful feature that allows customers to share their data easily and securely with other organizations without copying or moving it while maintaining full control over access and security. Note that the data must be in the same region as the consumer account to leverage data sharing.

Snowflake Time Travel, Fail-Safe, Replication, and Clones (BC/DR)

Time Travel: A feature that allows customers to access and query historical versions of their data. This feature allows customers to track changes to their data over time and to easily revert to earlier versions if needed.

Fail-safe: A powerful feature that provides automatic and transparent data recovery, ensuring high availability and data durability in the event of a failure or outage. This feature can help customers to protect their data and to ensure that their business can continue to operate without interruption in case of failure. Fail-safe requires Snowflake support and is *not* a method for accessing data outside the configured Time Travel period.

Replication: Enables users to replication databases between Snowflake accounts in a single Snowflake organization. Supports cross-region and cross-cloud replication.

Clones: In Snowflake, a clone is a fully independent and self-contained copy of a database, schema, or table. Clones can create multiple copies of an object for different purposes, such as testing, development, or analytics, while maintaining a single source of truth.

As we continue, we touch on these key features, explain how to implement them, and, most importantly, when to bring business value.

Additional Snowflake Documentation

`https://docs.snowflake.com/en/user-guide/intro-key-concepts.html`

`https://www.snowflake.com/en/data-cloud/workloads/unistore/`

`https://docs.snowflake.com/en/developer-guide/native-apps/native-apps-about`

CHAPTER 1

Introduction to Snowflake

This chapter is designed to familiarize you with Snowflake. It starts by walking through the setup of the most popular and common methods of interacting with Snowflake. It does not cover everything, so please check the Snowflake documentation to validate what is available. There are constantly new drivers being added by Snowflake and driven by the community.

The chapter also reviews how to navigate the differences and begin a conversation around selecting multi-cloud or multi-tenant environments leveraging Snowflake organizations and the need for those organizations within a business. By the end of the chapter, you'll have a general understanding of the Snowflake editions used in this book.

Recipe 1-1. Connecting to Snowflake

Snowflake is a cloud native Software as a Service (SaaS) platform. The ability to leverage Snowflake within the enterprise largely depends on what options are available for connectivity. Snowflake offers many industry-standard options in drivers, APIs, and native connectors. By leveraging several of these connection types, developers have the flexibility to build rich, robust, and efficient data environments, including data warehousing, security, DataOps, governance, app development, artificial intelligence/machine learning, and so forth. As with all our recipes, we encourage you to visit Snowflake's documentation.

Problem

Choose the appropriate method to connect and interact with Snowflake.

D. Dayton and J. Eipe, *Snowflake Recipes*, https://doi.org/10.1007/979-8-8688-0938-5_1

Solution

Snowflake provides a variety of drivers and connectors that allow customers to connect to and interact with their data cloud from a wide range of applications and platforms. The following are some of the most commonly used drivers and connectors.

Native Snowflake

Snowsight

Snowsight is a powerful and easy-to-use data visualization and analytics tool built into Snowflake. It allows users to easily explore, analyze, and visualize their data and collaborate on data analysis projects. Consider this the Snowflake user interface and the main interface to the Snowflake account. To access Snowsight, use the account URL.

SnowSQL

SnowSQL is a command-line client for Snowflake that allows customers to run SQL commands and queries against their Snowflake data warehouse. It is designed to be a simple and easy-to-use tool for running SQL commands and queries and is compatible with most SQL-based data analytics tools.

1. Download the SnowSQL client. The client is available for Windows, macOS, and Linux operations systems and can be found on the Snowflake website.

2. Install the SnowSQL client. The installation process is straightforward and depends on the operations system. For Windows, run the downloaded setup file. For macOS and Linux, extract the downloaded tar and run the installation script.

3. Connect to Snowflake. Run the following SnowSQL command in a command-line interface.

   ```
   snowsql -a <account_name> -u <username>
   ```

   The user is prompted to enter a password.

4. Configure SnowSQL. Use the following command to configure SnowSQL.

```
snowsql -c
```

In the configuration file, set default values for various parameters, such as the default database, schema, role, and warehouse.

5. Test the connection. Run the following query to test connectivity to Snowflake.

```
select current_timestamp();
```

If the query executes successfully, then SnowSQL is properly installed and configured.

Snowflake APIs

Snowflake APIs (application programming interfaces) allow customers to programmatically access and manage their data warehouse. These APIs can automate various tasks, such as creating and managing databases, schemas, and tables, and performing data operations, such as loading and querying data.

1. Get API credentials. To use the Snowflake API, you need to obtain API credentials, such as an API key or a user account with API access enabled.

2. Choose a programming language. Snowflake supports several programming languages, including Python, Java, Go, and Node.js. Choose the language that you are comfortable with or is supported by your organization.

3. Install the appropriate driver or library. Depending on the programming language you choose, you need to install the appropriate driver or library for Snowflake.

4. Connect to Snowflake. To connect to Snowflake, you must create a connection object and specify the connection parameters, such as the account name, username, password, database name, and authentication credentials.

5. Execute API requests. Once you have established a connection to Snowflake, you can execute API requests to perform various tasks, such as executing queries, creating tables, or loading data.

6. Handle responses. After executing an API request, you must handle the response, which typically includes a status code and a response body. The response body contains the result of the API request.

7. Handle errors. You should also implement error handling in your code to handle errors that may occur during API execution, such as connection errors or query execution errors.

Note The exact steps to use the Snowflake API depend on the programming language and the driver or library you use. The general steps provide a high-level overview of the process.

Snowpark API

The Snowpark API is an intuitive API for building data pipelines and applications directly in Snowflake, leveraging existing storage and computing. The following example uses Python, but Scala and Java are also supported.

1. Install Snowpark. First, you need to install the Snowpark library in your environment. This library provides an intuitive API for querying and processing data in Snowflake. At the time of this writing, Python 3.8 is required leveraging the Anaconda distribution platform.

   ```
   conda install snowflake-snowpark-python
   ```

2. Connect to Snowflake. To connect to Snowflake, you must create a connection object and specify the connection parameters, such as the account name, username, password, and database name. Common methods of connection are via Jupyter Notebooks or an integrated development environment (IDE). As an example, run the following to create a notebook session.

```
conda install notebook
jupyter notebook
```

3. Create a Snowpark context. To use Snowpark, you need to create a Snowpark context, which connects your application and Snowflake. This context contains the connection information and settings required to access Snowflake data.

```
from snowflake.snowpark import Session

conn = {
      "account": "<snowflake account>"
      "user":"<snowflake user>"
      "password": "<snowflake password>"
      "role": "<snowflake role>"
      "warehouse": "<snowflake warehouse>"
      "database": "<snowflake database>"
      "schema": "<snowflake schema>"
}

new_session = Session.buildcr.configs(conn).create()
```

4. Execute Snowpark code. Once you have created a Snowpark context, you can write Snowpark code to process and query data in Snowflake. Snowpark provides several functions and methods to perform data processing and analysis tasks.

```
session.sql('select current_timestamp()').collect()
```

5. Use Snowpark functions and methods. You can use Snowpark functions and methods to perform various tasks, such as executing SQL queries, transforming data, and aggregating data.

6. Store results. After processing data with Snowpark, you can store the results in Snowflake or another data storage solution and close the session.

Snowflake Connector(s)

Python

The Snowflake Python Connector is a Python library that allows customers to connect to and interact with their Snowflake data warehouse from within a Python application. It provides a simple, easy-to-use interface for running SQL commands and queries and loading and unloading data.

1. Install the Snowflake connector for Python. Open a command prompt and use pip, the Python package manager.

   ```
   pip install snowflake-connector-python
   ```

2. Establish a connection. To connect to Snowflake from Python, you need to provide the following information.

   a. Account name

   b. Username

   c. Password/Keypair

   d. Role

   e. Warehouse

3. Run a query. Once you have established a connection to Snowflake, you can query using the following example code.

   ```
   import snowflake.connector

   conn = snowflake.connector.connect(
       user='user',
       password='password',
       account='account',
       role='role',
       warehouse='warehouse'
   )

   cursor = conn.cursor()
   cursor.execute("SELECT * FROM my_table")
   print(cursor.fetchall())

   conn.close()
   ```

Spark

The Snowflake Spark connector is a plugin that allows customers to read and write data to and from Snowflake using Apache Spark, an open source big data processing framework. The connector provides a simple and easy-to-use interface for connecting to Snowflake and running Spark SQL commands and queries against the data stored in Snowflake. Currently, Snowflake supports Spark 3.1, 3.2, and 3.3 and maintains a connector for each version.

1. Install the Snowflake Spark connector. The connector is available in Maven Central and can be included in your Spark project by adding the following line to your `build.sbt` file or equivalent.

   ```
   libraryDependencies += "net.snowflake" % "snowflake-spark-
   connector_2.12" % "2.4.14-spark_2.4"
   ```

2. Configure Spark with Snowflake. Once the connector is installed, you need to configure Spark to use it. Add the following lines to your Spark configuration file.

   ```
   spark.jars.packages=net.snowflake:snowflake-jdbc:3.12.15,net.
   snowflake:spark-snowflake_2.11:2.4.14-spark_2.4
   spark.sql.catalogImplementation=hive
   spark.sql.hive.metastore.jars=maven
   spark.sql.hive.metastore.version=2.3.7
   ```

3. Establish a connection. To connect to Snowflake from Spark, you must provide the following information.

   a. Account name

   b. Username

   c. Password

   d. Warehouse

4. Load data. Once a connection has been established to Snowflake, use the following example code to load from Snowflake into Spark.

```
val df = spark.read
  .format("snowflake")
  .options(Map(
    "sfUrl" -> "jdbc:snowflake://<account>.snowflakecomputing.com",
    "sfUser" -> "<user>",
    "sfPassword" -> "<password>",
    "sfDatabase" -> "<database>",
    "sfSchema" -> "<schema>",
    "sfWarehouse" -> "<warehouse>"
  ))
  .option("query", "SELECT * FROM my_table")
  .load()
```

Kafka

The Snowflake Kafka connector is an integration that allows customers to read and write data to and from Snowflake using Apache Kafka, an open source, distributed streaming platform. The connector integrates with the Kafka Connect framework, which is a library for building and running connectors between Kafka and other systems. The connector allows customers to easily stream data into Snowflake and read data from Snowflake using Kafka's messaging system. With Snowflake as the focus, the following steps are to set up the connector. End to end process of Snowflake and Kafka is covered later.

1. Download the Snowflake Kafka connector JAR from Maven.

```
wget https://repo1.maven.org/maven2/com/snowflake/snowflake-
kafka-connector/1.8.2/snowflake-kafka-connector-1.8.2.jar
```

2. Move the JAR file to the kafka/libs directory.

```
mv snowflake-kafka-connector-1.8.2.jar kafka/libs
```

3. Configure the Snowflake Kafka connector.

```
vi kafka/config/connect-snowflake-kafka-connector.properties
```

Key values to update include the following.

- `name`
- `topics`
- `snowflake.url.name`
- `snowflake.user.name`
- `snowflake.private.key`
- `snowflake.database.name`
- `snowflake.schema.name`

Once data has been published on the topic, it is loaded into Snowflake. By default, the connector creates a table per topic using the topic *name*.

Snowflake Driver(s)

Node.js

Node.js is an open source, cross-platform, back-end JavaScript runtime environment that executes JavaScript code outside a web browser. It is built on Chrome's V8 JavaScript engine and allows developers to create fast, scalable, high-performance network applications using JavaScript on the server side. Snowflake offers a native driver, allowing seamless integration between Snowflake and any Node.js application or environment.

1. Install the node.js driver. Install the Snowflake node.js driver using the npm package manager and run the following code in a command-line prompt.

```
npm install snowflake-sdk
```

2. Import the driver. In the node.js script, import the Snowflake driver using the following statement.

```
var snowflake = require('snowflake-sdk');
```

3. Establish a connection. Use the following template to create a connection to Snowflake. Variables need to be updated with the appropriate connection information.

```
var connection = snowflake.createConnection({
  account: '<account name>',
  username: '<username>',
  password: '<password>',
  warehouse: '<warehouse name>'
});
```

4. Authenticate and execute. Once a connection has been established, use the following as an example to authenticate and execute a query.

```
connection.connect(function(err, conn) {
  if (err) {
    console.error('Unable to connect: ' + err.message);
  } else {
    conn.execute({
      sqlText: 'select current_timestamp()',
      complete: function(err, stmt, rows) {
        if (err) {
          console.error('Failed to execute statement: '
          + err.message);
        } else {
          console.log('Successful execution: ' + JSON.
          stringify(rows));
        }
      }
    });
  }
});

connection.destroy(function(err) {
  if (err) {
    console.error('Failed to close connection: ' + err.message);
```

```
    } else {
      console.log('Successful connection closure');
    }
  });
```

JDBC

Java Database Connectivity (JDBC) is a Java-based application programming interface (API) that allows Java applications to connect to and interact with relational databases. It provides a set of standard interfaces and classes that can be used to connect to a wide variety of relational databases. The Snowflake JDBC driver is a frequently used connection protocol, especially in modern cloud environments.

1. Download the Snowflake JDBC driver. You can download the latest version of the Snowflake JDBC driver from the Snowflake website.

2. Install the JDBC driver. You can install the JDBC driver by adding the JAR file to your classpath.

3. Set the connection URL and parameters. You need to set the connection URL that specifies the Snowflake account and the required connection parameters.

   ```
   jdbc:snowflake://myorganization-myaccount.snowflakecomputing.com/?
   user=username&warehouse=mywh&db=mydb&schema=public
   ```

4. Load the JDBC driver. Load the driver following best practices for the particular languages being used; for Java use the following.

   ```
   import java.sql.Connection;
   Import java.sql.DriverManager;

   private static Connection getConnection() throws SQLException {

           // build connection properties
           Properties properties = new Properties();
           properties.put("user", "snowflake user");
           properties.put("password", "snowflake password");
           properties.put("warehouse", "snowflake warehouse");
   ```

```
properties.put("db", "snowflake database");
properties.put("schema", "snowflake schema");

String connectStr = "jdbc:snowflake://<account_
identifier>.snowflakecomputing.com";

return DriverManager.getConnection(connectStr,
properties);
```

5. Open a connection. You can open a connection to Snowflake using the DriverManager.getConnection() method.

   ```
   Connection connection = getConnection();
   ```

   Execute SQL statements. You can execute SQL statements using the Statement and PreparedStatement objects in the JDBC API.

   ```
   Statement statement = connection.createStatement();
   ResultSet resultSet = statement.executeQuery("select current_
   timestamp()");
   System.out.println(ResultSet
   ```

6. Close the connection. You need to close the connection when you are done using it.

   ```
   resultSet.close();
   statement.close();
   connection.close();
   ```

ODBC

Open Database Connectivity (ODBC) is a widely used, platform-independent application programming interface (API) that allows applications to connect to and interact with relational databases. Again, this frequently used connection protocol is generally seen when connecting to database source systems like a transactional sales database, which feeds into the data warehouse.

1. Download the latest version of the Snowflake ODBC driver from the Snowflake website.

2. Install the ODBC driver on your system by running the installation program.

3. Create a system data source name (DSN) that identifies the Snowflake ODBC driver and contains the connection information. You can create a system DSN using the ODBC Data Source Administrator tool (Figure 1-1).

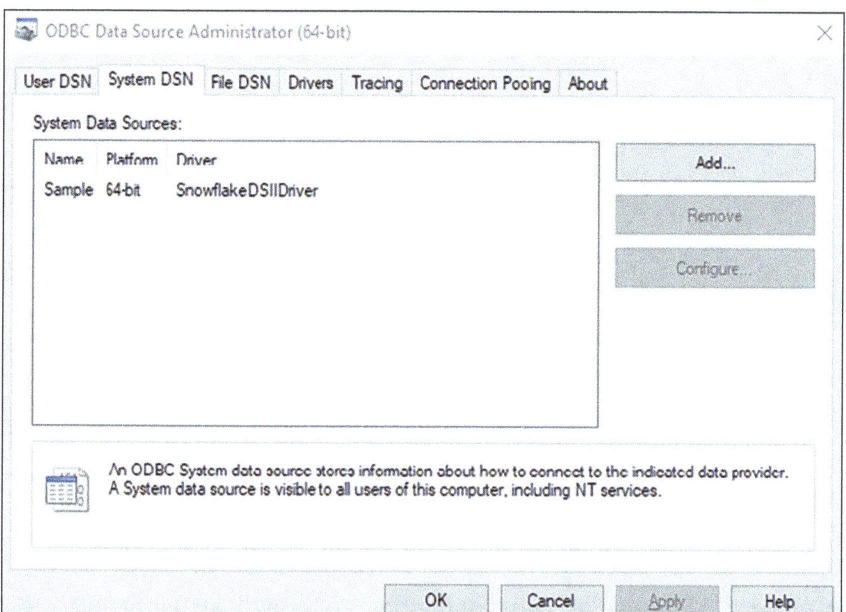

Figure 1-1. *ODBC Data Source Administrator*

4. Set the connection properties such as user, password, account, and warehouse that control the connection to Snowflake (Figure 1-2).

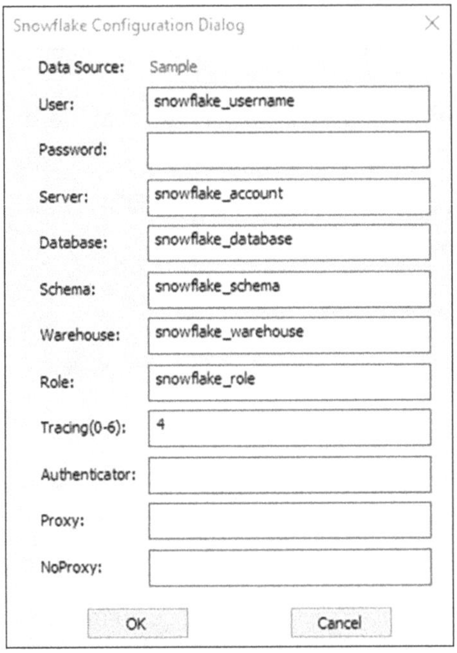

Figure 1-2. *ODBC setup*

5. Test the connection. You can test the connection using the Test Connection button in the ODBC Data Source Administrator tool.

6. Connect to Snowflake. You can connect to Snowflake using an application that supports ODBC connections.

Please visit the Snowflake Developers site to review all available drivers and connectors.

Note These steps are just basic examples to get you started. You may need to modify the code to meet your specific requirements and use case.

Additional Snowflake Documentation

`https://docs.snowflake.com/en/user-guide/conns-drivers.html`

Recipe 1-2. Selecting an Appropriate Cloud Provider

As previously discussed, Snowflake can operate in all major cloud providers—Amazon Web Services (AWS), Microsoft Azure, and Google Cloud Platform (GCP). While the core functionality of Snowflake remains the same in all clouds, there are minor differences in operations and best practices.

Problem

Is there a requirement for multi-tenant or multi-cloud when choosing the right cloud provider for my business?

Solution

Ultimately, all three clouds provide a robust ecosystem to run Snowflake and other aspects of your business, so deciding on the best cloud takes time and investment. Hopefully, the following sections provide enough context to assist in that decision.

Amazon Web Services (AWS)

Amazon Web Services (AWS) is a collection of remote computing services (also called web services) that make up a cloud computing platform offered by Amazon.com. These services operate from 12 geographic regions across the world. They provide a variety of services, including computing, storage, databases, and analytics. AWS is a leading cloud computing platform widely used by organizations of all sizes for hosting their applications, websites, and databases. With AWS, customers can build, deploy, and manage their applications and services in a highly scalable and secure environment without worrying about the underlying infrastructure.

In the context of Snowflake, AWS is one of the main cloud platforms on which Snowflake's cloud data warehouse can be deployed.

Microsoft Azure

Microsoft Azure is a cloud computing platform and infrastructure created by Microsoft for building, deploying, and managing applications and services through a global network of Microsoft-managed data centers. Azure provides a range of services, including virtual machines, storage, databases, and web applications, as well as a variety of tools and services for developing, deploying, and managing applications. With Azure, customers can build and run their applications in a secure and scalable environment without worrying about the underlying infrastructure.

Google Cloud Platform (GCP)

Google Cloud Platform (GCP) is a cloud computing platform and infrastructure created by Google for building, deploying, and managing applications and services. It provides a range of services, including virtual machines, storage, databases, and machine learning, as well as tools for developing and managing applications. GCP allows customers to build and run their applications on Google's infrastructure, eliminating the need to invest in their hardware and maintenance. With GCP, customers can take advantage of Google's global network, data centers, and expertise in machine learning and big data to build and scale their applications.

Key Differentiators

AWS, GCP, and Azure each have their own strengths and weaknesses. The following are some of their key differences.

- **Scale**: AWS is the largest cloud platform, followed by Azure and GCP. AWS has a larger user base and a wider range of services and offerings.
- **Geographic presence**: AWS has the largest geographic footprint, with data centers located all over the world, while Azure and GCP have a more limited global presence.
- **Pricing**: AWS, Azure, and GCP all offer a pay-as-you-go pricing model, but the specific costs can vary depending on the services and the region.
- **Services**: Each platform has its own set of core services and strengths. AWS is strong in computing, storage, and databases, while GCP is known for its machine learning and big data capabilities. Azure is strong in areas such as virtual machines and managed services.

- **Integration**: AWS and Azure strongly focus on integrating with existing enterprise IT systems, while GCP is known for its focus on innovation and new technologies.
- **Ease of use**: AWS and Azure offer a range of tools and services to make it easier to build and manage applications, while GCP has a steeper learning curve but is known for its innovation and cutting-edge technologies.

Additional Considerations

The choice of cloud provider in most businesses or applications is often based on factors other than Snowflake, including but not limited to the following.

- **Geographic locations**: Each platform has different geographic locations available for deploying Snowflake, so it is important to consider which locations are most suitable for your organization (Figure 1-3). Location can affect ingress/egress costs, timeliness, regulations, and more.

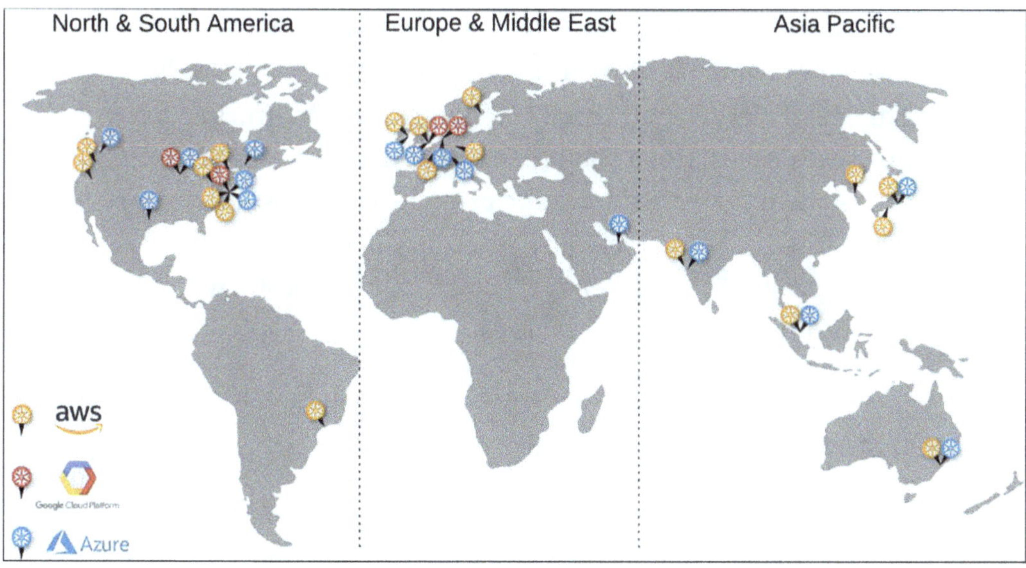

***Figure 1-3.** Global cloud availability at the time of writing* *https://docs.snowflake.com/en/user-guide/intro-regions.html*

- **Existing investments**: In a business, existing investments can play a significant role in determining the cloud provider. Companies may already have invested in a specific cloud platform or have existing licenses and agreements that they need to consider before making a new investment in a different platform. This can include factors such as the cost of switching, the compatibility of existing systems with the new platform, and the training required to use a new system. Additionally, the company may have existing security or data privacy requirements that it must consider when choosing a cloud provider.
- **Regulatory compliance**: Snowflake implements various measures and controls to meet compliance and regulatory requirements for its customers. These requirements can vary based on the industry, geography, and data being stored and processed. Some compliance certifications and standards Snowflake supports include SOC 1 Type II, SOC 2 Type II, PCI-DSS, HITRUST, ISO, and others. Certain regulators, such as FedRAMP and StateRAMP are only available in particular clouds and regions. To get the latest information, visit Snowflake's website to view their security compliance reports.
- **Cost**: The cost of Snowflake on AWS, Azure, or GCP can vary depending on the specific Snowflake edition and region, as well as the volume of data processed and stored. The cost structure for Snowflake on each cloud platform may also differ based on factors such as pricing for data storage and compute resources and network egress costs. It is recommended to consult with each cloud provider and Snowflake to comprehensively understand the costs and compare pricing based on specific use cases.

Additional Snowflake Documentation

`https://docs.snowflake.com/en/user-guide/intro-regions.html,`
`https://docs.snowflake.com/en/user-guide/intro-cloud-platforms.html`

Recipe 1-3. Snowflake Organizations

In Snowflake, an organization is an object that acts as an account manager or *umbrella* account. This object groups one or more accounts and provides billing and account management functions for them. If you sign up for Snowflake directly, the organization is created automatically for you; it has a randomly generated name. If you work with a sales engineer or resell partner when signing up with Snowflake, there is generally an opportunity to provide a user-friendly organization name. Accounts can also be renamed through an easy request to Snowflake support. While all Snowflake customers get an organization, it doesn't mean they will use or need it.

Problem

Consider the need for Snowflake organizations within your business and what the advantages are.

Solution

Snowflake organizations are a way to manage access and resources within a Snowflake account. An organization can represent a specific business grouping or a hierarchical structure for managing these resources. Customers can use an organization to take advantage of many key functionalities that allow them to transform Snowflake from a simple analytics database into a comprehensive data platform.

- Manage utilization
- Manage billing
- Create accounts
    - any cloud (AWS, GCP, Azure)
    - any region
    - any Snowflake edition

- Enables Snowflake replication
- Supports multi-cloud architecture
- Supports internal chargeback monetization
- Supports advanced architectures like data mesh and data lakehouse

One of the main benefits of using an organization is the ability to manage utilization, billing, and the creation of accounts. These accounts can be created in any cloud (AWS, GCP, Azure), region, or Snowflake edition. This isolates different environments, such as development, staging, production, or specific projects, possible.

Snowflake organizations also support multi-cloud architecture and enable Snowflake replication, which can help customers better manage their data and resources. Additionally, organizations support internal chargeback monetization capabilities and advanced architectures like data mesh, making it easier for customers to control costs and manage their data effectively.

Snowflake organizations are a powerful and flexible way to manage access to data and resources within a Snowflake environment. Using organizations, customers can easily manage access, isolate environments, and control costs, making it an essential tool for businesses looking to leverage Snowflake effectively. We always recommend that customers leverage Snowflake organizations for the best results now and in the future.

Additional Snowflake Documentation

`https://docs.snowflake.com/en/user-guide/organizations.html`

Recipe 1-4. Snowflake Editions

Snowflake Data Cloud offers four different editions: Standard, Enterprise, Business Critical, and Virtual Private Snowflake. These editions provide a range of features and functionality, so not all the recipes in this book can be completed using all the editions.

Problem

Select the appropriate Snowflake edition for your business or enterprise.

Solution

Snowflake's following editions allow customers to choose the option that offers the best balance of cost, functionality, and support for their needs.

Standard Edition

Snowflake's Standard Edition provides a basic level of service. It does not include many advanced features, so it is generally not recommended for most customers. All Snowflake editions come with support—Standard Edition customers have access to Premier Support 24 hours a day, 365 days a year.

Enterprise Edition

The Enterprise Edition includes all the Standard Edition's features and several additional features designed to support large environments that require enhanced security, management, and functionality. Some key features of the Enterprise Edition include multi-cluster warehouses, extended Time Travel, object tagging, and many others. These features are critical for large environments that need to ensure maximum security, manageability, and functionality. To see a full feature set comparison between editions, please visit Snowflake's Feature Matrix.

Business Critical Edition

The Business Critical Edition offers next-level security, governance, and data protection. It is commonly used by companies that handle sensitive data such as protected health information (PHI) and personally identifiable information (PII) and is subject to regulatory controls. This edition includes support for private connectivity, such as AWS PrivateLink, and is certified for various compliance standards, such as PHI, PCI DSS, and FedRAMP. It is important to note that a Business Associate Agreement (BAA) is required to store PHI data in Snowflake. Based on our experience, most enterprises choose the Business Critical Edition.

Virtual Private Snowflake (VPS)

Finally, there is the Virtual Private Snowflake (VPS) edition, which provides the highest level of security available from Snowflake. It is essentially the Business Critical Edition, except that the entire Snowflake environment is completely isolated from any other account and does not share any resources. This edition is typically reserved for the financial sector and other highly sensitive industries. As one might expect, this is also the most expensive edition.

Additional Snowflake Documentation

`https://docs.snowflake.com/en/user-guide/intro-editions.html#working-with-editions`

`https://docs.snowflake.com/en/user-guide/intro-editions.html#feature-edition-matrix`

CHAPTER 2

Bringing Your Data into Snowflake

With more and more data being moved into the cloud, services like AWS S3 and Azure Storage are becoming the de facto standard for data lakes. If you haven't yet moved your data to the cloud, the following are the driving factors.

- **Scalability**: Cloud object storage services offer virtually limitless scalability, allowing organizations to store massive volumes of data without worrying about capacity constraints.
- **Flexibility**: AWS S3 and Azure Object Storage support multiple data types, including structured, semi-structured, and unstructured data, making them suitable for data lakes.
- **Data security and compliance**: Cloud services provide robust security features for their object storage services, including encryption, access control, and compliance certifications.

The following recipe focuses on AWS S3 and Azure Storage, though Snowflake also supports Google Cloud Storage.

Recipe 2-1. Ready in AWS Cloud

Consider an e-commerce organization that stores its data in AWS S3 object store as flat files.

The organization now wants to bring its data residing in AWS S3 storage into Snowflake to utilize the compute and features provided by the Snowflake Data Cloud.

D. Dayton and J. Eipe, *Snowflake Recipes*, https://doi.org/10.1007/979-8-8688-0938-5_2

Now, some files receding in the storage bucket use the default server-side encryption (SSE). However, a few others use custom server-side encryption (SSE-KMS), and a few are encrypted using a client-side encryption (CSE) method.

Each of these problem statements is explored in the upcoming sections. In each solution, we stop at the stage creation process but note that once you have your data available in the Snowflake stage (a.k.a external stage), you could create external tables with their column definitions over these external stages or copy it into a regular Snowflake table using the "copy into" functionality.

Although there are several ways to configure and secure access to AWS S3 storage from Snowflake, we are focusing on the recommended approach of using Snowflake's ability to delegate authentication responsibility for external storage to an identity and access management (IAM) entity.

Prerequisites

This demonstration creates four buckets in S3 with the names and settings described as follows.

The first two buckets demonstrate the situation when data is server-side encrypted either using default or via KMS, and the next two are for demonstrating reading from client-side encrypted S3 data.

Step 1: Create Buckets with Appropriate Configuration

1. Create a bucket named **sfdemo-sse-test-1**. During the creation process, enable server-side encryption, and for encryption type, choose the default, which is the AWS S3 managed key (SSE-S3) as shown in Figure 2-1.

Default encryption
Automatically encrypt new objects stored in this bucket. Learn more
Server-side encryption
Disable
Enable
Encryption key type
To upload an object with a customer-provided encryption key (SSE-C), use the AWS CLI, AWS SDK, or Amazon S3 REST API.
Amazon S3-managed keys (SSE-S3)
An encryption key that Amazon S3 creates, manages, and uses for you. Learn more
AWS Key Management Service key (SSE-KMS)
An encryption key protected by AWS Key Management Service (AWS KMS). Learn more

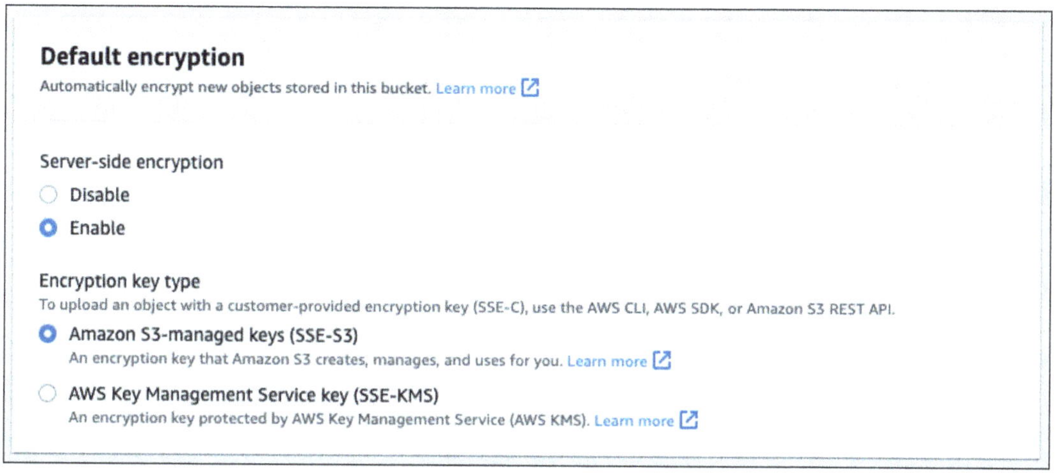

Figure 2-1. *Selecting encryption for s3 storage*

2. Create a bucket named **sfdemo-sse-test-2**. During the creation process, enable server-side encryption, and for encryption type, choose the AWS Key Management Service key (SSE-KMS) and then choose a KMS key (Create a KMS Key if not created before). The Figure 2-2 shown below highlights the section where this selection is made.

Default encryption
Automatically encrypt new objects stored in this bucket. Learn more
Server-side encryption
Disable
Enable
Encryption key type
To upload an object with a customer-provided encryption key (SSE-C), use the AWS CLI, AWS SDK, or Amazon S3 REST API.
Amazon S3-managed keys (SSE-S3)
An encryption key that Amazon S3 creates, manages, and uses for you. Learn more
AWS Key Management Service key (SSE-KMS)
An encryption key protected by AWS Key Management Service (AWS KMS). Learn more
AWS KMS key
AWS managed key (aws/s3)
arn:aws:kms:us-east-2:296080767349:alias/aws/s3
Choose from your AWS KMS keys
Enter AWS KMS key ARN
Available AWS KMS keys
arn:aws:kms:us-east-2:296080767349:key/1bb2d...
Create a KMS key
Bucket Key
When KMS encryption is used to encrypt new objects in this bucket, the bucket key reduces encryption costs by lowering calls to AWS KMS.
Learn more
Disable
Enable

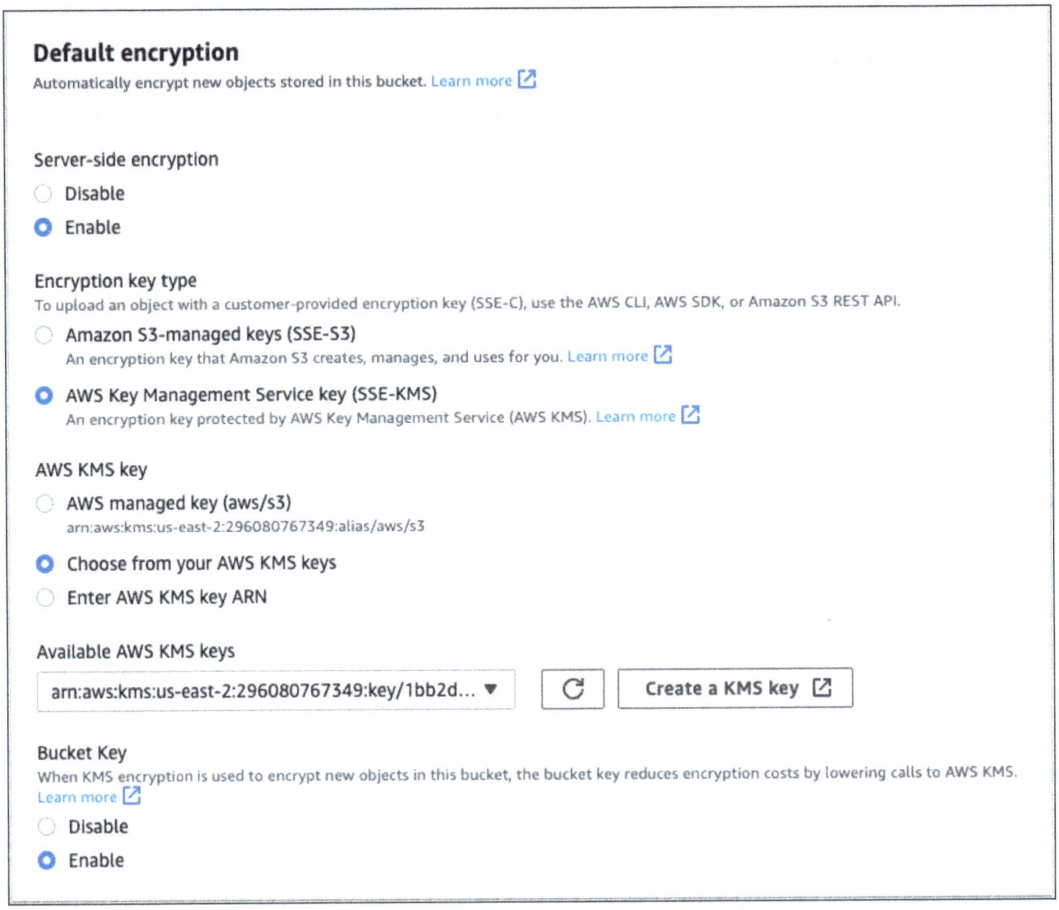

Figure 2-2. *Selecting AWS KMS key for s3 encryption*

3. Create a bucket named **sfdemo-cse-test-1**. Do not enable server-side encryption. It is used to demonstrate client-side encryption using KMS keys. You need to create the KMS key if none exist.

4. Create a bucket named **sfdemo-cse-test-2**. Do not enable service-side encryption. It is used to demonstrate client-side encryption using master keys created within the application.

Step 2: Activate Security Token Services

Next, activate security token services if you haven't already done so.

1. Log in to the AWS Management Console and choose Identity & Access Management (IAM).
2. Choose "Account settings" from the left navigation pane. Expand the Security Token Service Regions list, find the AWS region corresponding to the region where your account is located, and choose Activate it if the status is inactive as shown in the Figure 2-3.

- Access management
  - User groups
  - Users
  - Roles
  - Policies
  - Identity providers
  - Account settings
- Access reports
  - Access analyzer
    - Archive rules
    - Analyzers
    - Settings
  - Credential report

Region	Endpoint	Status
Asia Pacific (Mumbai)	https://sts.ap-south-1.amazonaws.com	Active
Asia Pacific (Singapore)	https://sts.ap-southeast-1.amazonaws.com	Active
Asia Pacific (Sydney)	https://sts.ap-southeast-2.amazonaws.com	Active
Canada (Central)	https://sts.ca-central-1.amazonaws.com	Active
Europe (Frankfurt)	https://sts.eu-central-1.amazonaws.com	Active
Europe (Stockholm)	https://sts.eu-north-1.amazonaws.com	Active
Europe (Ireland)	https://sts.eu-west-1.amazonaws.com	Active
Europe (London)	https://sts.eu-west-2.amazonaws.com	Active
Europe (Paris)	https://sts.eu-west-3.amazonaws.com	Active
South America (São Paulo)	https://sts.sa-east-1.amazonaws.com	Active

Figure 2-3. *Security Token Service Regions*

Reading S3 Data When Encryption Is SSE-S3 or SSE-KMS

Problem

Consider the first scenario when the organization's data is in S3 and uses default server-side encryption (SSE) or a custom server-side encryption (SSE-KMS).

SSE-KMS is similar to SSE with the difference being the keys used to encrypt the data are configured and used from the AWS Key Management Service (KMS) by the user.

Solution

This solution explores how to create the required permissions and use the built-in Snowflake capabilities to authenticate and use the data stored in S3, which is encrypted using SSE-S3 or SSE-KMS.

Step 1: Configure Access Permissions for the S3 Bucket

Snowflake requires the following permissions on an S3 bucket and folder to access files in the folder (and subfolders).

- s3:GetBucketLocation
- s3:GetObject
- s3:GetObjectVersion
- s3:ListBucket

The following instructions describe how to configure access permissions for Snowflake in your AWS Management Console so that Snowflake can read S3 bucket data.

1. Log in to the AWS Management Console and choose Identity & Access Management (IAM).
2. Choose Policies from the left navigation pane and click Create Policy.
3. Click the JSON tab. Add a policy document that allows Snowflake to access the S3 bucket and folder. The following policy provides Snowflake with the required permissions to load data from a single read-only bucket and folder path.

```
{
    "Version": "2012-10-17",
    "Statement": [
        {
            "Effect": "Allow",
            "Action": [
                "s3:GetObject",
                "s3:GetObjectVersion"
            ],
            "Resource": "arn:aws:s3:::<bucket>/<prefix>/*"
        },
        {
            "Effect": "Allow",
            "Action": [
```

```
                "s3:ListBucket",
                "s3:GetBucketLocation"
            ],
            "Resource": "arn:aws:s3:::<bucket>",
            "Condition": {
                "StringLike": {
                    "s3:prefix": [
                        "<prefix>/*"
                    ]
                }
            }
        }
    ]
}
```

Setting the "s3:prefix": condition to either "*" or "<path>/*" grants access to all prefixes in the specified bucket or path in the bucket, respectively.

Be sure to replace the bucket and prefix with your actual bucket name and folder path prefix.
The preceding example translates as follows.

```
{
    "Version": "2012-10-17",
    "Statement": [
        {
            "Effect": "Allow",
            "Action": [
                "s3:GetObject",
                "s3:GetObjectVersion"
            ],
            "Resource": [
                "arn:aws:s3:::sfdemo-sse-test-1/*",
                "arn:aws:s3:::sfdemo-sse-test-2/*",
                "arn:aws:s3:::sfdemo-cse-test-1/*",
                "arn:aws:s3:::sfdemo-cse-test-2/*"
```

```
                ]
            },
            {
                "Effect": "Allow",
                "Action": [
                    "s3:ListBucket",
                    "s3:GetBucketLocation"
                ],
                "Resource": [
                    "arn:aws:s3:::sfdemo-sse-test-1",
                    "arn:aws:s3:::sfdemo-sse-test-2",
                    "arn:aws:s3:::sfdemo-cse-test-1",
                    "arn:aws:s3:::sfdemo-cse-test-2"
                ],
                "Condition": {
                    "StringLike": {
                        "s3:prefix": [
                            "*"
                        ]
                    }
                }
            }
        ]
    }
```

4. Click Review policy. Enter the policy name (e.g., “snowflake_<actno>_s3_access” where actno is your account name) and an optional description. Click Create policy.

For this example, we created a policy named snowflake_vx15608_s3_access.

Step 2: Create IAM Policies for KMS

Since the second bucket is encrypted using a KMS, you need to provide snowflake access to the KMS.

Create another policy for the KMS you created earlier.

```
{
    "Version": "2012-10-17",
    "Statement": [
        {
            "Effect": "Allow",
            "Action": [
                "kms:GetParametersForImport",
                "kms:GetPublicKey",
                "kms:ListKeyPolicies",
                "kms:ListRetirableGrants",
                "kms:GetKeyRotationStatus",
                "kms:GetKeyPolicy",
                "kms:DescribeKey",
                "kms:ListResourceTags",
                "kms:ListGrants"
            ],
            "Resource": "arn:aws:kms:us-east-2:296080767349:key/1bb2ddd9-
            dc79-4493-b965-c1640d2f079a"
        },
        {
            "Effect": "Allow",
            "Action": [
                "kms:DescribeCustomKeyStores",
                "kms:ListKeys",
                "kms:ListAliases"
            ],
            "Resource": "*"
        }
    ]
}
```

For this example, it is named snowflake_vx15608_kms_access.

Step 3: Create the IAM Role in AWS

In the AWS Management Console, create an AWS IAM role to grant privileges on the S3 bucket containing your data files.

1. Log in to the AWS Management Console and choose Identity & Access Management (IAM).
2. Choose Roles from the left-hand navigation pane.
3. Click the "Create role" button.
4. Select "Another AWS account" as the trusted entity type.
5. In the Account ID field, enter your own AWS account ID temporarily as shown in Figure 2-4.
6. Later, you modify the trusted relationship and grant access to Snowflake.

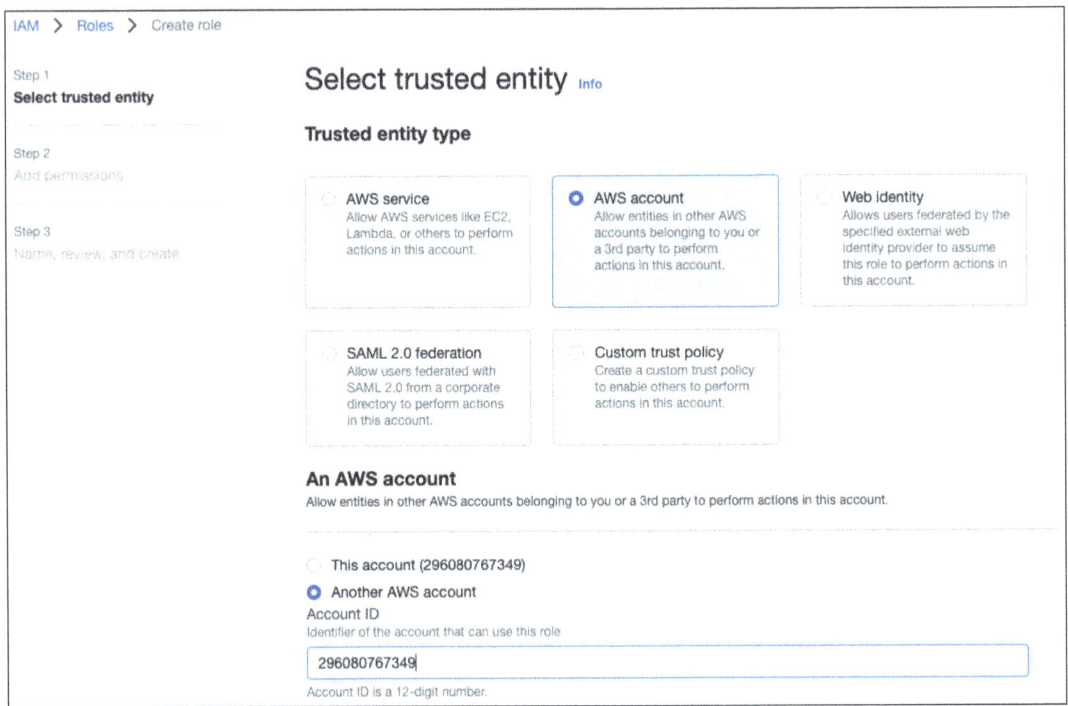

Figure 2-4. Creating IAM Role

7. Select the Require external ID option.

8. Enter a dummy ID such as "0000" as shown in below Figure 2-5.

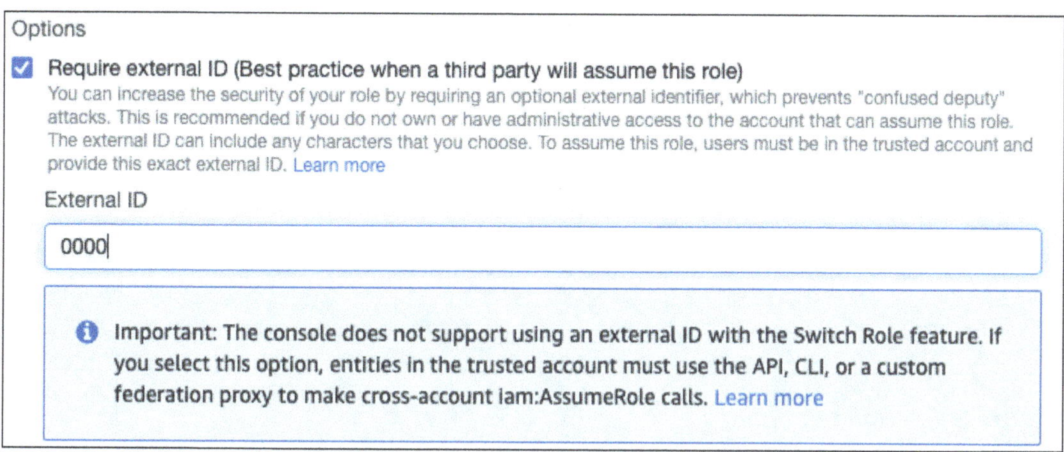

Figure 2-5. *Using External ID for the IAM role*

Later, you modify the trusted relationship and specify the external ID for your Snowflake stage. An external ID is required to grant access to your AWS resources (i.e., S3) to a third party (i.e., Snowflake).

9. Click the Next button.

10. Locate the policies you created previously and select the policies. In this example, it is snowflake_vx15608_s3_access and snowflake_vx15608_kms_access.

11. Click the Next button.

12. Enter a name (e.g., snowflake_<actno>_role) and description for the role, and click the Create role button.

You could attach policies later, as shown in Figure 2-6.

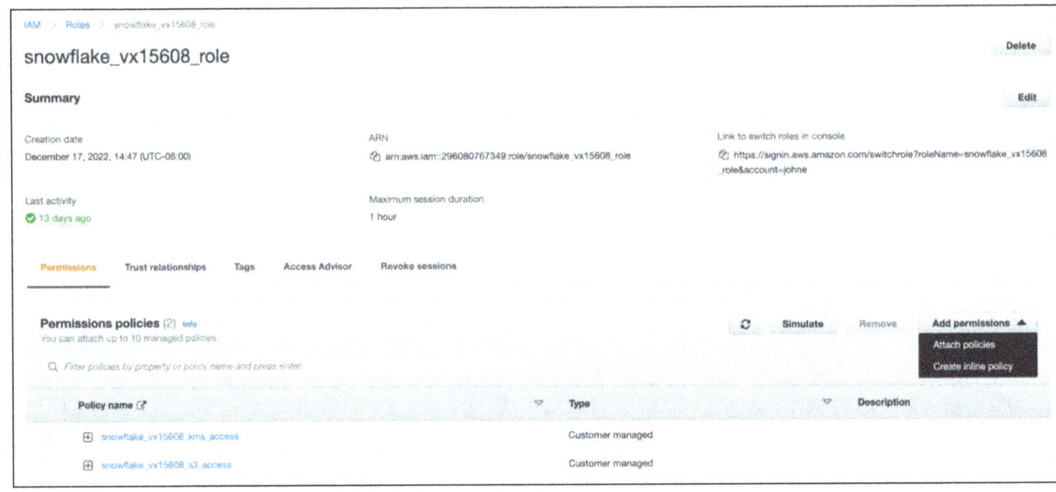

Figure 2-6. Roles Created

13. Record the Role ARN value located on the role summary page as shown in Figure 2-7. In the next step, you create a Snowflake integration that references this role.

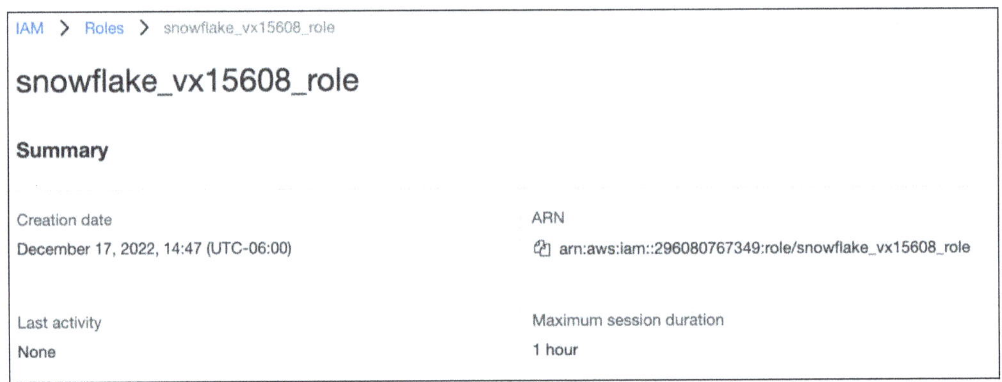

Figure 2-7. Summary and ARN of a selected role

14. Grant access to role to use key (this is needed if you are using SSE-KMS)

    a. Go to AWS Management Console and KMS service.

    b. Select the customer managed keys and the key you previously created.

    c. Under Key users grant the new role access to use the key. After the new role is granted access, it should show up in the list of Key users as shown in Figure 2-8.

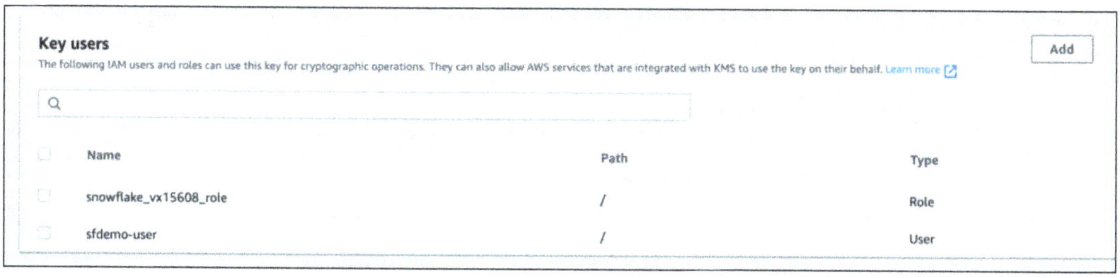

Key users

The following IAM users and roles can use this key for cryptographic operations. They can also allow AWS services that are integrated with KMS to use the key on their behalf. Learn more

Add

Name	Path	Type
snowflake_vx15608_role	/	Role
sfdemo-user	/	User

Figure 2-8. *Users of the role*

Step 4: Create a Cloud Storage Integration in Snowflake

A storage integration is a Snowflake object that stores a generated IAM user for your S3 cloud storage, along with an optional set of allowed or blocked storage locations

This option prevents users from supplying credentials when creating stages or loading data.

- Create a storage integration using the CREATE STORAGE INTEGRATION command, which is run using the ACCOUNTADMIN role.
- A single storage integration can support multiple external (i.e., S3) stages.

Here is the general syntax.

```
create storage integration <integration_name>
  type = external_stage
  storage_provider = 'S3'
  enabled = true
  storage_aws_role_arn = '<iam_role>'
  storage_allowed_locations = ('s3://<bucket>/<path>/',
  's3://<bucket>/<path>/')
  [ storage_blocked_locations = ('s3://<bucket>/<path>/',
  's3://<bucket>/<path>/') ]
In this example
create storage integration sfdemo_storage_intg
    type = external_stage
    storage_provider = 'S3'
    enabled = true
```

```
    storage_aws_role_arn = 'arn:aws:iam::296080767349:role/snowflake_
    vx15608_role'
    storage_allowed_locations = (
's3://sfdemo-cse-test-1/',
's3://sfdemo-cse-test-2/',
's3://sfdemo-sse-test-1',
's3://sfdemo-sse-test-2');
```

- Note that re-creating storage integration creates a different external ID.
- When storage integration is created Snowflake behind the scenes provisions a single IAM user for your entire Snowflake account. All S3 storage integrations use that IAM user.
- You could add and modify the allowed or disallowed locations later.

The following is an example.

```
alter STORAGE INTEGRATION sfdemo_storage_intg
set STORAGE_ALLOWED_LOCATIONS = ('s3://sfdemo-cse-test-1', 's3://sfdemo-
cse-test-2');
```

Step 5: Retrieve the AWS IAM User for your Snowflake Account

Execute the DESCRIBE INTEGRATION command to retrieve the ARN for the AWS IAM user that was created automatically for your Snowflake account.

```
desc storage integration sfdemo_storage_intg;
```

Record the following values.

- STORAGE_AWS_IAM_USER_ARN - The AWS IAM user created for your Snowflake account. In this example, it is `arn:aws:iam::119873109848:user/oz630000-s`, but it is a different value for your integration.
- STORAGE_AWS_EXTERNAL_ID - The external ID that is needed to establish a trust relationship.

Step 6: Grant the IAM User Permissions to Access Bucket Objects

1. Log into the AWS Management Console.
2. Choose Roles from the left-hand navigation pane.
3. Click on the role you created in the previous step.
4. Click on the Trust relationships tab.
5. Click the Edit trust policy button.
6. Modify the policy document with the values recorded earlier.

For this example, the values were changed as followed.

```
{
    "Version": "2012-10-17",
    "Statement": [
        {
            "Effect": "Allow",
            "Principal": {
                "AWS": "arn:aws:iam::119873109848:user/oz630000-s"
            },
            "Action": "sts:AssumeRole",
            "Condition": {
                "StringEquals": {
                    "sts:ExternalId": "VX15608_SFCRole=2_aWDC2YVrng/
                    dOllvQygZUE/DcYI="
                }
            }
        }
    ]
}
```

7. Click the Update Trust Policy button.
8. Make sure the changes are saved.

Step 7: Create External Stages in Snowflake

While creating the stage to access the sfdemo-sse-test-1 bucket, you need to provide just the storage integration name, and while creating the stage for sfdemo-sse-test-2, you need to additionally provide the KMS details.

```
create stage sfdemo_sse_1_stg
  storage_integration = sfdemo_storage_intg
  url = 's3://sfdemo-sse-test-1'
  file_format =( TYPE = CSV );

select $1 from @sfdemo_sse_1_stg; --displays the first column of the
csv file
list @sfdemo_sse_1_stg; --lists the files within the stage

create or replace stage sfdemo_sse_2_stg
  storage_integration = sfdemo_storage_intg
  url = 's3://sfdemo-sse-test-2'
  file_format =( TYPE = CSV )
ENCRYPTION =( type = 'AWS_SSE_KMS'  KMS_KEY_ID = '1bb2ddd9-dc79-4493-
****-******' );

select $1 from @sfdemo_sse_2_stg; --displays the first column of the
csv file
list @sfdemo_sse_2_stg; --lists the files within the stage
```

Reading S3 Data When Encryption Is SSE-CSE

Problem

Consider the second scenario when the organization's data is in S3 but uses client-side encryption.

Client-side encryption secures your data locally to ensure it is encrypted before transferring into S3. In this approach, Amazon S3 service receives your encrypted data, but it does not play a role in encrypting or decrypting it. The following example demonstrates client-side encrypted data residing in S3.

Solution

Like the previous solution, we explore how to create the required permissions and use the built-in Snowflake capabilities to authenticate and use the data stored in S3, which is encrypted using SSE-CSE.

Step 1: Create an IAM User and Grant Access to S3

We need to create an IAM user to generate secret keys and access keys for programmatic read/write access into S3 buckets.

1. Log into the AWS Management Console and choose Identity & Access Management (IAM).
2. Click "Add users", provide a username, and select the AWS Access type as programmatic access.
3. Click the Next button to attach permissions. You can attach the user to existing groups, use existing policies, or create and attach new policies.

For this example, specific policies are created to read/write to S3 and to access KMS keys.

Step 2: Write a Client Program to Put Files in S3

Write a client program to put files in S3 using both approaches.

For examples refer to the programs CSEWithKMS.java and CSEWithLocalKey.java in the source code repository.

Step 3: Alter or Create Storage Integration

This step is only needed if you haven't included these locations in the storage integration.

The following alter statement edits the already created storage integration to add the S3 storage locations.

```
alter STORAGE INTEGRATION sfdemo_storage_intg
set STORAGE_ALLOWED_LOCATIONS = ('s3://sfdemo-cse-test-1', 's3://sfdemo-
cse-test-2');
```

Step 4: Test Decryption

Test decryption and reconstruction of the file data using a client program that runs on your local machine. This is left to you as an exercise.

Step 5: Create a Snowflake Stage to Access the S3 Data

```
create or replace stage sfdemo_cse_1_stg
  storage_integration = sfdemo_storage_intg
  url = 's3://sfdemo-cse-test-1'
  file_format =( TYPE = CSV )
ENCRYPTION =( type = 'AWS_CSE'  MASTER_KEY = '1kCA8xM************' );

select $1 from @sfdemo_cse_1_stg; --displays the first column of the
csv file
list @sfdemo_cse_1_stg; --lists the files within the stage

create or replace stage sfdemo_cse_2_stg
  storage_integration = sfdemo_storage_intg
  url = 's3://sfdemo-cse-test-2'
  file_format =( TYPE = CSV )
ENCRYPTION =( type = 'AWS_CSE'  MASTER_KEY =
'nBCuxvZbju4j7S4cIja6GM+****************' );
```

Reading S3 Data Using Access Keys

Problem

Let's consider a scenario when the e-commerce organization wants to do a quick POC in Snowflake but needs to access the data in AWS S3. The idea is to avoid setting an integration via a Snowflake-managed IAM but be able to quickly access the S3 data via a Snowflake stage.

Solution

If you have an AWS IAM user with the appropriate permissions to your S3 bucket, you could utilize the AWS access key ID and secret to connect to the buckets from Snowflake.

Once you have generated the AWS access key and secret for the selected IAM user, you can directly use them in a Snowflake stage, as shown in the following example.

```
create or replace stage sfdemo_sse_3_stg
   credentials = (aws_key_id='AKIAUJ36J********' aws_secret_key='rRKEOIxJY
   6bAKKWMVl4W***********' )
   url = 's3://sfdemo-sse-test-3'
   file_format =( TYPE = CSV );
```

Additional Snowflake Documentation

https://docs.snowflake.com/en/user-guide/data-load-s3-config

https://docs.snowflake.com/en/sql-reference/sql/create-storage-integration

---

Recipe 2-2. Ready in Azure Cloud

Consider an e-commerce organization that stores its data in Azure object store as flat files.

The organization now wants to bring its data residing in the cloud storage into Snowflake to utilize the compute and features provided by the Snowflake Data Cloud.

The organization has some files encrypted using the Microsoft managed keys (SSE-MMK) and others using a customer managed key (SSE-CMK). There are also some files encrypted using CSE using your master key.

We explore these problem statements in the upcoming sections. Like the previous recipe, each solution for this recipe stops at the stage creation process, but you could create external tables with their column definitions over the external stages.

Prerequisites

This demonstration creates storage accounts and containers in Azure with appropriate configurations.

Two storage accounts are created.

- A storage account named **sfdemo1mmk** with default Microsoft encryption
- Another storage account named **sfdemo2cmk** with customer managed SSE

Creating storage accounts is straightforward. Log in to the Azure portal and select or search for "storage accounts". Click the Create button to create a new storage account.

Do note that all new storage accounts are created are created with these defaults,

- The storage v2 type is general-purpose storage v2.
- The redundancy is set to Geo-Redundant Storage.
- The hierarchical namespace is not enabled.
- MMK is the default encryption.

To use ADLS Gen2, select a general-purpose v2 account (if not selected by default), and then tick the hierarchical namespace check box. The selection screen will look as it is shown in the picture below (Figure 2-9).

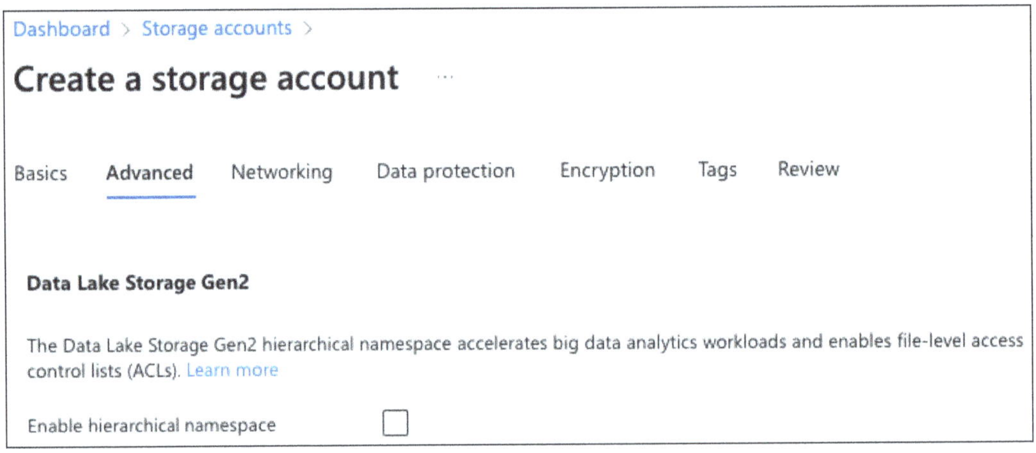

Figure 2-9. *Creating a ADLS Gen2 storage*

Does it matter which type of storage is used? Yes, it does. Snowflake supports the following types of blob storage accounts.

- Blob storage
- General-purpose v1
- General-purpose v2
- General-purpose v2 with Data Lake Storage Gen2 (when hierarchical namespaces are enabled)

For all new data projects, choosing general-purpose v2 with Data Lake Storage Gen2 enabled is preferred.

Azure Data Lake Storage Gen2 (**ADLS Gen2**) is a superset of Blob storage that is optimized for big data analytics by providing hierarchical namespace support, which means it supports directories and paths.

Unlike Blob storage, which provides pseudo directory operations via namespaces, ADLS Gen2 provides real support for directories with POSIX compliance and Access Control List (ACL) support in addition to the already existing RBAC controls. This makes operations such as renaming and deleting directories atomic and quick.

ADL Gen2 is also a Hadoop-compatible filesystem.

Step 1: Create a Storage Account with MMK Encryption

Create a storage account named **sfdemo1mmk** of type ADLS Gen 2 with MMK encryption.

The encryption type can be selected in the Encryption tab, as shown in Figure 2-10.

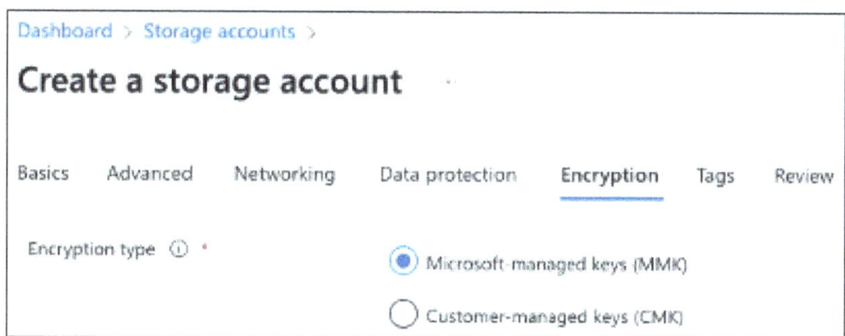

Figure 2-10. *Selecting encryption for Azure storage account*

Step 2: Create a Storage Account with CMK Encryption

CMK requires selecting an already created key from within the Azure Key Vault and specifying a user-assigned managed identity for the key.

You need to create a managed identity, as shown in Figure 2-11.

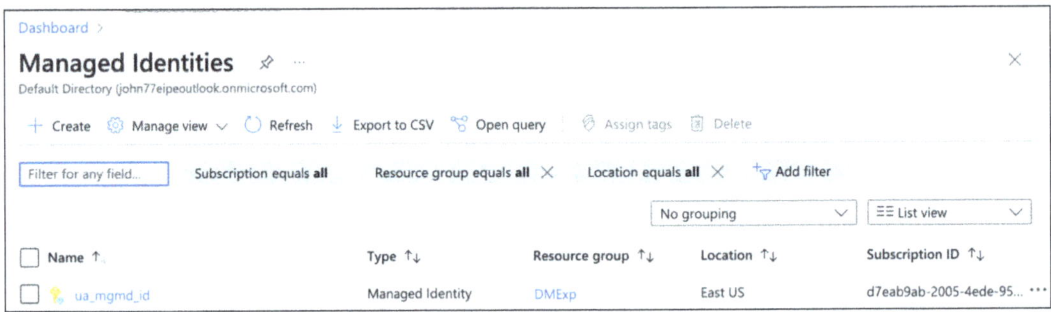

Figure 2-11. *Azure Managed Identities*

When you configure customer managed keys with a user-assigned managed identity, the user-assigned managed identity is then used to authorize access to the key vault that contains the key. Make sure you create the user-assigned identity before you configure customer managed keys.

You also should have purge protection enabled on the key.

Utilize the managed identity in the access policies for storage account and the customer managed key as shown in Figure 2-12.

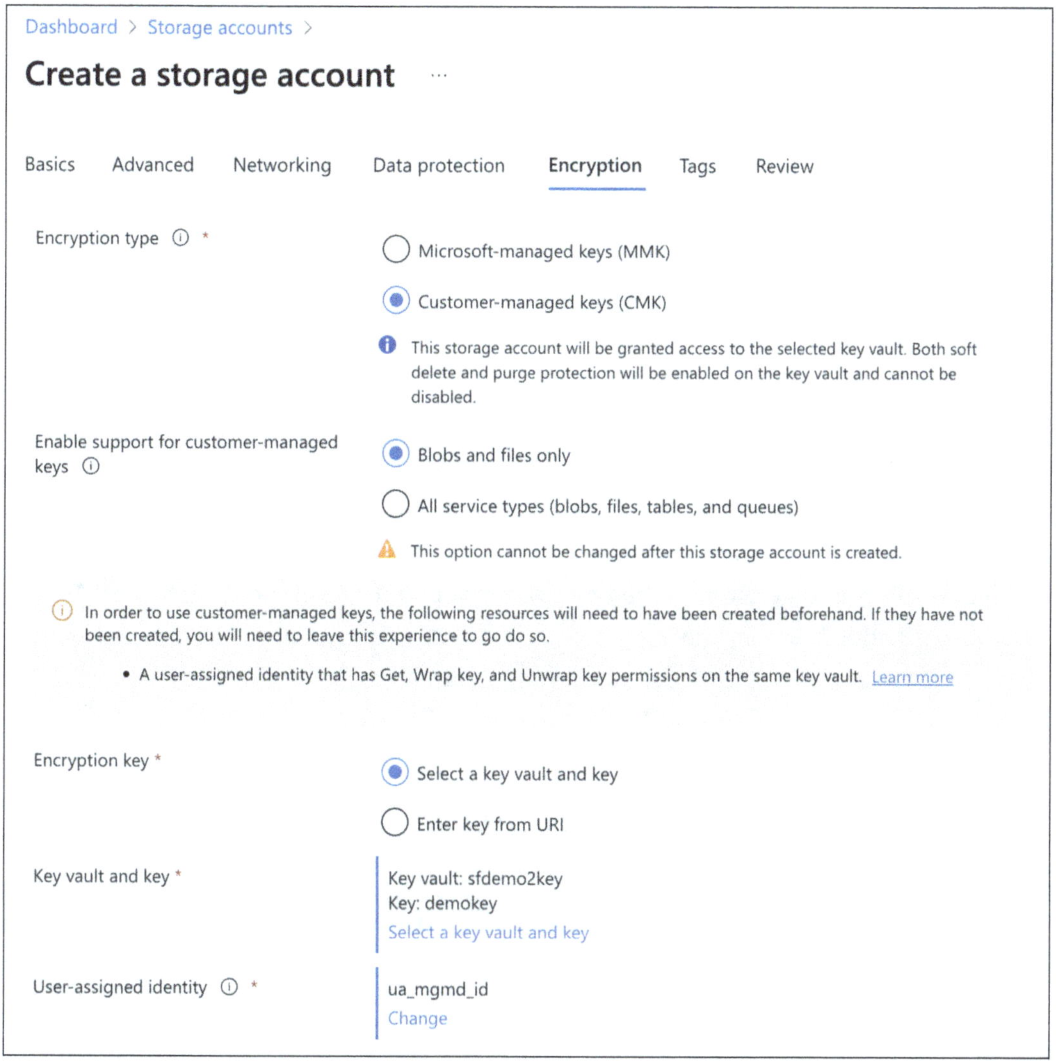

Figure 2-12. *Azure storage account with CMK encryption*

Figure 2-13 is a screenshot of the access policies assigned to the key.

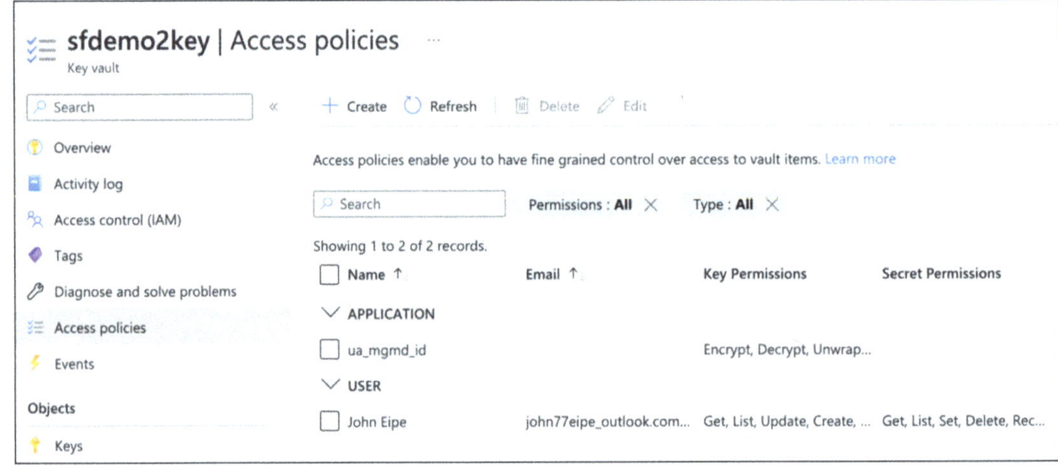

Figure 2-13. *Access policies*

Figure 2-14 is a screenshot of the two storage accounts we created—one with MMK and another with CMK encryption.

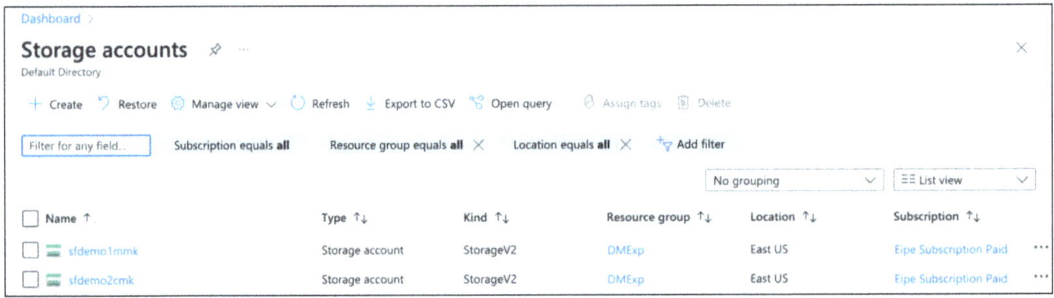

Figure 2-14. *Storage accounts*

Reading Azure Storage When Encryption Is SSE-MMK or SSE-CMK

Problem

Consider the first scenario when the organization's data is Azure storage and is SSE-MMK or SSE-CMK encrypted.

The following steps work under the premise that the prerequisites were completed.

Solution

This solution explores how to create the required permissions and use Snowflake native features to authenticate and use the data stored in Azure encrypted using SSE-MMK or SSE-CMK.

Step 1: Create a Storage Integration in Snowflake

Here is the general syntax to create a storage integration in Snowflake.

```
create storage integration <integration_name>
  type = external_stage
  storage_provider = 'AZURE'
  enabled = true
  azure_tenant_id = '<tenant id>'
  storage_allowed_locations = ('azure://<account>.blob.core.windows.
  net/<container>/<path>/', 'azure://<account>.blob.core.windows.
  net/<container>/<path>/')
  [ storage_blocked_locations = (''azure://<account>.blob.core.windows.
  net/<container>/<path>/", ''azure://<account>.blob.core.windows.
  net/<container>/<path>/") ]
```

The following pertains to this example.

```
create or replace storage integration sfdemo_az_storage_intg
    type = external_stage
    storage_provider = 'AZURE'
    enabled = true
    azure_tenant_id = 'fa170e8c-91aa-4b53-8348-6a25d76677f1'
    storage_allowed_locations = ('azure://sfdemo1mmk.blob.core.windows.net/
    customers', 'azure://sfdemo2cmk.blob.core.windows.net/customers');
```

Step 2: Retrieve the Azure App Name for Your Snowflake Integration

Execute the DESCRIBE INTEGRATION command to retrieve the details for providing Snowflake access to the Azure environment.

```
desc storage integration sfdemo_az_storage_intg;
```

Record the following values.

- AZURE_CONSENT_URL: the URL to the Microsoft permissions request page
- AZURE_MULTI_TENANT_APP_NAME: the name of the Snowflake client application created for your account.

Step 3: Grant Snowflake Access to Storage Location and Key Vault

1. Open the AZURE_CONSENT_URL in a web browser and click the Accept button. This action allows the Azure service principal created for your Snowflake account to obtain an access token on any resource inside your tenant. However the access token is only obtained if appropriate permissions are granted on the resources like storage containers.
2. To grant access to the storage account, log in to the Microsoft Azure portal and open the storage account to which you are granting the Snowflake service principal access.
   a. Click Access Control (IAM) and then Add Role Assignment.
   b. Select the Storage Blob Data Reader role to grant the Snowflake service principal read-only access. Use Storage Blob Data Contributor if you plan to read and write into the storage.
   c. In the next screen, search for the Snowflake service principal. You may search for the keyword "Snowflake" or get the exact identity value from the AZURE_MULTI_TENANT_APP_NAME property in the DESC STORAGE INTEGRATION output in the previous step.
3. To grant access to the key vault, log into the Microsoft Azure portal and open the key vault to which you are granting the Snowflake service principal access.
   a. Click Access Control (IAM) » Add role assignment.
   b. Select the Key Vault Reader role to grant the Snowflake service principal read-only access.

c. Search for the Snowflake service principal. This is the identity in the AZURE_MULTI_TENANT_APP_NAME property in the DESC STORAGE INTEGRATION output in the previous step.

Step 4: Create External Stages in Snowflake

Create two stages: one for sfdemo1mmk and the other for sfdemo2cmk.

```
Create or replace stage sfdemo_mmk_1_az_stg
  storage_integration = sfdemo_az_storage_intg
  url = 'azure://sfdemo1mmk.blob.core.windows.net/customers'
  file_format =( TYPE = CSV );

select $1 from @sfdemo_mmk_1_az_stg;


create or replace stage sfdemo_cmk_2_az_stg
  storage_integration = sfdemo_az_storage_intg
  url = 'azure://sfdemo2cmk.blob.core.windows.net/customers'
  file_format =( TYPE = CSV);

  select $1 from @sfdemo_cmk_2_az_stg;
```

Reading Azure Storage When Your Encryption Is Using CSE

Problem

Consider the case when an organization uses CSE to secure it's data locally and has uploaded the same data in Azure storage.

We won't be going into the details, but the steps are straightforward, assuming you have a storage account named sfdemocsk and Snowflake provided access to this storage.

Solution

The first step is to set up a project where you could use the azure-storage-blob and azure-storage-blob-cryptography libraries to encrypt your files using your selected encryption key.

Next, create the stage in Snowflake with the encryption and master key specified.

```
create or replace stage sfdemo_csk_az_stg
  storage_integration = sfdemo_az_storage_intg
  url = 'azure://sfdemocsk.blob.core.windows.net/customers'
  file_format =( TYPE = CSV )
  ENCRYPTION =( type = 'AZURE_CSE'  MASTER_KEY =
  'PGVuY3J5cHRpb*********' );
```

Reading Azure Storage Using SAS tokens

Problem

Let's consider a scenario when the e-commerce organization wants a quick POC in Snowflake but needs to access the data in Azure object storage. The idea is to avoid setting an integration via Snowflake-managed IAM but be able to quickly access the data via a Snowflake stage.

Solution

This approach doesn't need Snowflake principles created in Azure or any storage integrations created in Snowflake.

The only two steps involved are generating a shared access signature (SAS) token under your storage account and then using it during the creation of a Snowflake stage.

Step 1: Generate a SAS Token

1. Log into your storage account, and under Security + networking, choose "Shared access signature (SAS)".

   a) Select the following allowed services.

      - Blob

   b) Select the following allowed resource types.

      - Container (required to list objects in the storage account)
      - Object (required to read/write objects from/to the storage account)

c) Select the following allowed permissions to load data files from Azure resources.

- Read
- List

d) Select additional permissions if you plan to write/purge from the container.

e) Specify the expiry dates for the SAS token.

f) Make sure to leave the allowed IP addresses field blank.

g) Click the "Generate SAS and connection string" button. Record the full value in the SAS token field, starting with and including the "?".

Step 2: Create a Stage Referencing the SAS Token

Create an external (Azure) stage that references the SAS token you generated in step 1.

```
create or replace stage sfdemo_test_az_stg
  URL='azure://sfdemocsk.blob.core.windows.net/customers'
  CREDENTIALS=(AZURE_SAS_TOKEN='?sv=2016-05-31&ss=b&srt=sco&sp=rwdl&se=2018-
  06-27T10:05:50Z&st=2017-06-27T02:05:50Z&spr=https,http&sig=***************
  **8OFRLQ%3D')
  FILE_FORMAT =( TYPE = CSV );
```

Additional Snowflake Documentation

https://docs.snowflake.com/en/user-guide/data-load-azure-create-stage

https://docs.snowflake.com/en/sql-reference/sql/create-storage-integration

---

Recipe 2-3. Snowflake Object Types

Problem

Let's say you are tasked with setting up various storage objects in Snowflake, like tables or views, but you need to know the various options and when to use which type.

Though this seems like a simple choice, there are multiple native objects provided in Snowflake that act as containers to store data, and as of this writing, there are 15 types available. The list grows each year as new functionalities are added.

- Persisted query results (cache)
- Permanent or regular table
- External table
- Regular view
- Materialized view
- Secure view
- Secure materialized view
- Temporary table
- Transient table
- Directory table
- Dynamic table
- Hybrid table
- Streams
- Event table
- Iceberg table
- Dynamic Iceberg able

Solution

Let's briefly look at each native object and then how to make appropriate choices.

Persisted Query Results

This is an object that Snowflake internally uses and cannot be directly created or altered by the user, but Snowflake provides a way to utilize it.

By default, Snowflake persists (caches) the query result for a period of time. The cache expiry is set at 24 hours for query results of all sizes.

The query reuse is automatic, and multiple conditions need to be met, as explained in the documentation at `https://docs.snowflake.com/en/user-guide/querying-persisted-results.html`.

But do note that Snowflake does not guarantee the query reuse.

An elegant way to reuse query results is to capture the query ID and use the result_scan table function.

```
select last_query_id(); -- retrieve query id
select * from table(result_scan('<query id>')) -- use the query id here
```

Permanent Tables

A permanent table is the default table type.

- You create a permanent table using the CREATE TABLE syntax.
- Do note that regular tables incur storage but support clustering. The data is retained until you explicitly drop them.
- A permanent table can be guarded by Time Travel as per the edition of choice. Standard edition (0 or 1) and Enterprise edition (1 to 90).
- Fail-safe is set at seven days.

In a table definition, along with column names and their data types, you could optionally add whether the column

- requires a value (not null)
- has a default value
- has any referential integrity constraints (primary key, foreign key, etc.) (Note that this is not enforced in Snowflake but is just informational.)

Apart from the usual create table definition, you could create tables as follows.

- CREATE TABLE ... AS SELECT
    - creates a populated table, also referred to as CTAS
    - is useful for quickly constructing a table based on data from another query
    - allows you to enforce data types or let the query infer from the data

  The following is an example.

  ```
  create or replace table test (id int) as select 1 union all
  select 2.5;
  create or replace table test as select 1 as id union all select
  2.5 as id;
  ```

- CREATE TABLE ... USING TEMPLATE
    - creates a table with the column definitions derived from a set of staged files
    - is useful to infer from semi-structured data but limited to Apache Parquet, Apache Avro, ORC, JSON and CSV files.

    The following is an example.

    ```
    create table test
      using template (
        select array_agg(object_construct(*))
           from table(
            infer_schema(
              location=>'@mystage',
              file_format=>'my_parquet_format'
            )
          ));
    ```

- CREATE TABLE ... LIKE
    - creates an empty copy of an existing table

      The following is an example.

      ```
      create table test2 like abc;
      ```

- CREATE TABLE ... CLONE
    - creates a clone of an existing table
    - can be used at/before clause to utilize Time Travel to create the table as of a specified time in the past

      ```
      create table test3 clone test;
      ```

External Tables

As seen in the previous recipes, an external table helps maintain storage in any supported cloud object storage like S3.

- External tables are read-only tables.
- Since data is maintained outside Snowflake, it does not incur any storage charges within Snowflake, but the performance would not be great.
- Performance can be improved by partitioning your external data, which requires that your underlying data is organized using logical paths that include date, time, country, or similar dimensions in the path.
- The performance of the external table can be improved further by creating a materialized view on top of the external table.

  The following is an example.

  ```
  create external table extable(
   col1 int as (value:col1::integer),
   col2 varchar AS (value:col2::varchar))
   location=@stage/files/
  ```

  Handling different semi-structured data types is covered in Chapter 3.

- An external stage is a prerequisite for creating an external table.
- An external table can be refreshed automatically via event notifications.

Directory Tables

A directory table is an implicit object created when a stage is created and doesn't have grantable privileges. If a role has access to a stage, it should have access to the directory table.

- A directory table stores a catalog of staged files in cloud storage.
- A directory table can be added explicitly to a stage when the stage is created (using CREATE STAGE) or later (using ALTER STAGE).

Here is an example of querying a directory table available as part of the named stage.

```
select * from directory( @<stage_name> );
```

For the directory table to reflect the latest information, you need to initiate a refresh that synchronizes the Snowflake metadata with the latest files in the external stage.

Like external tables, directory tables can be refreshed automatically via event notifications, as follows.

```
alter stage @<stage_name> REFRESH;
```

Snowflake streams can be created on top of the directory tables to keep track of files added and dropped from the stage.

Regular Views

You use a view when you need to save on storage (by avoiding creating another table) and hence data duplication but at the same time achieve query reuse and open options for restrictive data access.

- Do note that you get the benefits of a view when the results change frequently but are not used often and when the query is not too complex and resource-intensive.
- Keep in mind that views are read-only, and schema changes to the underlying table are not propagated to a view.

- Never create views by doing a select * on the underlying table. Any changes to the schema break the view.

The following is an example.

```
create or replace view test_v as select id from test;
```

Materialized Views

Materialized views are like tables but with data pre-computed. This incurs storage and compute costs.

Because the data is pre-computed, querying a materialized view is faster than executing a query against the base table of the view.

You could also enable automatic clustering by specifying the clustering keys on the materialized view.

The following is an example.

```
create materialized view test_mv as
    select id, name from test;
```

There are several limitations to materialized views, namely,

- A materialized view can query only a single table or external table.
- Joins, including self-joins, are not supported.
- UDFs and UDTFs are not supported.
- Not all aggregate functions are supported. Refer to Snowflake documentation for the complete list [https://docs.snowflake.com/en/user-guide/views-materialized#limitations-on-creating-materialized-views].
- The order by clause is not supported.
- Window, Having, and Limit clauses are not supported.
- Functions used in a materialized view must be deterministic.

Secure Views

A secure view is similar to a regular view except that the details of the view definition or the internal optimizations on the view are not exposed to the users.

The following is an example.

```
create or replace secure view test_sv as select id from test;
```

Secure Materialized Views

Materialized views can be secure views.

The following is an example.

```
create secure materialized view test_smv as
    select id, name from test;
```

Temporary Tables

Temporary tables are session-specific, which means they are not visible outside the session of the user who created it.

- Time Travel for the temporary table is 0–1 day or the remainder of the session, whichever is shorter.
- Fail-safe is not available.
- Temporary table does *not* require the CREATE TABLE privilege on the schema in which the object is created.
- When you create a temporary table with the same name as an existing table in the same schema, the temporary table takes precedence and hides the existing non-temporary table.
- When you create a table that has the same name as an existing temporary table in the same schema, the newly created table gets hidden by the temporary table.

  ```
  create temporary table <table name> (<column definitions>);
  ```

Transient Tables

A transient table is like a permanent table with no Fail-safe, and Time Travel has a maximum of one day. It needs to be explicitly dropped like the permanent table.

The following is an example.

```
create transient table <table name> (<column definitions>);
```

Streams

A stream is a native object in Snowflake that looks like a table but captures underlying changes to a source object like a table, external table, view, or directory table.

- Once a stream is created it takes a snapshot of the table and tracks every change (DML) applied on the table.
- The change information provided within the stream mirrors the column structure of the tracked object but includes additional metadata columns to specify the type of the change event.
- Stream doesn't store or duplicate the table (or the underlying source) data; instead, it only stores an offset for the source object plus additional metadata for the change events.

```
create or replace stream cdc_customers on table customers;
```

Streams are covered in more depth in Chapter 5.

Dynamic Tables

Dynamic tables were initially called materialized tables and are currently in private preview as of this writing. They help developers use simple SQL statements to declaratively define the result of the data pipeline or the resultant table. These tables provide the collective feature of streams and tasks in a single create-table definition.

As of this writing, dynamic tables do not support the following.

- Certain non-deterministic functions (but functions like SEQ, CURRENT_ROLE, CURRENT_DATE, etc., are supported)
- Any UDF/UDTF function defined as VOLATILE (which marks it as non-deterministic)
- External functions
- Snowpark transformations written in Python, Java, or Scala
- Sources that are external tables, streams, and materialized
- Sources that are views on dynamic tables

You can create streams over dynamic tables but not the other way around.

```
create or replace dynamic table <table name>
  lag = ' 1 hour'
  warehouse = COMPUTE_WH
  as <select... query>;
```

Dynamic tables are covered in more depth in Chapter 5.

Hybrid Tables

Hybrid tables are in public preview as of this writing, but they enable you to build transactional applications on top of Snowflake.

- The query syntax is similar to permanent tables and can be managed using the same Snowflake web interface.
- It supports additional features, as shown in Table 2-1.

Table 2-1. *Hybrid Table Features vs. Standard Table Features*

Feature	Hybrid Tables	Standard Tables
Primary data layout	Row-oriented, with secondary columnar storage	Columnar micro-partitions
Locking	Row-level locking	Partition or table locking
PRIMARY KEY constraints	Required, enforced	Optional, not enforced
FOREIGN KEY constraints	Optional, enforced (referential integrity)	Optional, not enforced
UNIQUE constraints	Optional, enforced	Optional, not enforced
NOT NULL constraints	Optional, enforced	Optional, enforced
Indexes	Supported for performance; updated synchronously on writes	The search optimization service indexes columns for better point-lookup performance; batch updated/maintained asynchronously

Hybrid tables power Snowflake's Unistore capability to provide transactional and analytical data in a single unified platform. This means you can join hybrid and normal tables, and it works fine all together without any change in your query structure.

Event Tables

Before the advent of event tables, logging in Snowflake from various procedures was used to take different forms and strategies.

One common methodology is to create a logging table with at least a log message, dag/process/job ID, timestamp columns and then create a utility stored procedure to perform the logging (insert) operation on this table. This utility procedure would then be granted to execute all the appropriate roles. Based on your use case, you could have a column for severity, source, and so forth.

But now, the following applies to event tables.

- A central location for all your log entries and trace events.
- You can create multiple event tables, but you can associate only one with the account.
- Event tables can be associated only with the account. The active table associated with the account is called the *active* event table.
- You cannot add or modify the columns of the event table. The columns provided by Snowflake can be observed using the `describe event table <table name>` command.
- You can specify the severity level of log messages and the verbiage of the event message.
- You can only do a limited number of DDL operations on the table.
- There is a 1 MB limit for log and trace event payloads. If the payload is over the 1 MB threshold, the record in the event table is incomplete and only contains values for the following columns: TIMESTAMP, RECORD_TYPE, and RESOURCE_ATTRIBUTES.
- Event tables support both logging and tracing. Unlike log messages, trace events are more structured and automatically captured when something has happened in the system.

The following are log levels.

- TRACE
- DEBUG
- INFO
- WARN
- ERROR
- FATAL
- OFF

The following are trace levels.

- OFF
- ALWAYS
- ON_EVENT

Here is an example.

```
create event table cdw_event_log;
alter account set event_table = dev_raw.public.cdw_event_log;
desc event table cdw_event_log;
SHOW PARAMETERS LIKE 'event_table' IN ACCOUNT;

alter database DEV_RAW set LOG_LEVEL = TRACE;
alter database DEV_RAW set TRACE_LEVEL = ALWAYS;

create or replace procedure convert_meters(LENGTH_IN_FEET double)
 returns double
 language javascript
 as
 $$
   var length_in_meters =  LENGTH_IN_FEET/3.281;
   snowflake.log("info", "performing conversion");
   snowflake.addEvent('convert_meters');

   if(LENGTH_IN_FEET<=0) {
```

```
        length_in_meters = -1
        snowflake.log("warn", "invalid conversion hence defaulting to -1");
    }
    snowflake.addEvent('conversion', {'input': LENGTH_IN_FEET, 'output':
    length_in_meters});

    snowflake.log("info", "completed conversion");
    return length_in_meters;
  $$
  ;
Let's say we invoke the procedure and check the event table.
call convert_meters(3.0);
select * from cdw_event_log;
```

RECORD_TYPE ...	RECORD	RECORD_ATTRIBUTES	VALUE
LOG	{ "severity_text": "INFO" }	null	"performing convertion"
LOG	{ "severity_text": "INFO" }	null	"completed convertion"
SPAN	{ "kind": "SPAN_KIND_INTERNAL", "name": "snow.auto_instrumented",	null	null
SPAN_EVENT	{ "name": "convert_meters" }	null	null
SPAN_EVENT	{ "name": "convertion" }	{ "input": 3, "output": 0.9143553794	null

Figure 2-15. *Query Results*

The RECORD_TYPE column facilitates differentiating three types of event logs.

- LOG for a log message.
- SPAN for user-defined function invocations performed sequentially on the same thread.
- SPAN_EVENT for a single trace event. A single query can emit more than one SPAN_EVENT.

The RESOURCE_ATTRIBUTES column is automatically populated for every log or trace entry. It stores useful information about the execution context; for example, Figure 2-16 shows the values for the preceding log statement.

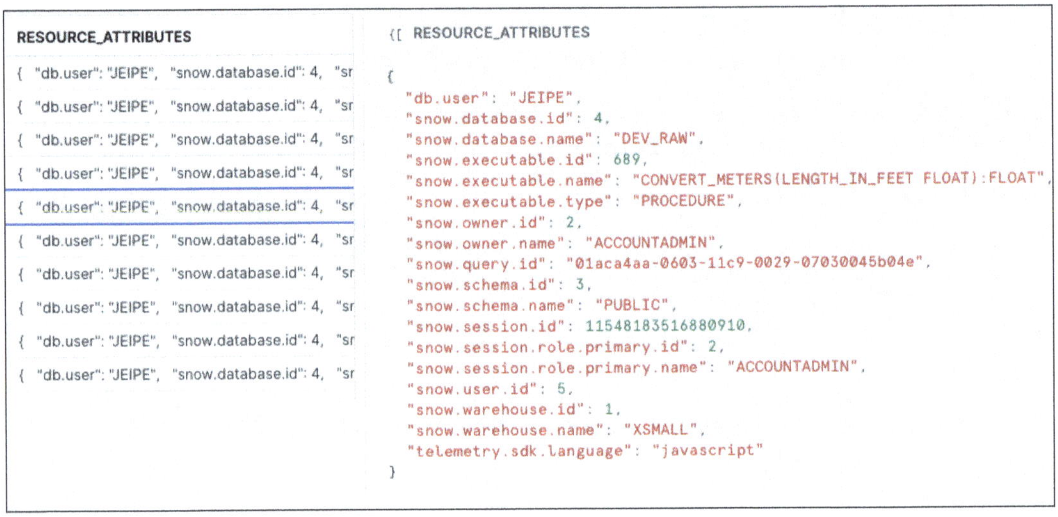

RESOURCE_ATTRIBUTES
{ "db.user": "JEIPE", "snow.database.id": 4, "sr
{ "db.user": "JEIPE", "snow.database.id": 4, "sr
{ "db.user": "JEIPE", "snow.database.id": 4, "sr
{ "db.user": "JEIPE", "snow.database.id": 4, "sr
{ "db.user": "JEIPE", "snow.database.id": 4, "sr
{ "db.user": "JEIPE", "snow.database.id": 4, "sr
{ "db.user": "JEIPE", "snow.database.id": 4, "sr
{ "db.user": "JEIPE", "snow.database.id": 4, "sr
{ "db.user": "JEIPE", "snow.database.id": 4, "sr
{ "db.user": "JEIPE", "snow.database.id": 4, "sr

{[RESOURCE_ATTRIBUTES


```
{
  "db.user": "JEIPE",
  "snow.database.id": 4,
  "snow.database.name": "DEV_RAW",
  "snow.executable.id": 689,
  "snow.executable.name": "CONVERT_METERS(LENGTH_IN_FEET FLOAT):FLOAT",
  "snow.executable.type": "PROCEDURE",
  "snow.owner.id": 2,
  "snow.owner.name": "ACCOUNTADMIN",
  "snow.query.id": "01aca4aa-0603-11c9-0029-07030045b04e",
  "snow.schema.id": 3,
  "snow.schema.name": "PUBLIC",
  "snow.session.id": 11548183516880910,
  "snow.session.role.primary.id": 2,
  "snow.session.role.primary.name": "ACCOUNTADMIN",
  "snow.user.id": 5,
  "snow.warehouse.id": 1,
  "snow.warehouse.name": "XSMALL",
  "telemetry.sdk.language": "javascript"
}
```


Figure 2-16. *Query Result*

The TIMESTAMP column is always populated but the OBSERVED_TIMESTAMP is populated only for log statements and START_TIMESTAMP for trace initialization.

Iceberg Tables

An Iceberg table uses the Apache Iceberg open table format specification as an abstraction on top of the files stored in the data lake.

If you are new to Iceberg, it is a distributed, community-driven, Apache 2.0-licensed, open source data table format that helps to simplify and add transactionality to data processing on large datasets stored in data lakes.

The idea is that the Iceberg metadata and abstraction on top of the data files help ensure that data is accurate and consistent and allows you to track how data and data structure changes over time.

- Note that Iceberg tables have a format similar to Hudi or Delta Lake.
- Apache Iceberg works on top of file formats like Parquet, ORC, and Avro, but Snowflake currently supports only Parquet.
- Apache Iceberg offers easy integrations with popular data processing frameworks such as Apache Spark, Apache Flink, Apache Hive, Presto, and more.
- Snowflake supports managing the Iceberg metadata and catalog in Snowflake and also let's you define the table in Snowflake.

Snowflake can have different types of Iceberg tables depending on where the catalog is managed for Iceberg tables, which are covered in Chapter 5.

Making the Right Choice

External Tables

Choose external tables when

- regulatory restrictions prevent data transfer between regions
- you *do not* need to perform DML operations on the table
- you have static reference data that doesn't change frequently
- performance is not a primary concern
- you have data in existing cloud object storages and want to apply schema-on-read to read a subset of the data
- you need to integrate Apache Hive metastores with Snowflake

Transient Tables

Choose transient tables when

- you need the performance of permanent tables but are only looking at a retention period of one day
- Fail-safe is not a concern
- you need to store the intermediate results of your data pipeline (ELT) process

Temporary Tables

Choose temporary tables when you need quick testing and validation.

Like transient tables, they can also be used for storing intermediate results of your ELT pipeline but keep in mind that the data is only available within the session.

Materialized Views

- Choose materialized views when creating visualizations in BI tools that require varying levels of aggregations (query rewrite)
- seeking to enhance the performance of external tables
- you have uncomplicated aggregation requirements on a single table
- you require the data to be refreshed as soon as possible
- the query results from the view don't change often, which means that the underlying base table for the view doesn't change often

Streams

Choose streams when

- you need change tracking on your source object (tables/external tables/directory tables/views)
- you don't mind managing additional objects (choose wisely - when you apply streams on 100 tables, you would have 100 stream objects)
- a stream is not consumed by Snowflake automatically but temporarily extends the data retention of the source table to 14 days; this incurs additional storage cost along with the compute cost (used by the virtual warehouse for querying streams)

Once a stream is consumed, it is removed from the stream; this might be a limitation in certain ELT pipelines.

Dynamic Tables

Choose dynamic tables when

- you are building SQL-based transformation pipelines and not planning on using Snowpark
- you require complex SQL, including joins, aggregates, and more
- you need more control over when tables are refreshed

Hybrid Tables

Choose hybrid tables when

- you plan to migrate or deploy a new OLTP database on Snowflake and utilize the best features of OLTP and OLAP systems and simply the overall architecture
- you have high concurrency random point reads vs. large-range reads
- you have high concurrency random writes versus large sequential writes
- hybrid tables are not available in Azure or Google Cloud Platform (GCP)
- hybrid tables are not available in SnowGov Regions and not available to Trial accounts
- cloning, materialized views, streams, replication, and data sharing are not supported
- there is a limitation on query throughput per Snowflake database of 1000 operations per second and 500 GB for data

Permanent Tables

Choose permanent tables when

- you need to benefit from data retention (Time Travel) and Fail-safe
- you need good performance

Permanent tables are the default option in most cases.

Iceberg Tables

Choose Iceberg tables in the following circumstances.

- Data is stored in a public cloud using an open file format like Parquet, which is compatible with Iceberg, and you need transactionality on top of your data.

- Data tables in data lakes require frequent deletes or record-level updates. Iceberg provides capabilities to update individual records without republishing the entire data set.
- Multiple teams use different computing engines like AWS Glue, Snowflake, or Spark to access the same data without storing data redundantly and not trading performance.
- Data is already stored in an Iceberg table format in the public cloud, and one does not want to migrate data to Snowflake native tables.
- Choose Snowflake-managed Iceberg tables when possible because they provide performance equal to Snowflake native tables in most cases as shown by the graph below (Figure 2-17).

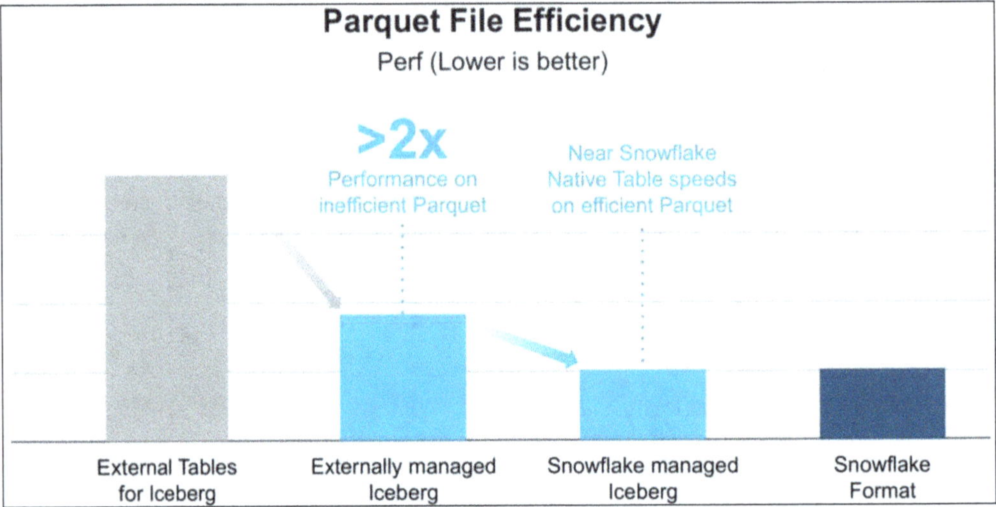

Figure 2-17. *Snowflake Blog*

Additional Snowflake Documentation

https://docs.snowflake.com/en/user-guide/views-materialized.html

https://docs.snowflake.com/en/user-guide/tables-iceberg

https://docs.snowflake.com/en/user-guide/tables-temp-transient

https://docs.snowflake.com/en/user-guide/tables-hybrid

https://docs.snowflake.com/en/user-guide/views-introduction

https://docs.snowflake.com/en/user-guide/views-secure

CHAPTER 3

Handling Atypical Data

Recipe 3-1. Handling Semi-Structured Data

Most databases and data warehousing solutions now support semi-structured data because of its increasing prevalence and the rapidly growing volume of sources such as social media, mobile devices, and the Internet of Things (IoT).

Snowflake provides built-in support for importing data from (and exporting data to) the following semi-structured data formats.

- Supported formats for importing: JSON, Avro, ORC, Parquet, XML
- Supported formats for exporting: JSON, Parquet, CSV

Snowflake provides three native data types (ARRAY, OBJECT, and VARIANT) for storing and managing these semi-structured data formats.

Handling JSON Data in Snowflake

Problem

Consider a scenario when an organization has its customers' data stored as JSON in AWS S3 as part of their data lake solution, and you want to read or move this data into Snowflake.

Refer to Chapter-3/customers.json in the code repository for a sample JSON file, which is used for illustration.

Prerequisites

This demonstration uses the sfdemo-data-formats S3 storage bucket with the customer data in the json/ path.

D. Dayton and J. Eipe, *Snowflake Recipes*, https://doi.org/10.1007/979-8-8688-0938-5_3

If you haven't done so, you need to create a storage integration. We created a storage integration called sfdemo_storage_intg.

Solution

Assuming the storage integration is already created, you must first construct a stage to point to the JSON files.

```
create stage sfdemo_customers_json
  storage_integration = sfdemo_storage_intg
  url = 's3://sfdemo-data-formats/json/'
  file_format = ( TYPE = JSON );
```

Now use the flatten and ":" operator to parse out the values from the data.

```
select
    e.value:C_CUSTKEY::integer       as C_CUSTKEY
    , e.value:C_NAME::varchar        as C_NAME
    , e.value:C_ADDRESS::varchar     as C_ADDRESS
    , e.value:C_NATIONKEY::integer   as C_NATIONKEY
    , e.value:C_PHONE::varchar       as C_PHONE
    , e.value:C_ACCTBAL::double      as C_ACCTBAL
    , e.value:C_MKTSEGMENT::varchar  as C_MKTSEGMENT
    , e.value:C_COMMENT::varchar     as C_COMMENT
from
    @sfdemo_customers_json as src,
    table(flatten(src.$1)) e;
```

The FLATTEN() function is used to get the nested structures and is not a feature just for JSON but for all VARIANT types where you encounter nested structures. The flatten function is further discussed in the upcoming sections.

Handling Avro Data in Snowflake

Problem

Consider a scenario when an organization has its customers' data stored as Avro files in AWS S3 as part of their data lake solution, and you want to read or move this data into Snowflake.

Refer to Chapter-3/customers.avro in the source code repository for a sample Avro file. You could also use the generate_avro_file.py to generate new Avro files from a CSV file.

Prerequisites

This demonstration uses the sfdemo-data-formats S3 storage bucket with the customer data in the avro/ path. If you haven't done so, you need to create a storage integration.

Solution

Assuming the storage integration is already created, you construct a stage to point to the Avro files.

```
create stage sfdemo_customers_avro
  storage_integration = sfdemo_storage_intg
  url = 's3://sfdemo-data-formats/avro/'
  file_format = ( TYPE = AVRO );
```

Next you can read from Avro without any issues.

```
select
    $1:C_CUSTKEY::integer      as C_CUSTKEY
    , $1:C_NAME::varchar       as C_NAME
    , $1:C_ADDRESS::varchar    as C_ADDRESS
    , $1:C_NATIONKEY::integer  as C_NATIONKEY
    , $1:C_PHONE::varchar      as C_PHONE
    , $1:C_ACCTBAL::double     as C_ACCTBAL
    , $1:C_MKTSEGMENT::varchar as C_MKTSEGMENT
    , $1:C_COMMENT::varchar    as C_COMMENT
from
    @sfdemo_customers_avro ;
```

Handling ORC Data in Snowflake

Problem

Consider a scenario when an organization has its customer's data stored as ORC files in AWS S3, and you want to read or move it into Snowflake.

Refer to Chapter-3/customers.orc in the source code repository for a sample ORC file. You could also use the generate_orc_file.py to generate new ORC files from CSV.

Prerequisites

This demonstration uses the sfdemo-data-formats S3 storage bucket with the customer data in the orc/ path.

Solution

Assuming the storage integration is already created, you construct a stage to point to the ORC files.

```
create stage sfdemo_customers_orc
  storage_integration = sfdemo_storage_intg
  url = 's3://sfdemo-data-formats/orc/'
  file_format = ( TYPE = ORC );
```

Now, you can parse and read with ease from this stage.

```
select
    $1:C_CUSTKEY::integer      as C_CUSTKEY
    , $1:C_NAME::varchar       as C_NAME
    , $1:C_ADDRESS::varchar    as C_ADDRESS
    , $1:C_NATIONKEY::integer  as C_NATIONKEY
    , $1:C_PHONE::varchar      as C_PHONE
    , $1:C_ACCTBAL::double     as C_ACCTBAL
    , $1:C_MKTSEGMENT::varchar as C_MKTSEGMENT
    , $1:C_COMMENT::varchar    as C_COMMENT
from
    @sfdemo_customers_orc ;
```

Handling Parquet Data in Snowflake

Problem

Consider a scenario when an organization has its customer's data stored as Parquet files in AWS S3, and you want to read or move this data into Snowflake.

Refer to Chapter-3/customers.parquet in the project folder for a sample parquet file. You could also use the generate_parquet_file.py to generate new Parquet files from CSV.

Prerequisites

This demonstration uses the sfdemo-data-formats S3 storage bucket with the customer data in the parquet/ path.

Solution

Assuming the storage integration is already created, you construct a stage to point to the parquet files.

```
create stage sfdemo_customers_parquet
  storage_integration = sfdemo_storage_intg
  url = 's3://sfdemo-data-formats/parquet/'
  file_format = ( TYPE = PARQUET );
```

Now, you can parse and read with ease from this stage.

```
select
    $1:C_CUSTKEY::integer      as C_CUSTKEY
    , $1:C_NAME::varchar       as C_NAME
    , $1:C_ADDRESS::varchar    as C_ADDRESS
    , $1:C_NATIONKEY::integer  as C_NATIONKEY
    , $1:C_PHONE::varchar      as C_PHONE
    , $1:C_ACCTBAL::double     as C_ACCTBAL
    , $1:C_MKTSEGMENT::varchar as C_MKTSEGMENT
    , $1:C_COMMENT::varchar    as C_COMMENT
from
    @sfdemo_customers_parquet ;
```

Handling XML Data in Snowflake

Problem

Consider a scenario when an organization has their customer's data stored as XML files in AWS S3, and you want to read or move this data into Snowflake.

Refer to Chapter-3/customers.xml in the project folder for a sample parquet file.

Prerequisites

This demonstration uses the sfdemo-data-formats S3 storage bucket with the customer data in the xml/ path.

Solution

Assuming the storage integration is already created, you construct a stage to point to the XML files.

```
create stage sfdemo_customers_xml
  storage_integration = sfdemo_storage_intg
  url = 's3://sfdemo-data-formats/xml/'
  file_format =( TYPE = XML );
```

Now, you can parse and read with ease from this stage.

```
select
    *
from
    @xml_stg src,
    lateral FLATTEN(src.$1:"$") customers
where GET( customers.value, '@') = 'CUSTOMER';

select
    XMLGET( customers.value, 'C_CUSTKEY' ):"$"::integer as C_CUSTKEY
    , XMLGET( customers.value, 'C_NAME' ):"$"::varchar as C_NAME
    , XMLGET( customers.value, 'C_ADDRESS'):"$"::varchar as C_ADDRESS
    , XMLGET( customers.value, 'C_NATIONKEY'):"$"::integer as C_NATIONKEY
    , XMLGET( customers.value, 'C_PHONE'):"$"::varchar as C_PHONE
    , XMLGET( customers.value, 'C_ACCTBAL'):"$"::double as C_ACCTBAL
    , XMLGET( customers.value, 'C_MKTSEGMENT'):"$"::varchar as C_MKTSEGMENT
    , XMLGET( customers.value, 'C_COMMENT'):"$"::varchar as C_COMMENT
from
    @xml_stg src,
    lateral FLATTEN(src.$1:"$") customers
where GET( customers.value, '@') = 'CUSTOMER';
```

Reading XML is slightly more involved than other formats like ORC or Parquet. We used the following functions to read the values from the file.

- The XMLGET() function gets the XML element or tag based on the outer XML element.

  Note that the result of XMLGET is not the tag's content (i.e., the text between the tags) but the entire element (the opening tag, content, and closing tag).

  It takes three arguments.

  - The expression from which to extract the element/tag (It should evaluate to an OBJECT/VARIANT type.)
  - The name of the XML element/tag (which exists within the expression)
  - An optional ordinal number that is used as an "index" in the XML structure
- Like XMLGET(), the GET() function helps extract values from an array or variant.
- The FLATTEN() function is used to get the nested structures.
- While parsing XML or using the GET function, there are three options for the location parameter.
  - The $ argument returns the element's value/tag contents.
  - The @ argument returns the element's name/tag name.
  - Passing @attrName returns the value of an attribute of the element/tag.

Additional Snowflake Documentation

`https://docs.snowflake.com/user-guide/semistructured-intro`

`https://docs.snowflake.com/en/sql-reference/data-types-semistructured`

Recipe 3-2. Schema Detection

In the data realm, schema detection refers to automatically determining the structure and properties of data, particularly when dealing with unstructured or semi-structured data.

The word *schema* here refers to the underlying organization and format of the data.

Schema Detection

Problem

Consider an organization that has its data infrastructure set up in Snowflake. The organization has stages and tables created in the landing layer, also known as its landing layer or raw layer. Most of this data is in a semi-structured format, and though most data feeds stay consistent with their schema, some are observed to have inconsistent schema. This means the columns or the keys can change at any time.

As a user of this data, you need the ability to query it or construct a table that accommodates this inconsistency.

Solution

This is where the schema detection feature of Snowflake comes into the picture. This feature helps when you have a very wide semi-structured file or Snowflake stages populated (with files) in real time but with unknown schema structures.

Snowflake's schema detection feature supports semi-structured file formats such as

- Parquet
- Avro
- ORC
- JSON
- CSV

Support for JSON and CSV was added during the 2023_06 Snowflake release.

The schema detection capability is made possible using three functions available within Snowflake.

- **INFER_SCHEMA** retrieves and returns the schema from a set of staged files.
- **GENERATE_COLUMN_DESCRIPTION** returns the columns necessary to create a table, external table, or view.
- **CREATE TABLE … USING TEMPLATE** expands upon Snowflake's CREATE TABLE functionality to automatically create the structured table using the detected schema from the staged files with no additional input.

The following is an example.

```
select * from table(
    infer_schema(
        location=>'@sfdemo_customers_parquet'
        , file_format => 'demo_parquet_format'
        , ignore_case => true
    )
);
```

It returns the output shown in Figure 3-1.

	COLUMN_NAME	TYPE	NULLABLE	EXPRESSION	FILENAMES	ORDER_ID
1	C_CUSTKEY	NUMBER(38, 0)	TRUE	GET_IGNORE_CASE($1, 'C_CUSTKEY')::NUMBER(38, 0)	parquet/customers.parquet	0
2	C_NAME	TEXT	TRUE	GET_IGNORE_CASE($1, 'C_NAME')::TEXT	parquet/customers.parquet	1
3	C_ADDRESS	TEXT	TRUE	GET_IGNORE_CASE($1, 'C_ADDRESS')::TEXT	parquet/customers.parquet	2
4	C_NATIONKEY	NUMBER(38, 0)	TRUE	GET_IGNORE_CASE($1, 'C_NATIONKEY')::NUMBER(38, 0)	parquet/customers.parquet	3
5	C_PHONE	TEXT	TRUE	GET_IGNORE_CASE($1, 'C_PHONE')::TEXT	parquet/customers.parquet	4
6	C_ACCTBAL	REAL	TRUE	GET_IGNORE_CASE($1, 'C_ACCTBAL')::REAL	parquet/customers.parquet	5

Figure 3-1. *Output*

```
select generate_column_description(array_agg(object_construct(*)), 'table')
as columns
    from table (
        infer_schema(
            location => '@sfdemo_customers_parquet'
            , file_format => 'demo_parquet_format'
```

```
            , ignore_case => true
        )
    );
```

This returns the column descriptions ready to be used for any automated script.

```
"C_CUSTKEY" NUMBER(38, 0),
"C_NAME" TEXT,
"C_ADDRESS" TEXT,
"C_NATIONKEY" NUMBER(38, 0),
"C_PHONE" TEXT,
"C_ACCTBAL" REAL,
"C_MKTSEGMENT" TEXT,
"C_COMMENT" TEXT
```

If the intention of inferring schema is to create a table, the quickest and easiest approach is to use the Snowflake feature of creating a table from a template.

```
create table demo_customer
    using template (select array_agg(object_construct(*))
    from table (
        infer_schema (
            location => '@sfdemo_customers_parquet'
            , file_format => 'demo_parquet_format'
            , ignore_case => true
        )
    )
);
```

Table Schema Evolution

Problem

Consider an organization that has semi-structured data from various sources stored in Snowflake. The data resides in Snowflake stages and the data engineers must build tables on top of this semi-structured data.

The challenge is that some of these sources don't follow a consistent schema and the engineers need to change or recreate the tables every time to keep it synchronized with the incoming keys/columns/features in the data files.

Table schema evolution refers to modifying or changing a database table's structure over time. It involves altering the schema when the underlying data stored in a table changes.

Solution

The structure of tables in Snowflake can evolve automatically to support the structure of new data received from the data sources.

Snowflake supports the following changes.

- adding new columns
- dropping the NOT NULL constraint from columns that are missing in new data files
- dropping existing columns
- changing the data type, length, or precision of existing columns.

To enable table schema evolution, set ENABLE_SCHEMA_EVOLUTION = TRUE on the table.

This feature also brings a new grant into the picture. If a role other than the owner needs to perform the schema evolution during insertion/copy into the table, then you must explicitly specify GRANT EVOLVE SCHEMA.

Let's create a similar table as before and then use another parquet file with additional column information. You may look at the customer_sample_with_age.csv for generating it.

```
create table demo_customer_raw
    using template (select array_agg(object_construct(*))
    from table (
        infer_schema (
            location => '@sfdemo_customers_parquet'
            , file_format => 'demo_parquet_format'
            , ignore_case => true
        )
    )
);
```

Now enable schema evolution on the table.

```
alter table demo_customer_raw set ENABLE_SCHEMA_EVOLUTION=TRUE;
```

Add the modified parquet files.

```
COPY INTO demo_customer_raw
  FROM @sfdemo_customers_parquet/customers_with_age.parquet
  FILE_FORMAT = demo_parquet_format
  MATCH_BY_COLUMN_NAME = CASE_INSENSITIVE;
```

Test the change.

```
show columns in demo_customer_raw;
select * from demo_customer_raw limit 10;
```

Additional Snowflake Documentation

https://docs.snowflake.com/en/sql-reference/sql/alter-schema

https://docs.snowflake.com/en/user-guide/data-load-schema-evolution

---

Recipe 3-3. Binary and Unstructured Data

Handling Binary Data in Snowflake Tables

Problem

Consider a scenario when you as a data engineer, encountered binary data in Snowflake when some data was imported from other systems that used BLOB data.

You need to decode this data and send back data encoded to an external integration.

Solution

Snowflake provides three ways to map the binary data to visual representation for easy human handling.

- hex (default)
- base64
- UTF-8

You might see inconsistencies when using UTF-8 as it works for text-to-binary encoding. But binary-to-text encoding may not work as expected because not all possible BINARY values can be converted to valid UTF-8 strings.

Let's briefly look at one example of the default hex encoding.

```
select hex_encode('😀',0) as reaction
union all
select hex_encode('😂',0) as reaction;
```

Figure 3-2 shown below is the result you would see for the above query.

	REACTION
1	f09f9880
2	f09f9882

Figure 3-2. *Query Result*

```
select hex_decode_string(reaction) from (
      select hex_encode('😀',0) as reaction
      union all
      select hex_encode('😂',0) as reaction
)
```

	HEX_DECODE_STRING(REACTION)
1	😀
2	😂

Figure 3-3. *Query Result*

Something to consider is the fact that the maximum length for the BINARY data type is 8 MB (8,388,608 bytes), and since it is binary data (0s and 1s), there is no notion of Unicode characters, and hence the length is always measured in bytes.

Note that if you have cases of CLOB data, you should use VARCHAR in Snowflake.

Handling Binary Data in Snowflake Stages

Problem

It is ideal for keeping binary data like images and PDFs in Snowflake stages, and as an organization, you have many binary files of the type HDF5 format in Snowflake stages.

The organization wants to read these files in stages using the computer provided by Snowflake.

Prerequisites

The HDF5 format is not natively supported in Snowflake. Let's say you have HDF5 files stored in S3 storage location.

You may use generate_hdf5_file.py to create an HDF5 file and upload it into the sfdemo-data-formats S3 storage bucket in the HDF5/ path.

Solution

There are various ways to approach this problem. One of them is to create a Python UDF or Java UDF and use the Snowflake warehouses to run the UDF and read the data.

First, create a stage if one doesn't exist already.

```
create stage sfdemo_customers_hdf5
  storage_integration = sfdemo_storage_intg
  url = 's3://sfdemo-data-formats/hdf5/';
```

Next, create a Python UDF as follows.

```
create or replace function python_read_hdf5(file string)
    returns string
    language python
    runtime_version = 3.8
    packages = ('snowflake-snowpark-python','h5py')
```

```
    handler = 'read_file'
as
$$
from snowflake.snowpark.files import SnowflakeFile
from io import BytesIO
import h5py
def read_file(file_path):
    response = ""
    with SnowflakeFile.open(file_path, 'rb') as file:
        f = BytesIO(file.readall())
        f =  h5py.File(f)
        d1 = f['C_CUSTKEY']
        d2 = f['C_MKTSEGMENT']

        response = str(d1[:]) + str(d2[:])
        return response
$$;
```

To test the function, invoke it by passing input to the function.

```
select python_read_hdf5(build_scoped_file_url(@sfdemo_customers_
hdf5,'customers.hdf5')) ;
```

The different file URLs are discussed in further detail later.

Additional Snowflake Documentation

https://docs.snowflake.com/en/sql-reference/binary-examples

https://docs.snowflake.com/en/sql-reference/binary-input-output

---

Recipe 3-4. Geospatial Data

Geospatial data, or geodata, includes information related to locations on the earth's surface. These representations of geospatial data can be used to map objects, events, and other real-world phenomena to a specific geographical area represented by latitude and longitude coordinates.

Snowflake provides native data types for representing geospatial data and specialized query functions that can be used to parse, construct, and run calculations over geospatial objects.

Problem

Consider an organization that wants to read and operate on geospatial data available in the Snowflake marketplace for New York.

Prerequisites

If you already don't have a geospatial dataset, you could add one from the marketplace. We're using the OpenStreetMap New York dataset from Sonra.

Solution

Once you have added the dataset from the marketplace, you notice in the object browser on the left the newly added database—a schema called NEW_YORK and several views under it, as shown in Figure 3-4.

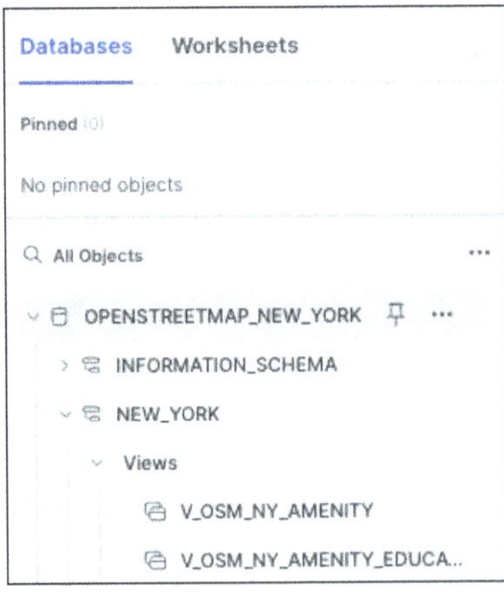

Figure 3-4. *Snowflake Data Sharing*

GEOGRAPHY Data Type

This GEOGRAPHY data type is similar to the GEOGRAPHY data type in other geospatial databases. It treats all points as longitude and latitude on a spherical earth instead of a flat plane.

You could pick a view like the V_OSM_NY_SHOP_FOOD_BEVERAGES to query bakeries around the Brooklyn bridge in a radius of 5,000 meters. (Use online tools like `https://geojson.io/` to capture the coordinates of Brooklyn Bridge)

```
select * from OPENSTREETMAP_NEW_YORK.NEW_YORK.V_OSM_NY_SHOP_FOOD_BEVERAGES
where st_dwithin(st_point( -74.00505479718738,
40.71234701524358),coordinates,5000);
```

There is a specific format in which the COORDINATES column is displayed. This is based on the GEOGRAPHY_OUTPUT_FORMAT value set for the account or session. The default is GeoJSON.

```
alter session set geography_output_format = 'GEOJSON';
```

A short description of the formats supported by Snowflake's GEOGRAPHY datatype is as follows.

- **GeoJSON** is a JSON-based standard for representing geospatial data.
- **WKT** and **EWKT** are short for a "well-known text" string format for representing geospatial data and the "extended" variation of that format.
- **WKB** and **EWKB** are the "well-known binary" format for representing geospatial data in binary and the "extended" variation of that format.

This format setting plays a role while files are ingested or unloaded from Snowflake and displaying query results.

Snowflake provides numerous functions for working the geography objects. The example uses two functions.

- **st_dwithin** returns TRUE if the minimum geodesic distance between two points (two GEOGRAPHY objects) is within the specified distance. Otherwise, returns FALSE.
- **st_point** constructs a GEOGRAPHY object representing a point with the specified longitude and latitude.

GEOMETRY Data Type

We won't be going into the details of using GEOMETRY data type, but here are a few key points about its usefulness and significance.

- Represents geometric spatial data in a flat (Euclidean) space.
- Uses traditional Euclidean coordinates, such as Cartesian (x, y) coordinates in a 2D plane.
- The coordinates are represented as pairs of real numbers (x, y).
- The units of the X and Y are determined by the spatial reference system (SRS) (`https://en.wikipedia.org/wiki/Spatial_reference_system`) associated with the GEOMETRY object. The spatial reference system is identified by the spatial reference system identifier (SRID) number. Unless the SRID is provided when creating the GEOMETRY object or by calling ST_SETSRID, the SRID is 0.

Choosing Between GEOGRAPHY and GEOMETRY Data Types

- The main difference between geography and geometry data types is how they represent spatial data. Geography focuses on representing data on a curved surface, specifically for earth-related information, while geometry deals with Euclidean coordinates and supports general geometric calculations.
- Geography is suitable for representing real-world features like cities, countries, rivers, and so forth. Geometry data types are commonly employed to handle geometric shapes like points, lines, polygons, or more complex structures.
- Using GEOMETRY with its multiple SRIDs can be particularly beneficial in areas like land surveying and urban planning, where precise measurements are essential. It also allows better compatibility among different data sources that use specific SRIDs.

Additional Snowflake Documentation

https://www.snowflake.com/blog/geometry-data-type-advancing-geospatial-analysis/

https://www.snowflake.com/blog/blog-getting-started-geometry-data/

Recipe 3-5. JSON Data Operations

Snowflake offers a wide range of built-in functions specifically designed for working with JSON data. These functions allow you to easily parse, extract, and manipulate JSON elements.

Problem

Consider a situation when you, as a data engineer, are tasked to handle semi-structured JSON data from various data sources.

You must use appropriate Snowflake data types to store the data and write code to parse, search and even reconstruct new data to be sent to downstream tables.

Solution

To mimic the scenario let's use the shared Snowflake sample database (SNOWFLAKE_SAMPLE_DATA) made available to every Snowflake account.

Let's look at Snowflake's data types that support semi-structured data.

- **ARRAY** is a language construct similar to an array in other languages.
- **OBJECT** is a dictionary, hash, or map in many languages, similar to a JSON object. This is a collection of key-value pairs.
- **VARIANT** is a data type that can hold any other data type's value (including ARRAY and OBJECT). VARIANT allows you to create and store hierarchical data.

Differences Between Array, Object, and Variant

Array

- It is a language construct similar to an array in other languages.
- In most other languages, an array is a collection of values of the same data type, but Snowflake supports multiple data types.
- Arrays can be nested, meaning that an array can contain other arrays.
- Arrays can be used to store related data, such as a list of customer names or product prices.
- The theoretical maximum combined size of all values in an ARRAY is 16 MB. However, ARRAYs have internal overhead. The practical maximum data size is usually smaller, depending on the number and values of the elements.

Object

- A Snowflake OBJECT is analogous to a JSON "object".
- In other programming languages, the corresponding data type is often called a *dictionary*, *hash*, or *map*.
- An object is a collection of key-value pairs.
- The key must be a varchar; the value could be any object type. Value could also be nested objects. Though Snowflake internally considers every value in the key-value pair as VARIANT.
- Objects can store related data, such as a customer record with a name, address, and phone number.
- The maximum length of an OBJECT is 16 MB.

Variant

- A variant is a data type that can hold any other data type.
- This means a variant can hold a string, an integer, a float, an array, an object, or even another variant.

- Variants are useful for storing not well-defined data, such as JSON data.
- The maximum uncompressed data size that a variant supports is 16 MB.

Which One Should You Use?

The best data type to use depends on the specific needs of your application.

- Choose ARRAY
    - if you need to store a collection of values of the same data type or data that is structured in the same way
    - if data has natural order (chronological order)
- Choose OBJECT
    - if you need to store a collection of key-value pairs
    - if you have multiple pieces of data that are identified by strings
    - if the information has no natural order or the order can be inferred solely from the keys
    - if the structure of the data varies or the data is incomplete
- Choose VARIANT
    - if the structure varies or if data is incomplete
    - if you are not sure what types of operations you want to perform on semi-structured data
- The following are performance considerations.
    - Non-native values (such as dates and timestamps in JSON) are stored as strings when loaded into a VARIANT column, so operations on these values could be slower and consume more space than when stored in a relational column with the

corresponding data type. If you anticipate frequent queries on these columns it is preferred to parse and store them as regular relational columns with appropriate data type.

- JSON "null" is stored as a string containing the word null. You must explicitly perform a string cast to verify if it is an SQL "NULL".

Exploring Various Functions for Parsing and Creating JSON Objects

- The array_construct function creates an array, and the array_construct_compact function creates an array after removing any NULL elements.

```
select array_construct(1, 'a', array_construct(1,2));
select array_append(array_construct(1, 2, 3), 'hi');
```

- Snowflake supports dense and sparse arrays.
    - In a dense array, each element consumes storage space, even if the element's value is NULL. In a dense array, the index values of the elements start at zero and are sequential (0, 1, 2, etc.).
    - In a sparse array, undefined elements do not directly consume storage space. In a sparse array, the index values can be non-sequential (e.g., 0, 2, 5).
    - Although sparse arrays are supposed to save space, the practical size limit in both cases is 16 MB.
    - You could create sparse arrays using the ARRAY_INSERT function to insert values at specific index points in an array (leaving other array elements undefined).
    - You could also convert sparse arrays into dense arrays using the ARRAY_COMPACT function, which returns a compacted array with missing and null values removed.

- You cannot use the array_construct function within a group by clause. If you need to pivot the values into an array based on a group by clause, use the array_agg() function. Note that the arrayagg() function is an alias of array_agg().

```
select
      year(O_ORDERDATE) as YEAR
      , O_ORDERSTATUS
      , arrayagg(O_CLERK) within group (order by O_
      TOTALPRICE desc)
      from SNOWFLAKE_SAMPLE_DATA.TPCH_SF1.ORDERS
where O_TOTALPRICE > 500000
group by YEAR, O_ORDERSTATUS
order by YEAR,O_ORDERSTATUS desc;
```

Figure 3-5 below shows the result of the this query.

	YEAR	O_ORDERSTATUS	ARRAYAGG(O_CLERK) WITHIN GROUP (ORDER BY O_TOTALPRICE DESC) ...
1	1,992	F	["Clerk#000000040", "Clerk#000000924", "Clerk#000000574", "Clerk#0000
2	1,993	F	["Clerk#000000105", "Clerk#000000020"]
3	1,994	F	["Clerk#000000230"]
4	1,995	O	["Clerk#000000630"]
5	1,996	O	["Clerk#000000650"]
6	1,997	O	["Clerk#000000699", "Clerk#000000336", "Clerk#000000245", "Clerk#0000
7	1,998	O	["Clerk#000000303"]

***Figure 3-5.** Query Result*

You could use the create table as the syntax to create a table from the SQL query.

```
create table order_status_year(year integer, O_ORDERSTATUS
varchar, CLERKS array) as
select
      year(O_ORDERDATE) as YEAR
      , O_ORDERSTATUS    as O_ORDERSTATUS
      , arrayagg(O_CLERK) within group (order by O_TOTALPRICE
      desc) as CLERKS
```

```
from SNOWFLAKE_SAMPLE_DATA.TPCH_SF1.ORDERS
where O_TOTALPRICE > 500000
group by YEAR, O_ORDERSTATUS;

select * from order_status_year where array_contains('Clerk
#000000693'::variant, CLERKS);
```

- The OBJECT_CONSTRUCT function is used to create OBJECT types.

```
select object_construct('user_id',1,'first_name','john');

create or replace table user_data(user_obj object) as
select object_construct('user_id',1,'first_name','john')
union all
select object_construct('user_id',2,'first_name','james')
union all
select object_construct('user_id',3,'first_name','dave');

select user_obj['user_id'] from user_data;
select substr(user_obj['first_name'], 0 ,1) from user_data;
```

- As with arrayagg, there is an object_agg function.

```
create or replace table NATION_REGION(N_NATIONKEY VARCHAR,
N_DETAILS object) as
select
N_NATIONKEY
, object_agg(N_NAME, N_REGIONKEY)  as N_DETAILS
from SNOWFLAKE_SAMPLE_DATA.TPCH_SF1.NATION
group by N_NATIONKEY;
```

- Snowflake supports object constant. That means you need not always use the object_construct function.

```
UPDATE user_data SET user_obj = { 'user_id',1,'first_
name','john' };
UPDATE user_data SET user_obj = object_construct( 'user_
id',1,'first_name','john' );
```

- The Snowflake VARIANT is a true JSON data type. You can create VARIANT data by parsing JSON strings.

```
create or replace table user_auto_data
(
  id integer, data variant
)
AS
SELECT column1 as id, PARSE_JSON(column2) AS src
FROM VALUES
(
    1,
    '{
        "first_name": "Antons",
        "last_name": "Yakushkev",
        "email": "ayakushkev0@sina.com.cn",
        "gender": "Male",
        "wallet": "$1681.18",
        "cars": [
            {
                "car_make": "Lexus",
                "car_age": 20,
                "car_year": 2011
            },
            {
                "car_make": "Jeep",
                "car_age": 27,
                "car_year": 2008
            },
            {
                "car_make": "Ford",
                "car_age": 9,
                "car_year": 1994
            }
        ]
    }'
),
```

```
    (
        2,
        '{
            "first_name": "Stacee",
            "last_name": "Geaves",
            "email": "sgeaves1@naver.com",
            "gender": "Agender",
            "wallet": "$88.82",
            "cars": [
                {
                    "car_make": "Buick",
                    "car_age": 26,
                    "car_year": 2004
                },
                {
                    "car_make": "Oldsmobile",
                    "car_age": 5,
                    "car_year": 1996
                }
            ]
        }'
    ),
    (
        3,
        '{
            "first_name": "Mickie",
            "last_name": "Crush",
            "email": "mcrush2@bing.com",
            "gender": "Male",
            "wallet": "$252.32",
            "cars": [
                {
                    "car_make": "Chevrolet",
                    "car_age": 19,
                    "car_year": 1995
```

```
            },
            {
                "car_make": "GMC",
                "car_age": 2,
                "car_year": 2008
            },
            {
                "car_make": "Dodge",
                "car_age": 13,
                "car_year": 2008
            },
            {
                "car_make": "Honda",
                "car_age": 6,
                "car_year": 1986
            }
        ]
    }'
);
```

- To create similar mock JSON data, use tools like Mockaroo (`www.mockaroo.com`).
- Use PARSE_JSON to convert a string to a variant type.
- VARIANT supports both dot and bracket notation.

```
select id, data:first_name from user_auto_data;
select id, data['first_name'] from user_auto_data;
```

- FLATTEN is a table function that converts a repeated field into a set of rows. FLATTEN removes one level of nesting. If you need to unwarp multiple levels of nesting, then you need to use FLATTEN multiple times

```
select
  id,
  data:first_name,
```

```
    value:car_make
    from
      user_auto_data
    , lateral flatten(INPUT => data:cars);

select
    id,
    data:first_name,
    arrayagg(value:car_make) within group (order by value:car_make
    desc) as car_brands
    from
      user_auto_data
    , lateral flatten(INPUT => data:cars)
    group by id, data:first_name ;
```

Additional Snowflake Documentation

https://docs.snowflake.com/en/user-guide/querying-semistructured

https://docs.snowflake.com/sql-reference/functions-semistructured

---

CHAPTER 4

Data Security and Privacy

In today's digital age, data security and privacy have become increasingly important issues for individuals, businesses, and governments. As more sensitive data is generated and stored in digital formats, the risk of data breaches and privacy violations has grown exponentially. From personal information such as health records, financial data, and online behavior to business-critical information such as intellectual property and trade secrets, the importance of protecting data from unauthorized access and theft has never been greater.

This chapter closely examines how Snowflake keeps your data secure and private in the cloud. It starts by examining how Snowflake encrypts and controls access to your data, including its support for multi-factor authentication and role-based access control. It also explores the network, row access, and masking policies that help protect your sensitive information.

We also dive into Snowflake's data privacy features, including support for data classification and masking. This understanding allows discussion on how Snowflake meets industry standards and regulations like HIPAA, GDPR, and SOC 2.

By the end of this chapter, you'll better understand how Snowflake keeps your data secure and private and how you can utilize its features to protect your sensitive information.

Recipe 4-1. Compliance Regulations

Data compliance regulations refer to laws and guidelines governing how organizations collect, store, process, and use personal and sensitive data. These regulations are designed to protect the privacy and security of individuals' data and prevent unauthorized access or misuse of data.

D. Dayton and J. Eipe, *Snowflake Recipes*, https://doi.org/10.1007/979-8-8688-0938-5_4

Some of the most well-known data compliance regulations include the General Data Protection Regulation (GDPR) in the European Union and the California Consumer Privacy Act (CCPA) in the United States. These regulations require organizations to obtain consent from individuals before collecting and processing their data, provide individuals with the right to access and control their data, and require organizations to take measures to ensure the security and confidentiality of the data they handle.

The importance of data compliance regulations in society cannot be overstated. With the increasing amount of personal data being generated and collected by organizations, there is a growing risk of data breaches, identity theft, and other forms of cybercrime. Data compliance regulations help to mitigate these risks by setting clear guidelines for organizations to follow and establishing penalties for non-compliance. They also help build trust between organizations and individuals, as individuals can feel more confident that their data is being handled responsibly and ethically.

Data compliance regulations play a critical role in protecting the privacy and security of individuals' data and ensuring that organizations are held accountable for handling that data.

Snowflake is designed to help organizations comply with various data compliance regulations. Here are some of the ways Snowflake treats data compliance regulations.

- **Compliance certifications**: Snowflake has achieved several compliance certifications, such as SOC 1 Type II, SOC 2 Type II, ISO 27001, HIPAA, and PCI-DSS. These certifications ensure that Snowflake has implemented the necessary controls and processes to protect the confidentiality, integrity, and availability of its customers' data. Be sure to check for the latest updates to existing and new certifications. Snowflake is continuously growing this program to ensure organizations are protected.

- **Data encryption**: Snowflake uses industry-standard encryption protocols to protect data at rest and in transit. Snowflake encrypts data using 256-bit Advanced Encryption Standard (AES) encryption and supports customer managed keys for added security.

- **Access controls**: Snowflake provides granular access controls that allow organizations to restrict access to data based on user roles and permissions. Snowflake also supports multi-factor authentication to ensure that only authorized users can access sensitive data.

- **Auditing and logging**: Snowflake provides detailed auditing and logging capabilities that allow organizations to track user activities and access data. Snowflake logs all user activity, including queries, logins, and modifications, to help organizations detect and prevent unauthorized access.
- **Compliance reporting**: Snowflake provides compliance reporting features that allow organizations to generate reports that demonstrate compliance with various regulations. These reports can help organizations prove their compliance to auditors and regulators. The latest information on compliance reports can be found at `www.snowflake.com/en/legal/snowflakes-security-and-compliance-reports/`.

Overall, Snowflake treats data compliance regulations seriously and has implemented various features and certifications to help organizations comply with these regulations.

Problem

Consider a financial institution that handles sensitive financial information, such as customer account details and transactions. The institution must comply with various data privacy and security regulations, such as PCI-DSS, to ensure that customer data is kept secure and confidential.

To meet these compliance requirements, the financial institution needs a data platform that provides robust security and privacy features, including compliance with various regulatory frameworks, such as PCI-DSS, GDPR, and SOC 2.

Furthermore, compliance regulations often require organizations to implement specific security measures, such as encryption, access control, and data retention policies. The data platform must provide various security and privacy features that meet these requirements, enabling the organization to implement secure best practices and solutions.

Solution

There are various compliance regulations that different types of organizations may be required to abide by, depending on the nature of their operations and the data they handle. The following are some ways a company can identify if it must abide by compliance regulations.

- **Identify the type of data they handle.** Different types of data, such as personal identifiable information (PII), protected health information (PHI), financial data, or confidential business information, may be subject to different regulations. A company should identify what types of data they handle to determine which regulations apply to them.
- **Determine the industry.** Certain industries, such as healthcare, finance, and government, have specific regulations that apply to them. Companies operating in these industries should know the specific regulations they must follow.
- **Evaluate the geographic location.** Different countries and regions have their own data privacy and security regulations. A company operating in multiple locations may be subject to multiple sets of regulations.
- **Review contracts and agreements.** Companies may also be required to comply with regulations based on contractual agreements with clients or partners. For example, a contract may require compliance with certain security standards or data handling practices.
- **Consult with legal experts.** If a company is unsure whether it must comply with certain regulations, it should consult with legal experts who are knowledgeable in the specific regulations and can provide guidance on compliance requirements.

In general, organizations need to stay up to date with the latest compliance regulations in their industry and geographic location, as non-compliance can result in significant legal and financial consequences.

Let's revisit our example where a financial institution is in the process of onboarding Snowflake. After following the necessary steps, the legal department has determined that proof of PCI-DSS certification is required before Snowflake can be approved and onboarded.

If the organization already has a Snowflake account, the most efficient way to obtain the report is by logging in to the support portal and submitting a case to request the PCI-DSS certification report or any other required reports.

The process is just as straightforward for organizations without an existing relationship with Snowflake. Simply visit Snowflake's Security Compliance Reports (`www.snowflake.com/snowflakes-security-compliance-reports/`) and complete the information on the Contact page. To receive appropriate support, select Security Information as the inquiry type.

Additional Snowflake Documentation

`https://www.snowflake.com/snowflakes-security-compliance-reports/`

Recipe 4-2. Security Best Practices

Data security is an increasingly important issue in modern society as the amount of digital information generated and stored grows exponentially. With the rise of digital technologies and the Internet of Things (IoT), organizations and individuals generate vast amounts of data daily. This data can be incredibly valuable, but it also represents a significant risk if it falls into the wrong hands. Data breaches can result in financial losses, reputational damage, and legal liability. As a result, data security has become a critical concern for businesses, governments, and individuals alike. In this context, robust data security measures are essential to protect sensitive data, prevent unauthorized access, and ensure compliance with industry regulations and standards.

Multi-Factor Authentication

Multi-factor Authentication (MFA) is an authentication mechanism that requires users to provide two or more forms of identification before they can access a system or application. Snowflake MFA adds an extra layer of security to the authentication process and helps to prevent unauthorized access to sensitive data.

Overall, Snowflake MFA provides an additional layer of security to help protect data warehouses from unauthorized access, helping to ensure the privacy and security of sensitive data.

Private Link

Private Link is a service offered by all three cloud providers. It allows users to access their VPC (virtual private cloud) resources using private IP addresses without needing internet gateways, network address translation (NAT) devices, or VPN connections.

- AWS: AWS PrivateLink
- Azure: Private Link
- Google Cloud Platform: Private Service Connect

Private Link creates a private endpoint for the resource, which can be accessed from within the VPC or from an on-premises network connected to the VPC using a direct connection or VPN.

Snowflake uses Private Link to enable secure, private access to its cloud data warehouse service from within a customer's VPC. This allows Snowflake customers to securely access their data without exposing it to the public internet or requiring complex network configurations.

To use Private Link with Snowflake, a customer creates a private endpoint in their VPC, which is then associated with the Snowflake service endpoint in the cloud account. The customer can then use the Snowflake service endpoint to connect to Snowflake using private IP addresses over the cloud network instead of using a public IP address and the internet.

By using Private Link, Snowflake customers can improve their data security posture by reducing their exposure to potential threats from the public internet. Private Link also provides a more reliable and performant connection to Snowflake by eliminating the need for a VPN connection or internet gateway. Overall, Private Link is a valuable tool for Snowflake customers who require secure, private access to their data warehouse service.

Network Policies

Snowflake's network policies are a security feature that allows customers to control the network traffic to access their Snowflake account. Network policies define which IP addresses and networks are allowed to connect to Snowflake and which are blocked.

Network policies provide granular control over network traffic by allowing customers to define rules based on Classless Inter-Domain Routing (CIDR) addresses. For example, a network policy could be set up to allow connections only from a specific set of IP addresses or subnets or to block all traffic except for a specific range of IP addresses.

Snowflake's network policies are a valuable security feature because they help to prevent unauthorized access to an organization's Snowflake account. By limiting network traffic to only authorized sources, network policies help to reduce the attack surface and minimize the risk of data breaches and other security incidents.

Network policies also help to meet compliance requirements by ensuring that access to Snowflake is restricted to authorized parties only. For example, certain regulations such as HIPAA require organizations to control access to sensitive data to only authorized personnel. By using Snowflake's network policies, organizations can demonstrate compliance with these regulations.

Overall, Snowflake's network policies are an important security feature that enables customers to have granular control over network traffic and helps ensure the security of their Snowflake account and data.

Role-Based Access Control

Role-based access control (RBAC) controls access to resources or features within a system or application based on the roles assigned to individual users or groups of users. In RBAC, access to resources or features is determined by user roles rather than their individual identities.

Under RBAC, roles are defined based on the specific responsibilities and permissions required for a given task or set of tasks. These roles are then assigned to individual users or groups of users based on their job functions or responsibilities within an organization.

RBAC provides several benefits over traditional access control methods, such as discretionary access control (DAC) or mandatory access control (MAC). RBAC helps to simplify access control management by allowing administrators to assign permissions based on roles rather than individually managing permissions for each user. RBAC also provides a more fine-grained control over access to resources, as permissions can be tailored to the specific needs of individual roles. This granularity makes RBAC valuable for modern data platforms like a Data Lakehouse architecture.

In addition, RBAC helps to improve security by reducing the risk of accidental or malicious access to sensitive resources. RBAC can help ensure that users only have access to the resources and features necessary for their job function and can prevent unauthorized access to sensitive resources.

Overall, RBAC is a widely used access control method that helps to simplify access control management, provide more fine-grained control over access to resources, and improve security by reducing the risk of accidental or malicious access.

Problem

Consider a financial institution that handles sensitive financial information, such as customer account details and transactions. The institution must comply with various data privacy and security regulations, such as PCI-DSS, to ensure that customer data is kept secure and confidential. While Snowflake is PCI-DSS compliant, there is a responsibility to enforce security within the organization's account.

The financial institution uses Snowflake as their data warehousing solution to meet these requirements, which provides robust security and privacy features, including MFA, Private Link, and network policies.

By implementing MFA, the financial institution can ensure that only authorized personnel can access sensitive financial information. This helps prevent unauthorized access to sensitive information, reducing the risk of data breaches and potential legal liabilities.

Private Link further enhances security by providing a secure, private communication channel between the financial institution's network and the Snowflake data cloud. This helps ensure that sensitive data is not exposed to the public internet, reducing the risk of data breaches and potential cyberattacks.

Finally, network policies allow the financial institution to define and enforce security rules that govern data access and network communication within the Snowflake data cloud. This helps ensure that data is accessed and communicated only by authorized personnel and devices, reducing the risk of data breaches and potential security incidents.

By implementing these features in Snowflake the financial institution can better protect their sensitive financial data, maintain regulatory compliance, and build trust with their customers.

For this recipe, it is assumed that PrivateLink and RBAC models have already been deployed. These features require collaboration across multiple teams, including security, DevOps, cloud, and business. If assistance is needed in these areas, opening a support case with Snowflake and following the documentation links provided is suggested.

Solution

This solution is a three-step process.

1. Create a local user with MFA enabled.

Tip Comments are variant columns to leverage them by using JSON.

```
CREATE
USER
IF NOT EXISTS
   'JDOE'
   PASSWORD = 'DemoP@55w0rd'
   LOGIN_NAME = 'JDOE'
   DISPLAY_NAME = 'Jane Doe'
   FIRST_NAME = 'Jane'
   LAST_NAME = 'Doe'
   EMAIL = 'jane.doe@fakemail.com'
   DEFAULT_ROLE = 'DEVELOPER'
   DEFAULT_WAREHOUSE = 'DEVELOPER_WH'
   TIMEZONE = 'UTC'
   MUST_CHANGE_PASSWORD = TRUE
   DISABLED = FALSE
   TIMESTAMP_DAY_IS_ALWAYS_24H = FALSE
   COMMENT = '[{"DEPARTMENT":"Data"}, {"ROLE":"Developer"}]'
;

-- enable MFA
ALTER USER JDOE SET DISABLE_MFA = FALSE;
```

2. Create a network policy that blocks all public WAN access for the account.

```
-- requires all users to connect via office or corporate VPN
-- PrivateLink creates a secured connection between corp
   LAN and Snowflake Account
CREATE NETWORK POLICY np_account_block_public ALLOWED_IP_
LIST=('192.168.1.0/24');

-- validate policy
DESC NETWORK POLICY mypolicy1;

-- apply network policy to ACCOUNT

ALTER ACCOUNT SET NETWORK_POLICY = np_account_block_public;
```

3. Create a network policy that restricts source access for an individual user account.

```
-- extra protection from stolen credentials
-- for example, ELT service account can only access SF from
   specific IP or host server
CREATE NETWORK POLICY np_elt_block ALLOWED_IP_
LIST=('192.168.1.58);

-- validate policy
DESC NETWORK POLICY np_elt_block;

-- apply network policy to USER (Service Acct for this example)
ALTER USER svc_elt_user SET NETWORK_POLICY = np_elt_block;
```

Additional Snowflake Documentation

https://docs.snowflake.com/en/user-guide/security-mfa

https://docs.snowflake.com/en/user-guide/network-policies

https://docs.snowflake.com/en/user-guide/private-snowflake-service

https://docs.snowflake.com/en/user-guide/security-access-control-overview

https://docs.aws.amazon.com/vpc/latest/privatelink/what-is-privatelink.html

https://azure.microsoft.com/en-us/products/private-link/

https://cloud.google.com/vpc/docs/private-service-connect

Recipe 4-3. Data Privacy

Data privacy is a fundamental right that protects an individual's personal information, including their name, address, email, phone number, and any other data that can be used to identify them. Data privacy has become increasingly important due to governments, businesses, and other organizations' widespread collection, storage, and use of personal data.

The social impact of data privacy is significant, as the misuse of personal data can have far-reaching consequences for individuals and society as a whole. For example, the Cambridge Analytica scandal in 2018 highlighted how personal data collected from social media platforms was used to manipulate public opinion during the US presidential election. This misuse of data can erode trust in democratic institutions and undermine the integrity of elections.

In addition to political implications, the misuse of personal data can also have economic consequences. Data breaches can result in financial losses for businesses and individuals, damage to reputations and loss of trust. For example, the Equifax data breach in 2017 resulted in the theft of personal information from 147 million individuals, causing significant financial losses and damage to the company's reputation.

Data privacy also has significant implications for individuals, particularly those in marginalized communities who may be at greater risk of discrimination or other harm. For example, the use of facial recognition technology has been criticized for its potential to perpetuate racial biases and further marginalize already vulnerable communities.

Despite the risks associated with the misuse of personal data, there are also many benefits to the collection and use of data. Personal data can improve public health, advance scientific research, and create more personalized and effective products and services.

Many governments have enacted laws and regulations to protect individuals' personal data to address the risks associated with data privacy. For example, the GDPR in the European Union sets out strict rules for the collection, storage, and use of personal data and provides individuals with the right to access and control their data.

Data privacy is a complex and evolving issue with significant social implications. While the collection and use of personal data can have many benefits, it is important to balance these benefits with the need to protect individuals' privacy and prevent the misuse of their data. For those reasons, Snowflake has enabled several features to help organizations protect their data.

Snowflake Row Access Policies (Row-Level Security)

Snowflake row access policies control access to rows within a table based on specific conditions or attributes. Row access policies enable more granular control over data access, allowing administrators to restrict access to sensitive data based on specific criteria.

With row access policies, administrators can create rules determining which users or roles can access certain rows of data within a table. These rules are based on conditions or attributes such as user roles or subqueries.

Snowflake Masking Policies (Column-Level Security)

Snowflake masking policies protect sensitive data by obscuring or replacing it with non-sensitive data. Masking policies allow administrators to control what data is visible to users or roles while still allowing them to perform their necessary work functions. This helps prevent sensitive data from being viewed by unauthorized users or roles while allowing authorized users to access the data they need to perform their work.

With masking policies, administrators can define rules that determine how data is masked or obscured. This can include partial masking, where only certain parts of the data are visible, or full masking, where the data is completely obscured or replaced with non-sensitive data. For example, a masking policy may hide or replace social security or credit card numbers in a database.

Snowflake Classification

Snowflake classification allows organizations to classify data based on sensitivity or confidentiality level. Snowflake classification is designed to help organizations protect sensitive data by providing an automated way to identify and categorize data according to its level of sensitivity.

Classification tags can be automatically applied to data based on predefined rules and policies or manually applied by administrators. These tags can then enforce security policies, such as access controls, masking policies, and retention policies. For example, a healthcare organization may classify patient medical records as highly sensitive data and apply security policies to ensure that only authorized users have access to the data and that the data is properly secured and retained.

Problem

Consider a healthcare organization that collects sensitive patient data, including medical history, lab results, and prescriptions. The organization must comply with various data privacy regulations, such as HIPAA, to ensure that patient data is kept secure and confidential.

To meet these requirements, the healthcare organization uses Snowflake as their data warehousing solution, which provides robust security and privacy features, including row-level security, data masking, and data classification.

With row-level security, the healthcare organization can ensure that only authorized personnel, such as doctors and nurses, can access patient data. This helps prevent unauthorized access to sensitive information, reducing the risk of data breaches and potential legal liabilities.

Data masking further enhances security by ensuring that even authorized users only see the minimum amount of sensitive data necessary for their job function. For example, a nurse may be able to view a patient's medical history but not their social security number or home address.

Finally, data classification allows the healthcare organization to label sensitive data, such as patient diagnoses or treatment plans, and apply appropriate security measures based on sensitivity. This helps ensure that sensitive data is treated with the appropriate level of security and confidentiality.

By using row-level security, data masking, and data classification in Snowflake, the healthcare organization can better protect their sensitive patient data, maintain regulatory compliance, and build trust with their patients and communities.

Solution

Taking our healthcare example, let's explore how to apply different levels of data protection to sensitive data stored in a Snowflake database. Figure 4-1 is an example CLAIMS table that the finance department uses for billing and insurance claims.

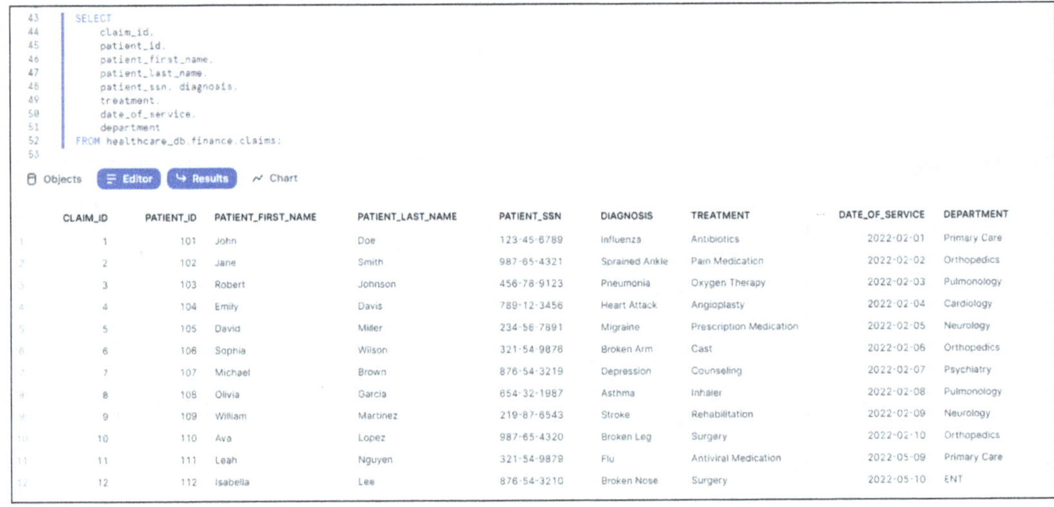

```
SELECT
    claim_id,
    patient_id,
    patient_first_name,
    patient_last_name,
    patient_ssn, diagnosis,
    treatment,
    date_of_service,
    department
FROM healthcare_db.finance.claims;
```


	CLAIM_ID	PATIENT_ID	PATIENT_FIRST_NAME	PATIENT_LAST_NAME	PATIENT_SSN	DIAGNOSIS	TREATMENT	DATE_OF_SERVICE	DEPARTMENT
1	1	101	John	Doe	123-45-6789	Influenza	Antibiotics	2022-02-01	Primary Care
2	2	102	Jane	Smith	987-65-4321	Sprained Ankle	Pain Medication	2022-02-02	Orthopedics
3	3	103	Robert	Johnson	456-78-9123	Pneumonia	Oxygen Therapy	2022-02-03	Pulmonology
4	4	104	Emily	Davis	789-12-3456	Heart Attack	Angioplasty	2022-02-04	Cardiology
5	5	105	David	Miller	234-56-7891	Migraine	Prescription Medication	2022-02-05	Neurology
6	6	106	Sophia	Wilson	321-54-9876	Broken Arm	Cast	2022-02-06	Orthopedics
7	7	107	Michael	Brown	876-54-3219	Depression	Counseling	2022-02-07	Psychiatry
8	8	108	Olivia	Garcia	654-32-1987	Asthma	Inhaler	2022-02-08	Pulmonology
9	9	109	William	Martinez	219-87-6543	Stroke	Rehabilitation	2022-02-09	Neurology
10	10	110	Ava	Lopez	987-65-4320	Broken Leg	Surgery	2022-02-10	Orthopedics
11	11	111	Leah	Nguyen	321-54-9879	Flu	Antiviral Medication	2022-05-09	Primary Care
12	12	112	Isabella	Lee	876-54-3210	Broken Nose	Surgery	2022-05-10	ENT

Figure 4-1. *Snowflake Snowsight Worksheet*

The CLAIMS table contains PII and PHI (example data). Consider two roles used by the finance team—FINANCE_ANALYST and FINANCE_ADMIN—currently have the same data access permissions against the CLAIMS table.

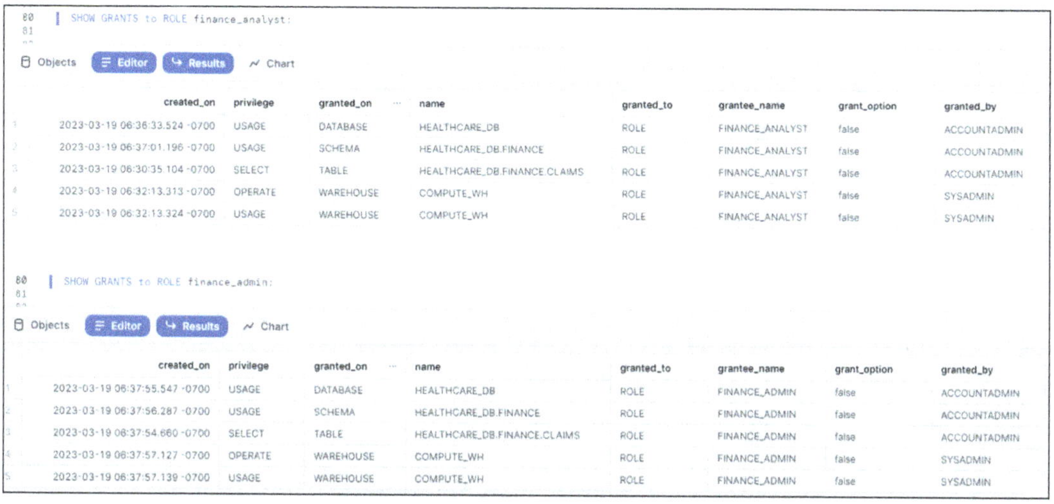

```
SHOW GRANTS to ROLE finance_analyst;
```

Objects Editor Results Chart

	created_on	privilege	granted_on	name	granted_to	grantee_name	grant_option	granted_by
1	2023-03-19 06:36:33.524 -0700	USAGE	DATABASE	HEALTHCARE_DB	ROLE	FINANCE_ANALYST	false	ACCOUNTADMIN
2	2023-03-19 06:37:01.196 -0700	USAGE	SCHEMA	HEALTHCARE_DB.FINANCE	ROLE	FINANCE_ANALYST	false	ACCOUNTADMIN
3	2023-03-19 06:30:35.104 -0700	SELECT	TABLE	HEALTHCARE_DB.FINANCE.CLAIMS	ROLE	FINANCE_ANALYST	false	ACCOUNTADMIN
4	2023-03-19 06:32:13.313 -0700	OPERATE	WAREHOUSE	COMPUTE_WH	ROLE	FINANCE_ANALYST	false	SYSADMIN
5	2023-03-19 06:32:13.324 -0700	USAGE	WAREHOUSE	COMPUTE_WH	ROLE	FINANCE_ANALYST	false	SYSADMIN

```
SHOW GRANTS to ROLE finance_admin;
```

Objects Editor Results Chart

	created_on	privilege	granted_on	name	granted_to	grantee_name	grant_option	granted_by
1	2023-03-19 06:37:55.547 -0700	USAGE	DATABASE	HEALTHCARE_DB	ROLE	FINANCE_ADMIN	false	ACCOUNTADMIN
2	2023-03-19 06:37:56.287 -0700	USAGE	SCHEMA	HEALTHCARE_DB.FINANCE	ROLE	FINANCE_ADMIN	false	ACCOUNTADMIN
3	2023-03-19 06:37:54.660 -0700	SELECT	TABLE	HEALTHCARE_DB.FINANCE.CLAIMS	ROLE	FINANCE_ADMIN	false	ACCOUNTADMIN
4	2023-03-19 06:37:57.127 -0700	OPERATE	WAREHOUSE	COMPUTE_WH	ROLE	FINANCE_ADMIN	false	SYSADMIN
5	2023-03-19 06:37:57.139 -0700	USAGE	WAREHOUSE	COMPUTE_WH	ROLE	FINANCE_ADMIN	false	SYSADMIN

Figure 4-2. *Snowflake Snowsight Worksheet*

The goal is to leverage Snowflake functionality to harden data security and leverage the principle of least privilege.

Adding Masking Polices

The analyst team runs monthly reports and does not need access to particular PII and PHI data. Let's build column-level security by adding masking policies to each patient's full name, SSN, diagnosis, and treatment.

Note Use appropriate role to build objects and assign grants for your organization.

---

1. Create a generic masking policy to mask string values.

```
CREATE OR REPLACE MASKING POLICY masked_string AS (val string)
returns string ->
   CASE
     WHEN current_role() in ('FINANCE_ADMIN') THEN val
     ELSE '********'
   END;
```

2. Add generic masking policy to patient_first_name, patient_last_name, diagnosis, and treatment columns.

```
alter table healthcare_db.finance.claims modify column patient_
first_name set masking policy masked_string;
alter table healthcare_db.finance.claims modify column patient_
last_name set masking policy masked_string;
alter table healthcare_db.finance.claims modify column diagnosis
set masking policy masked_string;
alter table healthcare_db.finance.claims modify column treatment
set masking policy masked_string;
```

3. Create an SSN masking policy to mask SSN or return the last four depending on the role.

```
CREATE OR REPLACE MASKING POLICY masked_ssn AS (val string)
returns string ->
 CASE
   WHEN current_role() in ('FINANCE_ADMIN') THEN REGEXP_
   REPLACE(val,'((\\d{3})-(\\d{2})-(\\d{4}))', '***-**-\\4')
   ELSE '*********'
 END;
```

4. Add SSN masking policy to PATIENT_SSN column.

```
alter table healthcare_db.finance.claims modify column patient_ssn
set masking policy masked_ssn;
```

5. Test analyst role.

```
use role finance_analyst;
select patient_first_name, patient_last_name, patient_ssn,
diagnosis, treatment from healthcare_db.finance.claims;
```

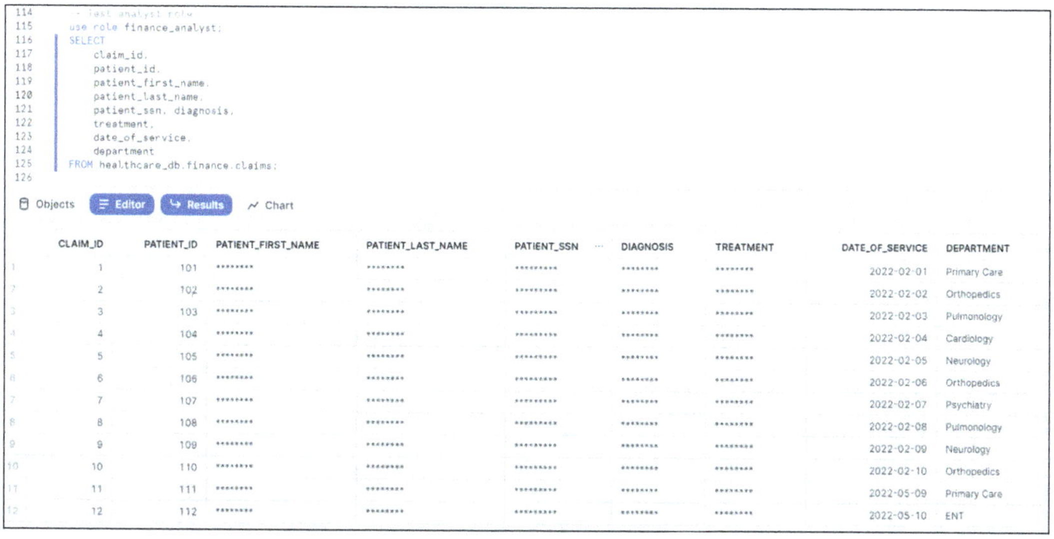

```
-- Test analyst role
use role finance_analyst;
SELECT
    claim_id,
    patient_id,
    patient_first_name,
    patient_last_name,
    patient_ssn, diagnosis,
    treatment,
    date_of_service,
    department
FROM healthcare_db.finance.claims;
```

Objects | Editor | Results | Chart

	CLAIM_ID	PATIENT_ID	PATIENT_FIRST_NAME	PATIENT_LAST_NAME	PATIENT_SSN	DIAGNOSIS	TREATMENT	DATE_OF_SERVICE	DEPARTMENT
1	1	101	********	********	*********	********	********	2022-02-01	Primary Care
2	2	102	********	********	*********	********	********	2022-02-02	Orthopedics
3	3	103	********	********	*********	********	********	2022-02-03	Pulmonology
4	4	104	********	********	*********	********	********	2022-02-04	Cardiology
5	5	105	********	********	*********	********	********	2022-02-05	Neurology
6	6	106	********	********	*********	********	********	2022-02-06	Orthopedics
7	7	107	********	********	*********	********	********	2022-02-07	Psychiatry
8	8	108	********	********	*********	********	********	2022-02-08	Pulmonology
9	9	109	********	********	*********	********	********	2022-02-09	Neurology
10	10	110	********	********	*********	********	********	2022-02-10	Orthopedics
11	11	111	********	********	*********	********	********	2022-05-09	Primary Care
12	12	112	********	********	*********	********	********	2022-05-10	ENT

***Figure 4-3.** Snowflake Snowsight Worksheet*

6. Test the admin role as seen in Figure 4-4.

```
use role finance_admin;
select patient_first_name, patient_last_name, patient_ssn,
diagnosis, treatment from healthcare_db.finance.claims;
```

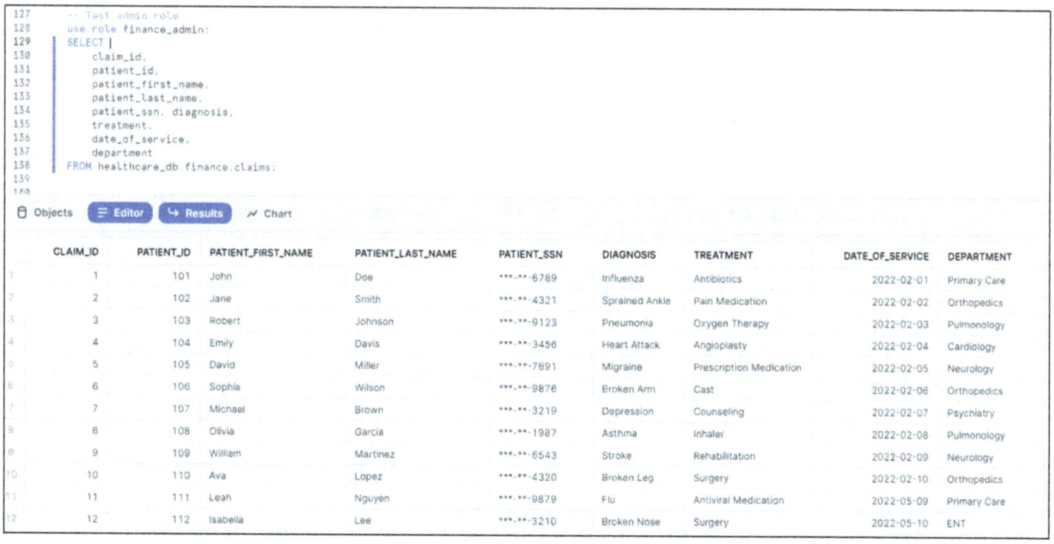

```
-- Test admin role
use role finance_admin;
SELECT
    claim_id,
    patient_id,
    patient_first_name,
    patient_last_name,
    patient_ssn, diagnosis,
    treatment,
    date_of_service,
    department
FROM healthcare_db.finance.claims;
```

Objects | Editor | Results | Chart

	CLAIM_ID	PATIENT_ID	PATIENT_FIRST_NAME	PATIENT_LAST_NAME	PATIENT_SSN	DIAGNOSIS	TREATMENT	DATE_OF_SERVICE	DEPARTMENT
1	1	101	John	Doe	***-**-6789	Influenza	Antibiotics	2022-02-01	Primary Care
2	2	102	Jane	Smith	***-**-4321	Sprained Ankle	Pain Medication	2022-02-02	Orthopedics
3	3	103	Robert	Johnson	***-**-9123	Pneumonia	Oxygen Therapy	2022-02-03	Pulmonology
4	4	104	Emily	Davis	***-**-3456	Heart Attack	Angioplasty	2022-02-04	Cardiology
5	5	105	David	Miller	***-**-7891	Migraine	Prescription Medication	2022-02-05	Neurology
6	6	106	Sophia	Wilson	***-**-9876	Broken Arm	Cast	2022-02-06	Orthopedics
7	7	107	Michael	Brown	***-**-3219	Depression	Counseling	2022-02-07	Psychiatry
8	8	108	Olivia	Garcia	***-**-1987	Asthma	Inhaler	2022-02-08	Pulmonology
9	9	109	William	Martinez	***-**-6543	Stroke	Rehabilitation	2022-02-09	Neurology
10	10	110	Ava	Lopez	***-**-4320	Broken Leg	Surgery	2022-02-10	Orthopedics
11	11	111	Leah	Nguyen	***-**-9879	Flu	Antiviral Medication	2022-05-09	Primary Care
12	12	112	Isabella	Lee	***-**-3210	Broken Nose	Surgery	2022-05-10	ENT

***Figure 4-4.** Snowflake Snowsight Worksheet*

Although we have taken the initial step of implementing Snowflake masking policies and column-level security for our tables, there is still room for improvement in data security.

Implementing Row-Level Security

We should consider implementing row-level security to further enhance our security measures within Snowflake.

For instance, let's consider Jane Doe, the finance controller for the pulmonology department, who has both FINANCE_ADMIN and FINANCE_ANALYST roles, giving her access to PII and PHI data. However, Jane does not require access to data from other departments. To address this, we can create a row access policy that utilizes the user profile to identify the department and restricts access accordingly.

1. Create a cross-reference table for users and departments.

```
create table department_user_xref (
   user varchar,
   department varchar,
   insert_date datetime,
   update_date datetime
);
```

2. Seed the newly created table with known users.

```
INSERT INTO department_user_xref (user, department, insert_date,
update_date)
VALUES
('JDOE', 'Pulmonology', current_timestamp(), current_timestamp());
```

3. Create the row-level access policy.

```
create or replace row access policy dept_policy as (department_
value varchar) returns boolean ->
   exists (select 1
           from healthcare_db.finance.department_user_xref
           where user = current_user()
             and department = department_value);
```

4. Add the row policy to the table.

```
alter table healthcare_db.finance.claims add row access policy
dept_policy on (department);
```

5. Test the row access policy and masking policies together.

```
use role finance_analyst;
select * from healthcare_db.finance.claims;
use role finance_admin;
select * from department_user_xref;
```

Now that functional row- and column-level security is available, let's dive into Snowflake data classification.

A Closer Look at the Data Classification

This can assist in automating processes such as masking policies and access revocation. Please note that the following examples are for demonstration purposes only. If these features can provide value to your organization, it is recommended to understand the necessary steps for productionalizing and automating them in your environment.

The following describes the manual process for setting classification system tags. However, if classification is implemented in a production environment, you would expect to see the entire process automated using Snowflake-provided stored procedures such as EXTRACT_SEMANTIC_CATEGORIES and ASSOCIATE_SEMANTIC_CATEGORY_TAGS.

1. Identify and apply privacy classification system tags.

```
select system$get_tag_allowed_values('snowflake.core.privacy_
category');

ALTER TABLE healthcare_db.finance.claims
 SET TAG SNOWFLAKE.CORE.PRIVACY_CATEGORY='SENSITIVE';

ALTER TABLE healthcare_db.finance.claims
 MODIFY COLUMN patient_ssn
 SET TAG SNOWFLAKE.CORE.PRIVACY_CATEGORY = 'IDENTIFIER';
```

2. Identify and apply semantic classification system tags.

```
select system$get_tag_allowed_values('snowflake.core.semantic_
category');

ALTER TABLE healthcare_db.finance.claims
 MODIFY COLUMN patient_first_name
 SET TAG SNOWFLAKE.CORE.SEMANTIC_CATEGORY='NAME';

ALTER TABLE healthcare_db.finance.claims
 MODIFY COLUMN patient_last_name
 SET TAG SNOWFLAKE.CORE.SEMANTIC_CATEGORY='NAME';

ALTER TABLE healthcare_db.finance.claims
 MODIFY COLUMN patient_ssn
 SET TAG SNOWFLAKE.CORE.SEMANTIC_CATEGORY='US_SSN';
```

3. Validate the classification tags have been applied.

```
SELECT object_database, object_schema, object_name, column_name,
level, tag_name, tag_value
from table(healthcare_db.information_schema.tag_references_all_
columns('claims','table'))
order by tag_name;
```

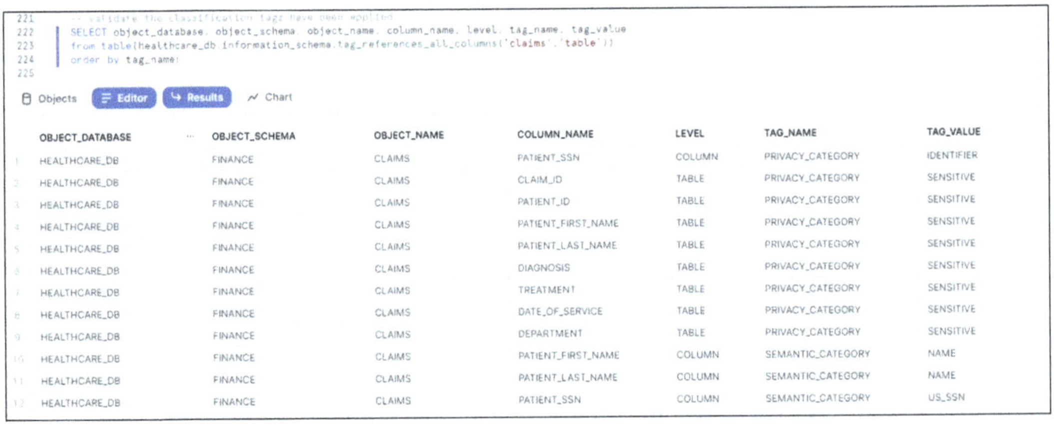

```
-- validate the classification tags have been applied
SELECT object_database, object_schema, object_name, column_name, level, tag_name, tag_value
from table(healthcare_db.information_schema.tag_references_all_columns('claims','table'))
order by tag_name;
```

Objects Editor Results Chart

	OBJECT_DATABASE	OBJECT_SCHEMA	OBJECT_NAME	COLUMN_NAME	LEVEL	TAG_NAME	TAG_VALUE
1	HEALTHCARE_DB	FINANCE	CLAIMS	PATIENT_SSN	COLUMN	PRIVACY_CATEGORY	IDENTIFIER
2	HEALTHCARE_DB	FINANCE	CLAIMS	CLAIM_ID	TABLE	PRIVACY_CATEGORY	SENSITIVE
3	HEALTHCARE_DB	FINANCE	CLAIMS	PATIENT_ID	TABLE	PRIVACY_CATEGORY	SENSITIVE
4	HEALTHCARE_DB	FINANCE	CLAIMS	PATIENT_FIRST_NAME	TABLE	PRIVACY_CATEGORY	SENSITIVE
5	HEALTHCARE_DB	FINANCE	CLAIMS	PATIENT_LAST_NAME	TABLE	PRIVACY_CATEGORY	SENSITIVE
6	HEALTHCARE_DB	FINANCE	CLAIMS	DIAGNOSIS	TABLE	PRIVACY_CATEGORY	SENSITIVE
7	HEALTHCARE_DB	FINANCE	CLAIMS	TREATMENT	TABLE	PRIVACY_CATEGORY	SENSITIVE
8	HEALTHCARE_DB	FINANCE	CLAIMS	DATE_OF_SERVICE	TABLE	PRIVACY_CATEGORY	SENSITIVE
9	HEALTHCARE_DB	FINANCE	CLAIMS	DEPARTMENT	TABLE	PRIVACY_CATEGORY	SENSITIVE
10	HEALTHCARE_DB	FINANCE	CLAIMS	PATIENT_FIRST_NAME	COLUMN	SEMANTIC_CATEGORY	NAME
11	HEALTHCARE_DB	FINANCE	CLAIMS	PATIENT_LAST_NAME	COLUMN	SEMANTIC_CATEGORY	NAME
12	HEALTHCARE_DB	FINANCE	CLAIMS	PATIENT_SSN	COLUMN	SEMANTIC_CATEGORY	US_SSN

Figure 4-5. *Snowflake Snowsight Worksheet*

Additional Snowflake Documentation

https://docs.snowflake.com/en/user-guide/security-column-ddm

https://docs.snowflake.com/en/user-guide/governance-classify-using

https://docs.snowflake.com/en/user-guide/security-column-intro

https://docs.snowflake.com/en/user-guide/security-row-intro

Recipe 4-4. Data Encryption

Data encryption is converting plain text into ciphertext, which is a scrambled version of the original text that can only be read by someone with the decryption key. Encryption is an important tool for protecting sensitive data, particularly in modern society, where data collection, storage, and transmission is ubiquitous.

One of the primary reasons why encryption is important is because it helps to ensure the confidentiality of sensitive data. For example, financial institutions use encryption to protect customer data, such as credit card numbers and bank account information. If this data were to be intercepted by an unauthorized party, it could be used for fraudulent activities or identity theft. Encryption helps prevent this by ensuring that authorized parties can only read the data with the appropriate decryption key.

Encryption is also important for protecting the privacy of individuals. With the widespread collection of data by governments, businesses, and other organizations, there is a risk that personal information could be accessed or misused. Encryption helps to mitigate this risk by ensuring that personal data is protected from unauthorized access or disclosure.

In addition to protecting data confidentiality and privacy, encryption is important for ensuring data integrity. For example, digital signatures use encryption to ensure that a message or document has not been tampered with during transmission. This helps to prevent fraud and ensures that the data remains accurate and reliable.

Another important application of encryption is in securing communications. Encryption is commonly used in email, messaging apps, and other forms of digital communication to protect the privacy of conversations and prevent unauthorized access. This is particularly important for individuals and organizations that need to communicate sensitive information, such as lawyers, journalists, and human rights activists.

Overall, encryption is a crucial tool for protecting sensitive data, ensuring data privacy and confidentiality, and maintaining data integrity. As the collection and transmission of data continue to expand in modern society, encryption becomes increasingly important for ensuring the security and privacy of individuals and organizations.

Problem

Consider an e-commerce organization that is building a new data platform to store customer data, such as purchase history, payment details, and shipping addresses. The organization recognizes that this data is highly sensitive and must be protected from potential security breaches.

To determine the required level of data encryption in the new data platform, the organization must consider various factors, such as the type of data being stored, the potential impact of a data breach, and the compliance requirements that apply to their business.

For example, if the organization operates in a region with strict data privacy regulations, such as the EU's GDPR, it may need to implement a higher level of encryption to comply with these regulations.

By identifying the required level of data encryption in the new data platform, the organization can ensure that their sensitive data is protected from potential security breaches, comply with applicable regulations, and maintain the trust of their customers.

Solution

Now that the requirements have been identified, our e-commerce organization has determined that Snowflake using Business Critical Edition meets all the regulatory requirements. However, due to the rise in social engineering and data breaches, a company standard requires the use of Tri-Secret Secured.

Key rotation and rekeying serve distinct purposes in key management. Key rotation transitions a key from its active state to a retired state after 30 days and is automatic in Snowflake. While rekeying transitions a key from its retired state to being permanently destroyed and is disabled by default.

When periodic rekeying is activated, Snowflake triggers the creation of a new encryption key and re-encryption of all previously protected data using the retired key if it has been inactive for over a year. The newly generated key then takes over as the decryption key for the table data.

To enable rekeying, run the following as ACCOUNTADMIN role.

```
ALTER ACCOUNT SET PERIODIC_DATA_REKEYING = true;
```

Note There are charges from the additional storage for Fail-safe protection of rekeyed data files.

Key rotation and rekeying are standard practices that provide a high level of security for most Snowflake accounts. However, in some cases, businesses may choose to enhance their security measures by implementing Tri-Secret Secure, which combines a Snowflake-managed key and a customer managed key in the cloud provider platform.

It's important to note that Snowflake does not support key rotation for customer managed keys and does not recommend automatic rekeying due to the risk of data decryption issues. If you're using Tri-Secret Secure, you should disable PERIODIC_DATA_REKEYING.

The composite master key resulting from Tri-Secret Secure acts as the account master key and wraps all keys in the hierarchy. If the customer managed key is revoked, Snowflake cannot decrypt the data, providing an additional layer of security and control. This approach, combined with Snowflake's integrated user authentication, creates three levels of data protection for your data.

Using Amazon Web Services' Key Management Service

This example uses the Amazon Web Services (AWS) Key Management Service (KMS) to create a customer managed key.

1. Log in to the organization's AWS account and navigate to KMS customer managed keys.

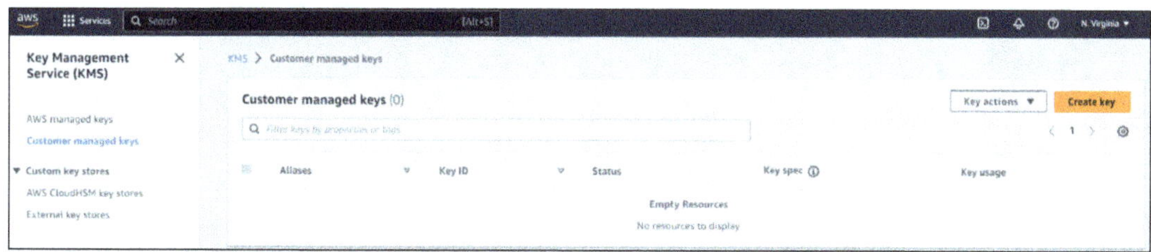

Figure 4-6. *AWS Key Management Service (KMS) dashboard*

2. Create a new key with the following configuration.

   a. Symmetric key type

   b. Encrypt and Decrypt key usage

   c. Default Advanced options

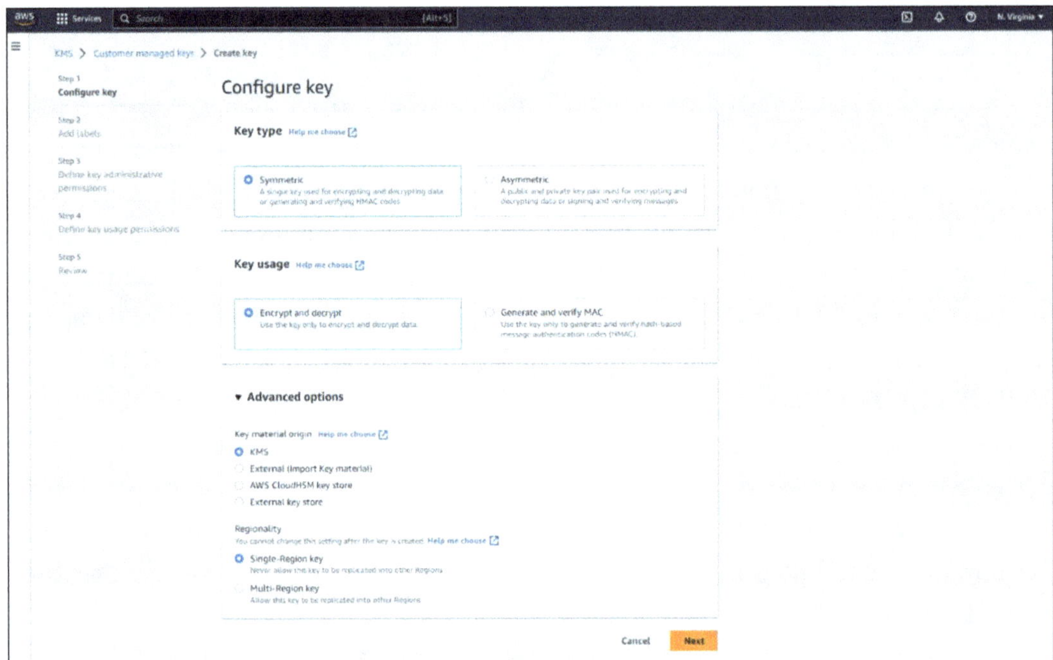

Figure 4-7. *AWS Key Management Service (KMS) key creation wizard*

3. Assign the alias, description, and tags to CMK.

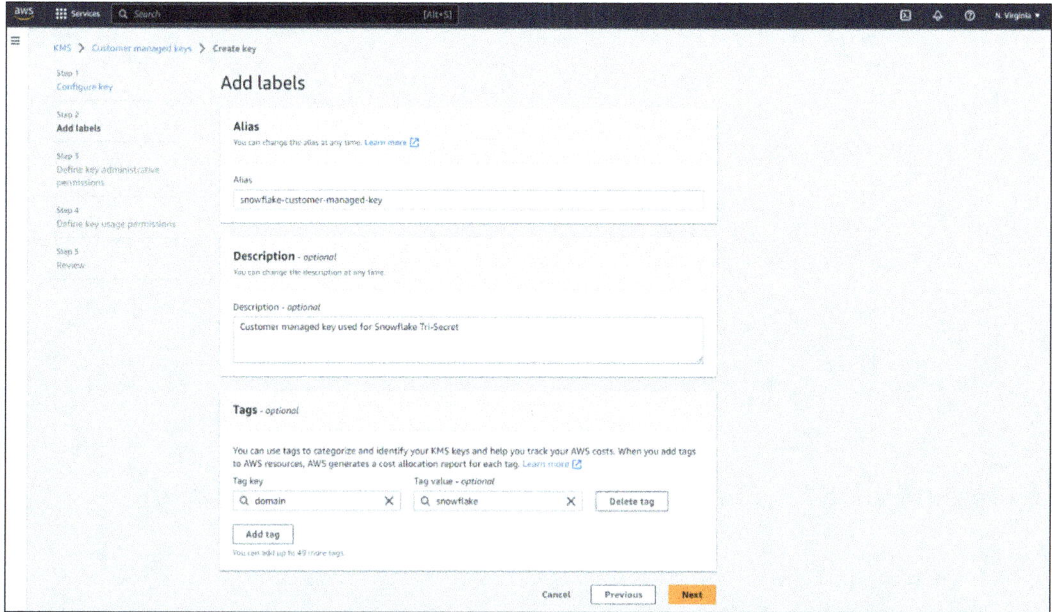

Figure 4-8. *AWS Key Management Service (KMS) key creation wizard*

4. Assign key administrators.

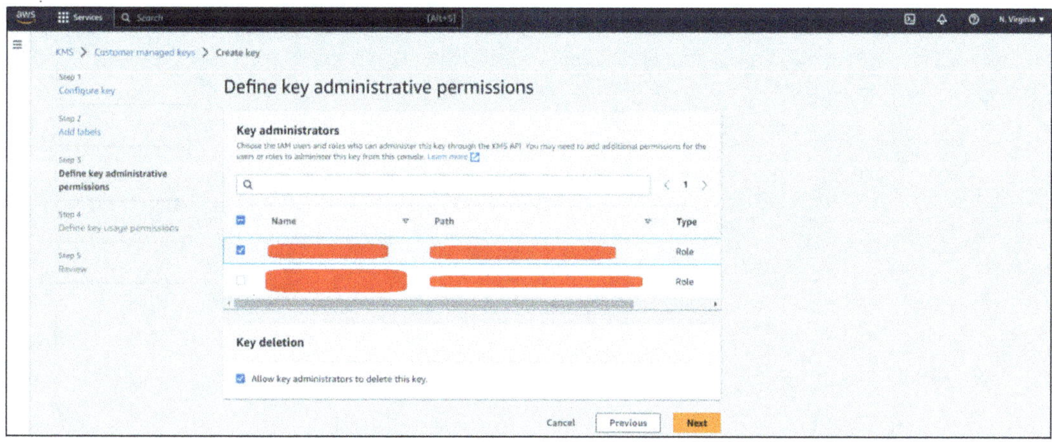

Figure 4-9. *AWS Key Management Service (KMS) key creation wizard*

5. Define usage permissions.

Figure 4-10. *AWS Key Management Service (KMS) key creation wizard*

6. Create the CMK.

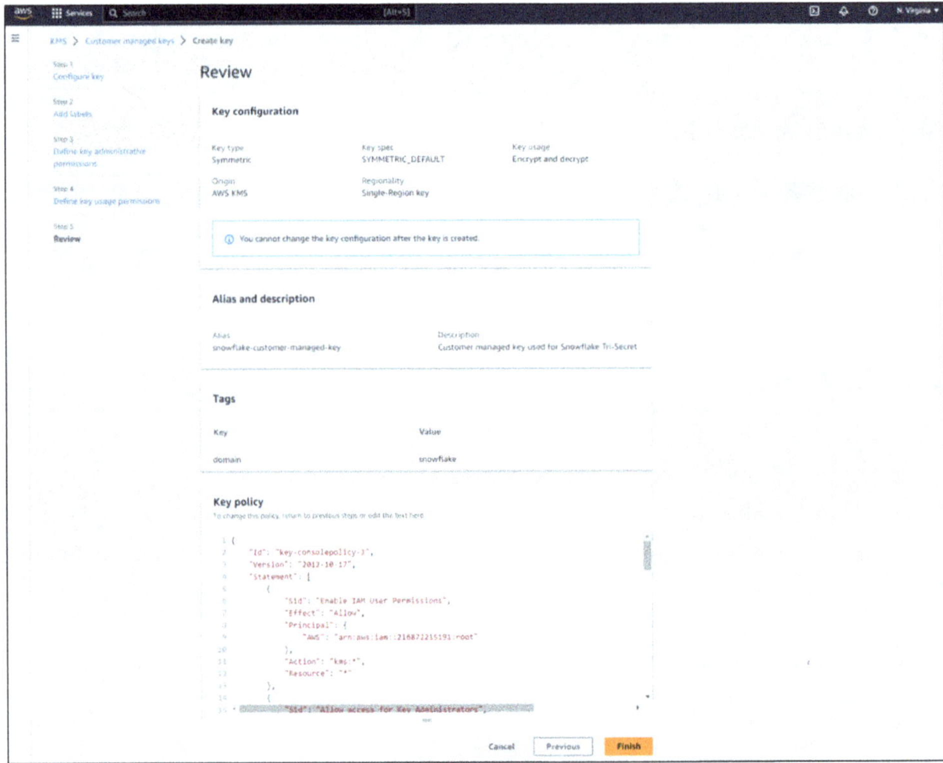

Figure 4-11. *AWS Key Management Service (KMS) key creation wizard*

To enable Tri-Secret Secure, you must have a Business Critical Edition or higher Snowflake account, and you need to create a support case requesting the change. In the support case, you are asked to provide the Amazon Resource Name (ARN) of the Customer Master Key (CMK) in AWS. Therefore, make sure to have this information available when submitting your request.

7. Get the ARN value of the CMK from KMS.

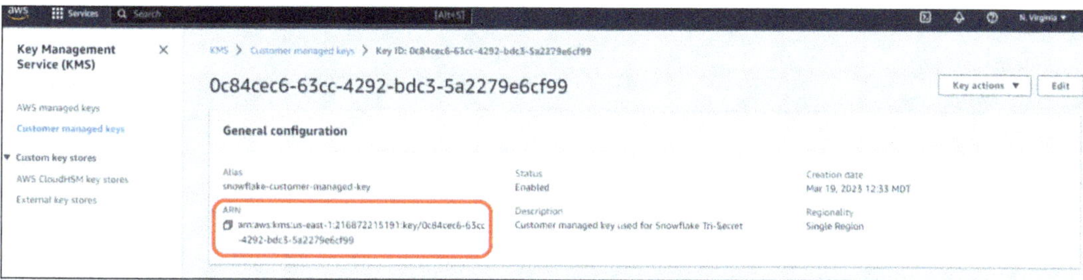

Figure 4-12. *AWS Key Management Service (KMS)*

Before initiating Tri-Secret Secure for your Snowflake account, it's important to carefully assess your responsibility for protecting your customer managed key.

Additional Snowflake Documentation

`https://docs.snowflake.com/en/user-guide/security-encryption-manage`

CHAPTER 5

Handling Near and Real-Time Data

The world of big data is no stranger to massive volumes of information. But when that data needs to be processed and analyzed close to real time, a whole new set of challenges emerge.

Fortunately, there are several solutions available to tackle these challenges.

- **Stream processing engines**: Tools like Apache Spark Streaming and Apache Flink are built specifically to handle high-velocity data streams. They process data as it arrives, enabling real-time analytics.
- **Message queues**: Services like Apache Kafka and RabbitMQ are examples of message queues that act as buffers for incoming data streams.
- **Real-time data warehousing**: Cloud-based data warehouses like Google BigQuery and Snowflake offer near real-time capabilities.

The solutions in this chapter focus on various options provided natively by Snowflake to handle and ingest near real-time data.

Recipe 5-1. Data Loading Using Snowpipe

Snowflake provides the COPY INTO command and a native object called Snowpipe. Before we delve into different problems and their solutions, let's look at what they are and how they differ.

The COPY INTO command enables loading batches of data available in external cloud storage or an internal stage within Snowflake. In fact, you could use COPY INTO for most of your use cases of data ingestion, provided you consider these aspects of the COPY command.

D. Dayton and J. Eipe, *Snowflake Recipes*, https://doi.org/10.1007/979-8-8688-0938-5_5

Snowpipe is an automated and serverless solution provided by Snowflake so that you don't have to worry about running and managing any extraction and loading jobs or your compute warehouses.

- Snowpipe is designed to load small volumes of data (i.e., mini-batches).
- Snowpipe loads data within minutes after files are added to a stage and submitted for ingestion.
- Snowpipe is an excellent choice when you have continuous streams of data, and you need Snowflake to manage the ingestion (serverless). This means you do not manage the compute warehouses.
- The COPY option enables loading batches of data from files already available in cloud storage or copying data files from a local machine to an internal Snowflake stage.
- COPY provides file-level transaction granularity and hence it provides different ON_ERROR handling options.
    - CONTINUE continues loading the file even if there are errors.
    - SKIP_FILE skips only the files with errors.
    - ABORT_STATEMENT aborts the load operation if an error is encountered in any of the data files.
- By default, the ON_ERROR parameter for COPY is set to ABORT_STATEMENT.
- When you use COPY INTO, it needs to be programmatically controlled or client-driven, whereas SNOWPIPE is a Snowflake-managed serverless feature.
- When you use COPY INTO, compute is determined by the user's selection of warehouse for the operation. But this means you must carefully consider choosing the appropriate warehouse size. The most critical aspect is the degree of parallelism because each (warehouse) node can handle eight threads in parallel. Therefore, if you have 32 files, a medium warehouse with four nodes can process all of them simultaneously, running 32 threads in parallel.

- Not all options that work with COPY INTO work with Snowpipe.
  - Since COPY INTO is more user/client code controlled, you could use the VALIDATION_MODE parameter to validate before loading the files using COPY INTO. This acts like a pre-load test, and files are not loaded even if the validation succeeds.
  - COPY INTO default for the ON_ERROR parameter is ABORT_STATEMEN, and for Snowpipe, it is SKIP_FILE. Snowpipe does not support ABORT_STATEMENT.
  - COPY INTO doesn't load files already loaded into the table but you can force it by using the FORCE load parameter. Snowpipe doesn't support FORCE load.
  - COPY INTO... PURGE = TRUE can be used to automatically delete the loaded files from the stage, or they can be deleted later using the REMOVE command. When Snowpipe is used, you do not have the option to purge during the copy into operation but can only use REMOVE later to delete the files.
  - You cannot pick files using the FILES parameter in Snowpipe, but it is possible in the COPY command.
  - You cannot set the SIZE_LIMIT parameter in Snowpipe, but it is possible in the COPY command.
- You could use the VALIDATE table function to validate files loaded as part of the COPY INTO command. This acts like a post-load test.

Problem

Consider a scenario when an organization's data team needs to provide an analytical solution that is refreshed multiple times a day and needs to reflect the latest data with only a lag of 5 to 10 minutes.

Assume that the data is available in AWS S3 storage, and the team needs to ingest it into Snowflake once it is available in S3.

Let's assume the organization has no S3 event notification set up, and everything must be done from scratch.

The following is a set of prerequisite steps for demonstrating this solution.

Prerequisites

1. Create a storage bucket named sfdemo-landing in S3 with the default SSE encryption.

2. Configure access permissions in AWS IAM for Snowflake to access the new bucket.

```
{
    "Version": "2012-10-17",
    "Statement": [
        {
            "Effect": "Allow",
            "Action": [
              "s3:GetObject",
              "s3:GetObjectVersion"
            ],
            "Resource": "arn:aws:s3:::sfdemo-landing/*"
        },
        {
            "Effect": "Allow",
            "Action": [
                "s3:ListBucket",
                "s3:GetBucketLocation"
            ],
            "Resource": "arn:aws:s3:::sfdemo-landing",
            "Condition": {
                "StringLike": {
                    "s3:prefix": [
                        "*"
                    ]
```

```
                }
            }
        }
    ]
}
```

Refer to Recipe 2-1 in Chapter 1 for the steps to create an IAM policy and attach an AWS role to it.

3. Modify an existing or create a new role in AWS IAM.
   We used the existing role (snowflake_vx15608_role) we created in Chapter 2 and attached the new policy we created in the previous step to this role.

4. Update existing storage integration or create a new storage integration.
   Since we used the same role and storage integration from Chapter 2, we updated this storage integration.

```
alter STORAGE INTEGRATION sfdemo_storage_intg
set STORAGE_ALLOWED_LOCATIONS = (
    <add all existing s3 buckets>,
    's3://sfdemo-landing/');
```

5. If you have created a new role in step 3, you need to update the role with the IAM user name and external ID generated by the storage integration. (Refer to Chapter 2.)

6. Test if the integration works fine by creating a folder in S3 and a stage.

   a) Add a folder called mini-batch tos the S3 bucket.

   b) Create a Snowflake stage.

   ```
   create or replace stage sfdemo_landing_minibatch_stg
   storage_integration = sfdemo_storage_intg
   url = 's3://sfdemo-landing/mini-batch/';
   ```

   c) Test the stage.

   ```
   list @sfdemo_landing_minibatch_stg;
   ```

Solution

Snowpipe relies on event notification and distribution, such as AWS SQS or SNS, Azure Event Grid, or GCP Pub/Sub, based on the cloud vendor that you are using.

This setup requires corresponding privileges configured in the cloud account to deliver event notifications from the source bucket to Snowpipe.

But before creating event notifications, ensure you do not have an existing event notification set for the S3 bucket.

Figure 5-1 shows various components and how they are related to each other.

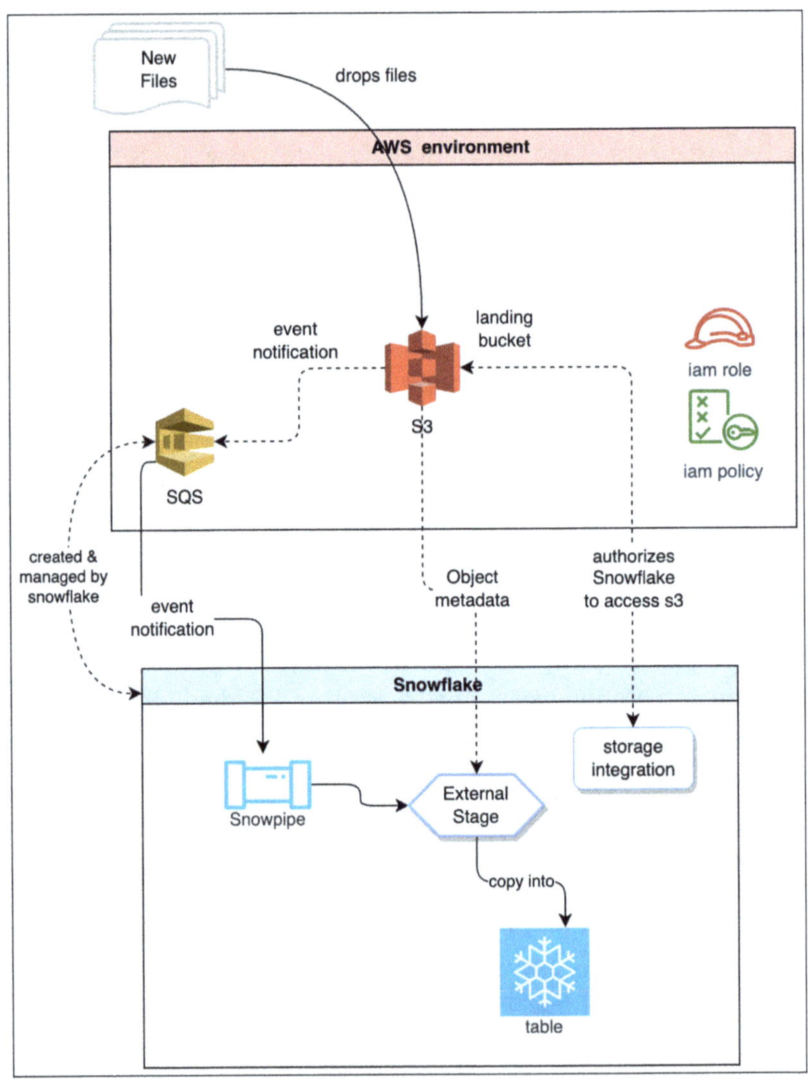

***Figure 5-1.** Snowpipe Ingestion from AWS s3*

Step 1: Create an external stage.

Create an external stage if you haven't created one already. We created a stage named sfdemo_landing_minibatch_stg as part of the prerequisite steps.

Step 2: Confirm file formats.

Make sure you have file formats defined (or use default file formats) for your data.

Step 3: Create a table.

If you haven't already done so, create a table for consuming the piped data.

```
create or replace TABLE RAW_CUSTOMER (
    C_CUSTKEY number(38,0),
    C_NAME varchar(16777216),
    C_ADDRESS varchar(16777216),
    C_NATIONKEY number(38,0),
    C_PHONE varchar(16777216),
    C_ACCTBAL float,
    C_MKTSEGMENT varchar(16777216),
    C_COMMENT varchar(16777216)
);
```

Step 4: Create a pipe.

Create a pipe using the following template.

```
create PIPE <name>
auto_ingest = true
  as <copy_statement>
```

For this specific demo, we planned to use the data set customer.csv used in the previous examples and with the default file format which works fine.

```
create pipe sfdemo_minibatch
auto_ingest=true as
    copy into RAW_CUSTOMER(
        C_CUSTKEY
```

```
        , C_NAME
        , C_ADDRESS
        , C_NATIONKEY
        , C_PHONE
        , C_ACCTBAL
        , C_MKTSEGMENT
        , C_COMMENT
    )
    from (
    select
        $1::number
        , $2::varchar
        , $3:: varchar
        , $4::number
        , $5:: varchar
        , $6:: float
        , $7:: varchar
        , $8:: varchar
    from
        @sfdemo_landing_minibatch_stg
    )
   file_format =    (
        type = CSV
        field_delimiter =  ','
        record_delimiter = '\n'
        field_optionally_enclosed_by= '"'
        skip_header = 1
);
```

Step 5: Configure the S3 bucket to use the event notification.

Notice that Snowflake created the AWS SQS queue on your behalf, and you can retrieve the Amazon Resource Name (ARN) of the SQS queue by running a `desc pipe` or `show pipe` command.

1. Look for the Notification Channel column in the show pipes output. Save the value of the ARN to a convenient location.
2. In the Buckets list, choose the name of the bucket that you want to enable events.
3. Choose Properties.
4. Navigate to the Event Notifications section and choose Create event notification.
   a) Provide an appropriate name and suffix (if required).
   b) Choose "All object create events" from the Event Types.
   c) In the destination section, select SQS Queue from the list.
   d) Paste the SQS queue name from the SHOW PIPES output.

Creating new pipes doesn't guarantee that new SQS queues are created. Snowflake tries to reuse the SQS queues as much as possible. If you create pipes with overlapping directory paths, you could get the same data delivered to both tables.

Step 6: Test the ingestion.

Check if the pipe is running fine. Running the PIPE_STATUS should show that the pipe is running.

```
select SYSTEM$PIPE_STATUS( 'SFDEMO_MINIBATCH' );
```

Now upload the customer_sample.csv file into the S3 location (s3://sfdemo-landing/mini-batch/), wait for few seconds and observe the copy history by running the following command.

```
select * from table(information_schema.copy_history(table_name=>'RAW_
CUSTOMER', start_time=> dateadd(hours, -1, current_timestamp())));
```

This command should show the files that were recently loaded into the table along with the stage location, last load time, row count, and so forth.

Problem

Consider the scenario similar to the previous problem statement when an organization's data team needs to ingest this data into Snowflake as soon as it is available in AWS S3 such that the lag is only 5 to 10 minutes but has S3 event notifications (SNS) already setup that was used with other solutions in AWS such as SQS queues and Lambda endpoints.

Assuming that the organization does have an SNS configured for its S3 bucket, the prerequisites for the demonstration are pretty straightforward.

Prerequisites

You need to have the SNS topic created and configured and access control granted to Snowflake to subscribe to the SNS topic.

For this demonstration, create a folder within the current S3 bucket (sfdemo-landing) called mini-batch-custom. Figure 5-2 shows the folders created within the bucket.

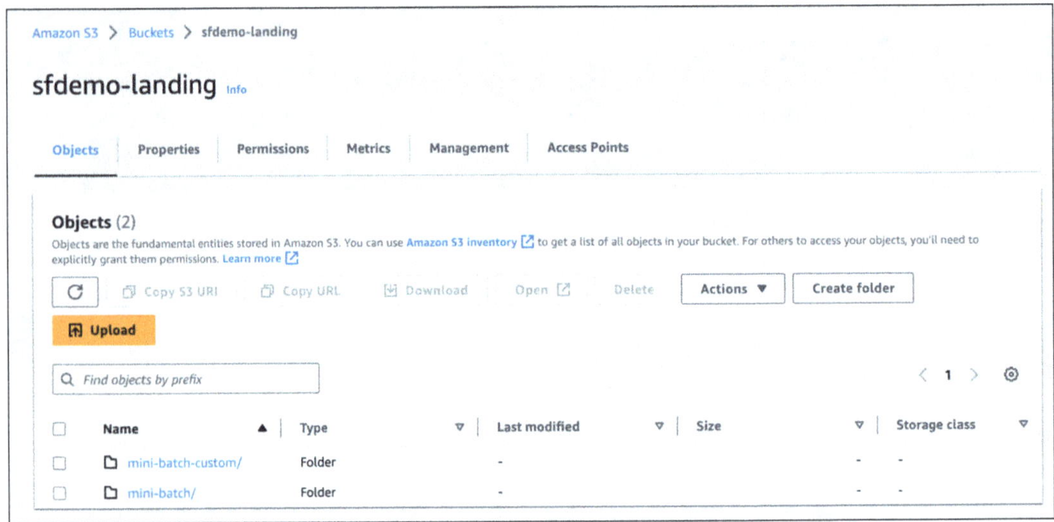

Figure 5-2. s3 buckets

The SNS topic should be created in the same region as the S3 bucket.

For this demonstration, we created an SNS topic named sfdemo-minibatchcustom-topic. Figure 5-3 shows the first step in creating an SNS Topic in AWS.

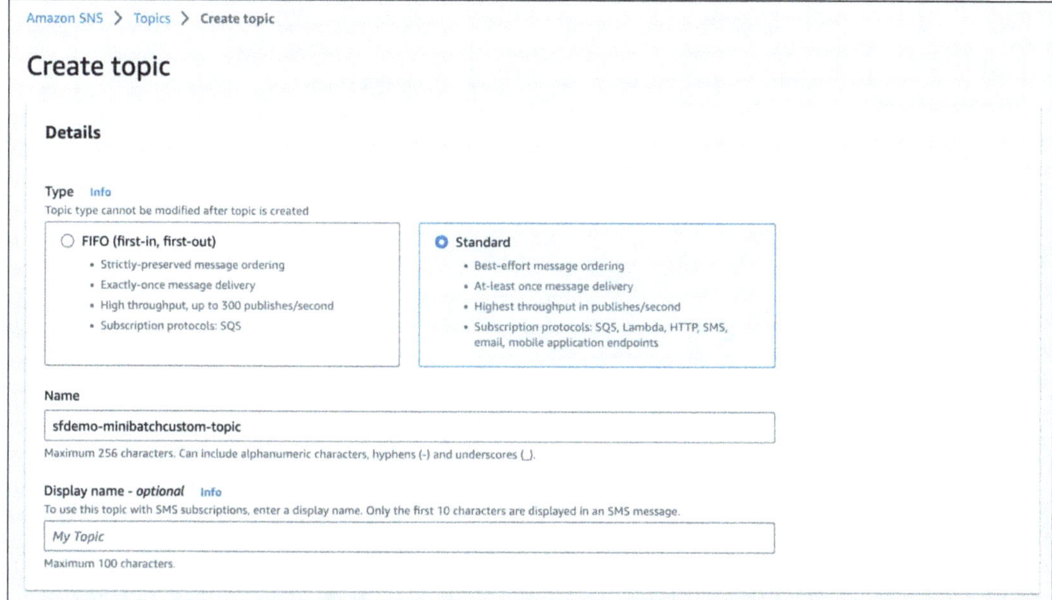

Figure 5-3. *Creating SNS topic*

Once the topic has been created, you must add the event notification in the S3 properties as shown in Figure 5-4.

General configuration

Event name

sfdemo_landing_minibatch_events_custom

Event name can contain up to 255 characters.

Prefix - *optional*

Limit the notifications to objects with key starting with specified characters.

mini-batch-custom

Suffix - *optional*

Limit the notifications to objects with key ending with specified characters.

.jpg

Event types

Specify at least one event for which you want to receive notifications. For each group, you can choose an event type for all events, or you can choose one or more individual events.

Object creation

- [x] All object create events
  s3:ObjectCreated:*
- [ ] Put
  s3:ObjectCreated:Put
- [ ] Post
  s3:ObjectCreated:Post
- [ ] Copy
  s3:ObjectCreated:Copy
- [ ] Multipart upload completed
  s3:ObjectCreated:CompleteMultipartUpload

Figure 5-4. *Creating event notification*

Make sure to select the SNS topic as your destination.

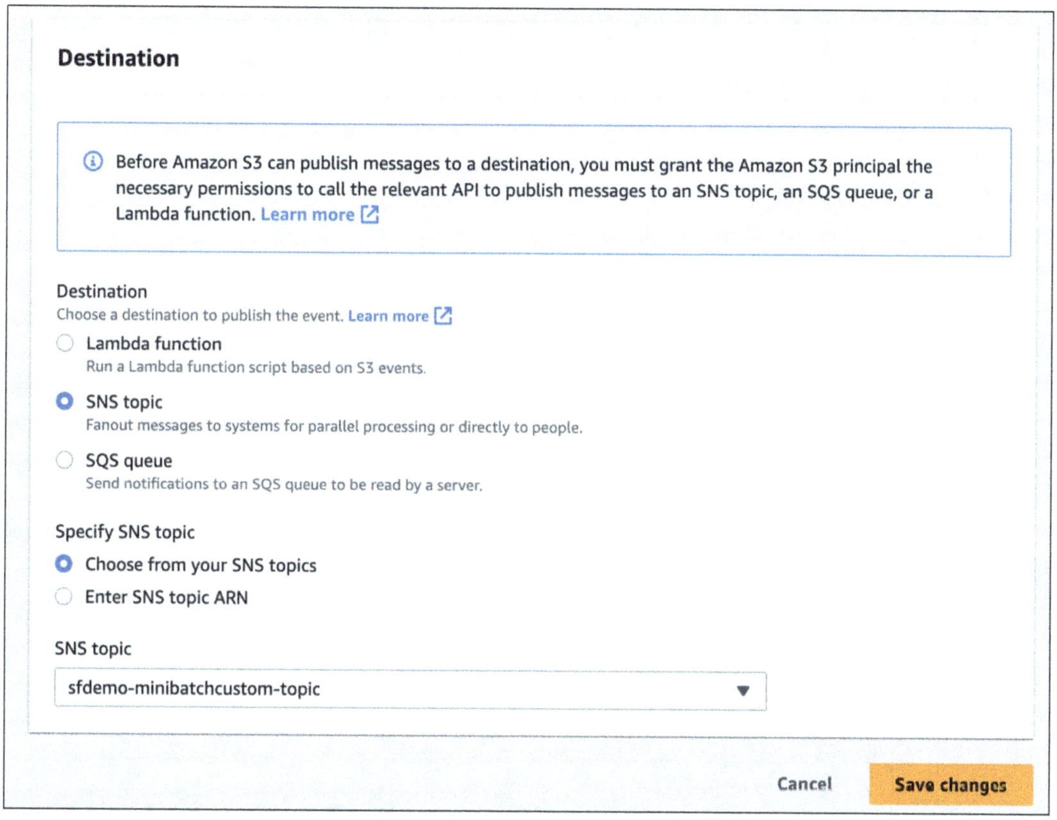

Destination

Before Amazon S3 can publish messages to a destination, you must grant the Amazon S3 principal the necessary permissions to call the relevant API to publish messages to an SNS topic, an SQS queue, or a Lambda function. Learn more

Destination
Choose a destination to publish the event. Learn more

- Lambda function
  Run a Lambda function script based on S3 events.
- SNS topic
  Fanout messages to systems for parallel processing or directly to people.
- SQS queue
  Send notifications to an SQS queue to be read by a server.

Specify SNS topic

- Choose from your SNS topics
- Enter SNS topic ARN

SNS topic

sfdemo-minibatchcustom-topic

Cancel Save changes

Figure 5-5. SNS Topic configuration

You could add any number of event notifications for an S3 bucket provided, each with different filters or criteria.

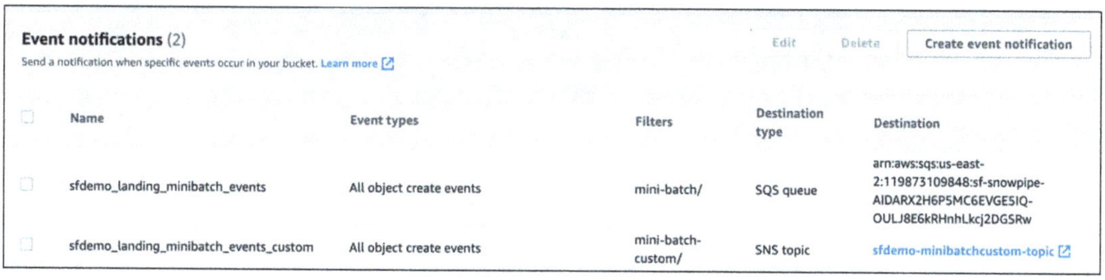

Event notifications (2)
Send a notification when specific events occur in your bucket. Learn more

Edit Delete Create event notification

Name	Event types	Filters	Destination type	Destination
sfdemo_landing_minibatch_events	All object create events	mini-batch/	SQS queue	arn:aws:sqs:us-east-2:119873109848:sf-snowpipe-AIDARX2H6P5MC6EVGE5IQ-OULJ8E6kRHnhLkcj2DG5Rw
sfdemo_landing_minibatch_events_custom	All object create events	mini-batch-custom/	SNS topic	sfdemo-minibatchcustom-topic

Figure 5-6. Event notifications created

Solution

This is similar to the previous solution with the only exception that we would utilize the existing SNS topic to listen for events by Snowpipe.

Figure 5-7 shows various components and how they are related to each other.

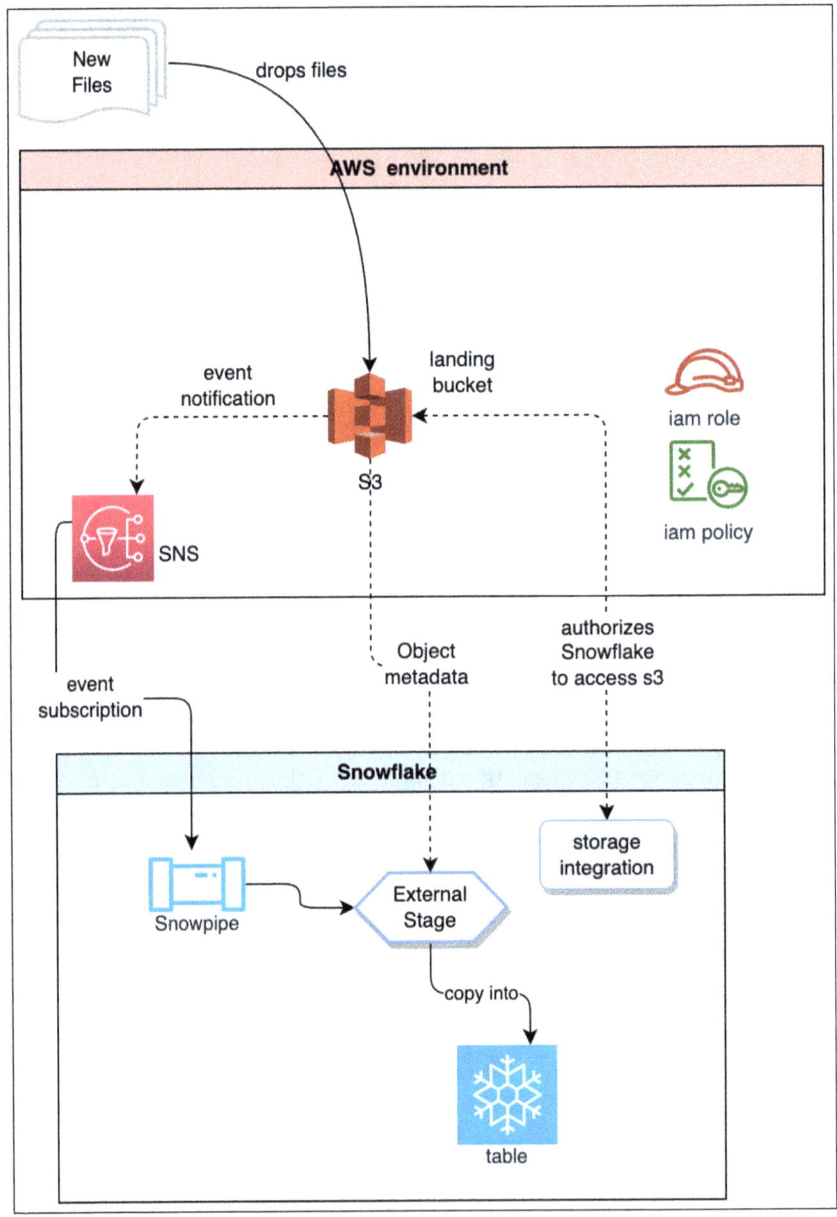

***Figure 5-7.** Snowpipe ingestion from AWS s3 using specific SNS topic*

1. Make a note of the ARN of your SNS topic. In this demonstration, it is arn:aws:sns:us-east-2:296080767349:sfdemo-minibatchcustom-topic.

2. In Snowflake, run the following system function.

```
select system$get_aws_sns_iam_policy('arn:aws:sns:us-
east-2:296080767349:sfdemo-minibatchcustom-topic');
```

The function returns an IAM policy that grants a Snowflake SQS queue permission to subscribe to the SNS topic.

3. Modify the Access policy of the SNS topic.

a) In AWS Management Console, choose Topics from the left navigation pane.

b) Select the topic for your S3 bucket, and click the Edit button. The Edit page opens.

c) Click Access policy is optional to expand this area of the page.

d) Merge the value obtained in step 2 by copying and adding only the highlighted section.

e) Note that the resource ARN and principal values you see would be different.

```
{
    "Version": "2012-10-17",
    "Statement": [
        {
            "Sid": "1",
            "Effect": "Allow",
            "Principal": {
                "AWS": "arn:aws:iam::119873109848:user/oz630000-s"
            },
            "Action": [
                "sns:Subscribe"
            ],
            "Resource": [
```

```
                    "arn:aws:sns:us-east-2:296080767349:sfdemo-
                    minibatchcustom-topic"
                ]
            }
        ]
    }
```

4. Create a stage and a pipe.

```
create or replace stage sfdemo_landing_minibatch_custom_stg
    storage_integration = sfdemo_storage_intg
    url = 's3://sfdemo-landing/mini-batch-custom/';

create or replace pipe sfdemo_minibatch_custom
auto_ingest=true
aws_sns_topic='arn:aws:sns:us-east-2:296080767349:sfdemo-
minibatchcustom-topic'
as
    copy into RAW_CUSTOMER(
        C_CUSTKEY
        , C_NAME
        , C_ADDRESS
        , C_NATIONKEY
        , C_PHONE
        , C_ACCTBAL
        , C_MKTSEGMENT
        , C_COMMENT
    )
    from (
    select
        $1::number
        , $2::varchar
        , $3:: varchar
        , $4::number
        , $5:: varchar
        , $6:: float
        , $7:: varchar
```

```
            , $8:: varchar
        from
            @sfdemo_landing_minibatch_custom_stg
        )
      FILE_FORMAT = (
            TYPE = CSV
            FIELD_DELIMITER = ','
            RECORD_DELIMITER = '\n'
            FIELD_OPTIONALLY_ENCLOSED_BY= '"'
            SKIP_HEADER = 1
    );
```

5. Verify and test the ingestion. After the pipe is created, you can check the status of the pipe, which should show as running.

```
select  SYSTEM$PIPE_STATUS( 'SFDEMO_MINIBATCH_CUSTOM' );
```

You could also return to the AWS SNS topic and look for subscriptions. The new subscription should automatically show up if your Snowpipe creation is successful. Figure 5-8 shows the screen after the successful creation of the subscription.

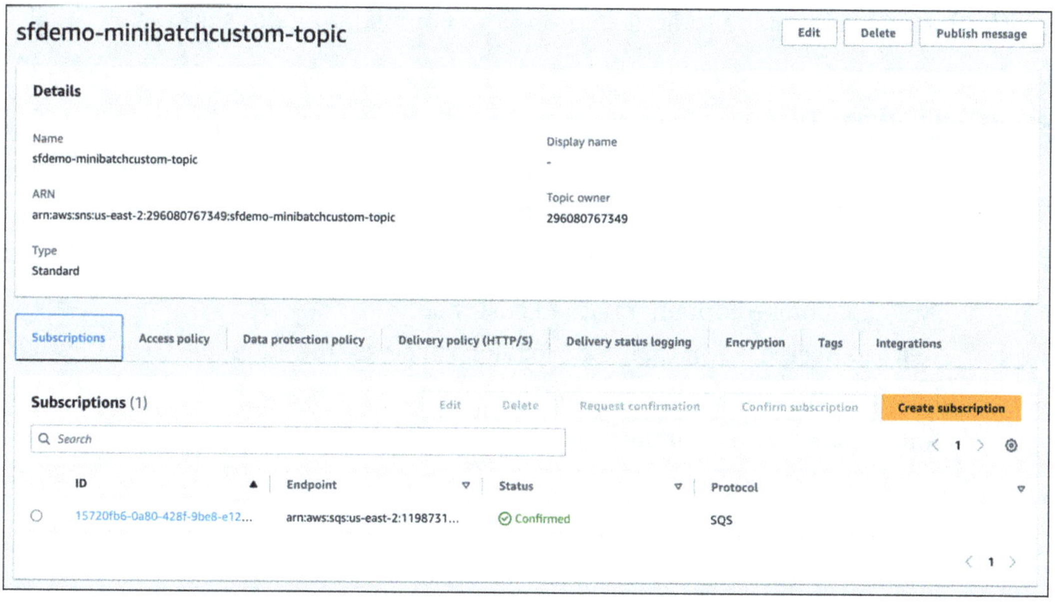

sfdemo-minibatchcustom-topic
Edit Delete Publish message
Details
Name
sfdemo-minibatchcustom-topic
Display name
-
ARN
arn:aws:sns:us-east-2:296080767349:sfdemo-minibatchcustom-topic
Topic owner
296080767349
Type
Standard
Subscriptions | Access policy | Data protection policy | Delivery policy (HTTP/S) | Delivery status logging | Encryption | Tags | Integrations
Subscriptions (1)
Edit Delete Request confirmation Confirm subscription Create subscription
Search

ID	Endpoint	Status	Protocol
15720fb6-0a80-428f-9be8-e12...	arn:aws:sqs:us-east-2:1198731...	Confirmed	SQS

Figure 5-8. *AWS SNS Topic subscriptions*

As before, check the copy history on the table.

```
select SYSTEM$PIPE_STATUS('SFDEMO_MINIBATCH');
```

Now upload the customer_sample.csv file into the S3 location (s3://sfdemo-landing/mini-batch-custom/), wait for a few seconds, and observe the copy history by running the following command.

```
select * from table(information_schema.copy_history(table_name=>'RAW_
CUSTOMER', start_time=> dateadd(hours, -1, current_timestamp())));
```

This command should show the files that were lately loaded into the table along with the stage location, last load time, row count, and so forth.

Query the RAW_CUSTOMER table to view the data.

Problem

Consider a scenario when an organization has data in Azure Storage and must make it available in Snowflake as soon as it is in the cloud storage. The overall lag is only expected to be not more than 10 minutes.

This is similar to the previous problem, except that the source is now Azure Storage.

The following is a set of prerequisite steps for demonstrating this solution.

Prerequisites

1. For this demonstration, create a storage account named sfdemolanding in Azure Cloud.
2. Create a storage container named customers.
   Refer to Recipe 2-2 in Chapter 2 for steps to create the storage integration for Azure Storage. We used the same Azure account and resource group for the following demonstration.

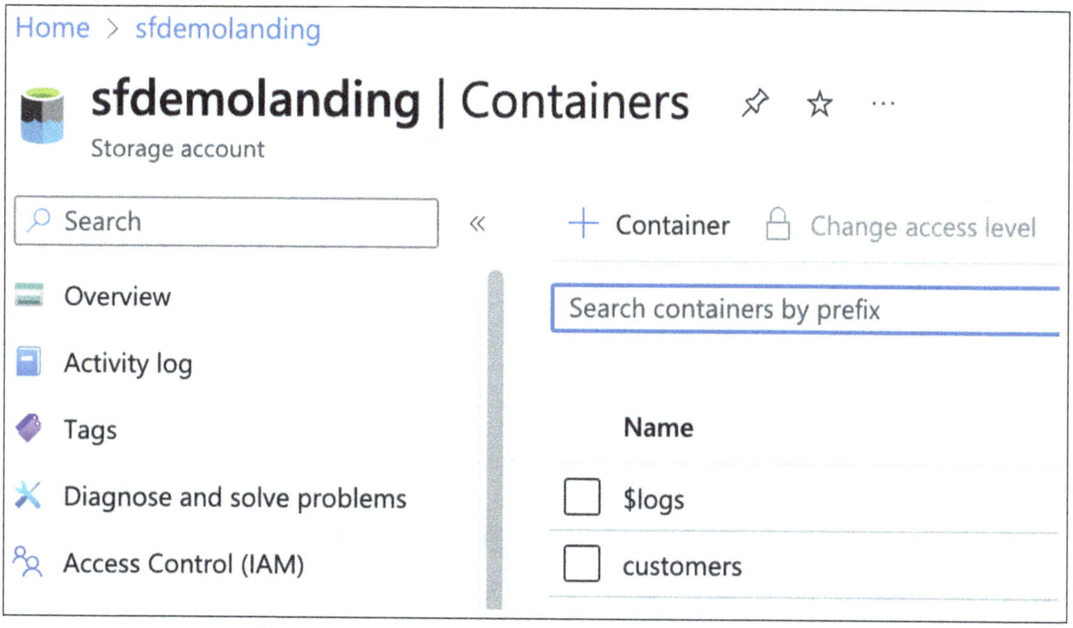

Figure 5-9. Azure Storage Account

You may update the existing storage integration or create a new storage integration.

Since we used the same role and storage integration from Chapter 2, we updated the storage integration.

```
alter STORAGE INTEGRATION sfdemo_az_storage_intg
set STORAGE_ALLOWED_LOCATIONS = (
    <add all existing azure storage locations>,
    ' azure://sfdemolanding.blob.core.windows.net/customers');
```

3. You need to grant the Storage Blob Data Reader role to the Snowflake service principal. (Refer to Recipe 2-2.)

4. Test if the integration is working fine by creating a stage.

```
create or replace stage sfdemo_landing_minibatch_az_stg
    storage_integration = sfdemo_az_storage_intg
    url = 'azure://sfdemolanding.blob.core.windows.net/customers';

list @sfdemo_landing_minibatch_az_stg;
```

Solution

Snowpipe relies on event notification and distribution, such as AWS SQS or SNS, Azure Event Grid, or GCP Pub/Sub, based on the cloud vendor that you are using.

This setup requires corresponding privileges configured in the cloud account to deliver event notifications from the storage container to Snowpipe.

Snowflake supports the following types of blob storage accounts.

- Blob storage
- Data Lake Storage Gen2
- General-purpose v2

Note that we are only interested in BlobCreated events.

Adding new objects to blob storage triggers these events but renaming a directory or object does not.

Figure 5-10 demonstrates the connection between various components of this integration.

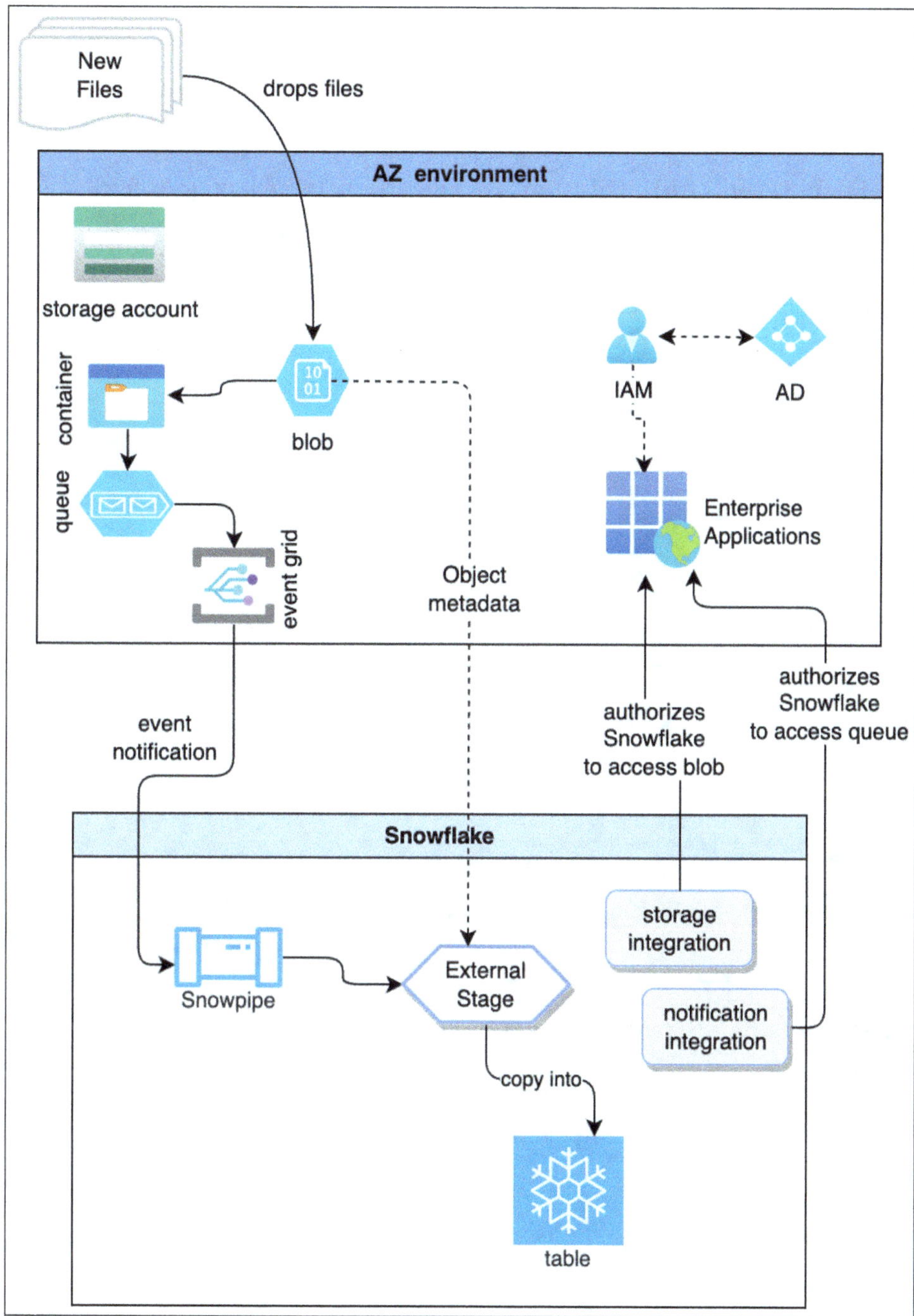

Figure 5-10. *Snowpipe Ingestion from Azure storage*

Configure the Event Grid Subscription

This section describes setting up an Event Grid subscription for Azure Storage events using the Azure CLI.

1. Enable Event Grid. (This step is needed if you use event grids for the first time in your Azure account.)

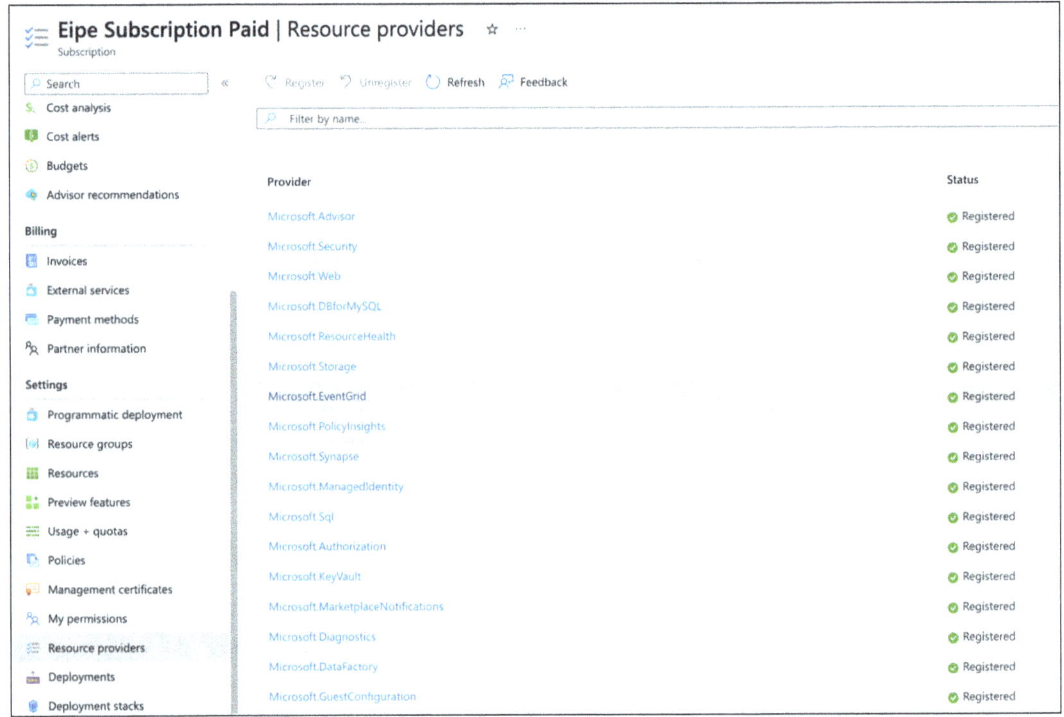

Figure 5-11. *Azure Resource Providers*

2. Create an event queue.

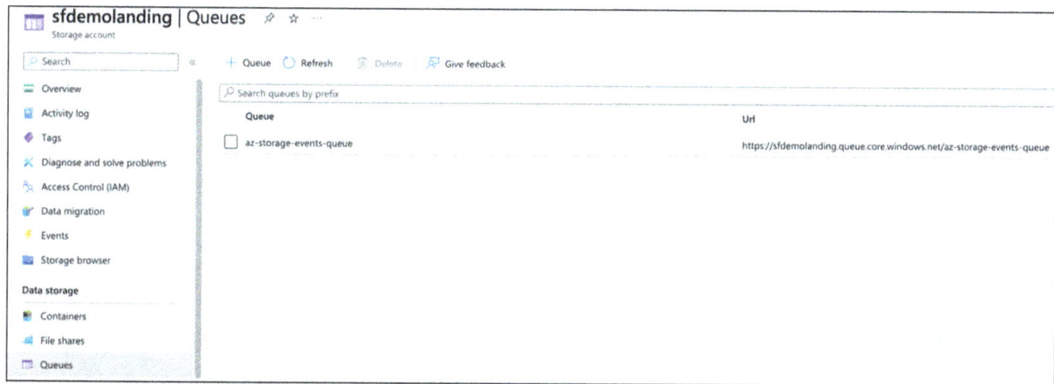

Figure 5-12. *Azure Queues*

3. Create an Event Grid subscription.

   This involves creating a topic and registering an Event Grid subscription.

   a. An Event Grid topic provides an endpoint where the source (i.e., Azure Storage) sends events. A topic is a logical container or a collection of events.

   b. Registering an event queue to receive the events.

Figure 5-13 shows the process of creating Event Subscriptions.

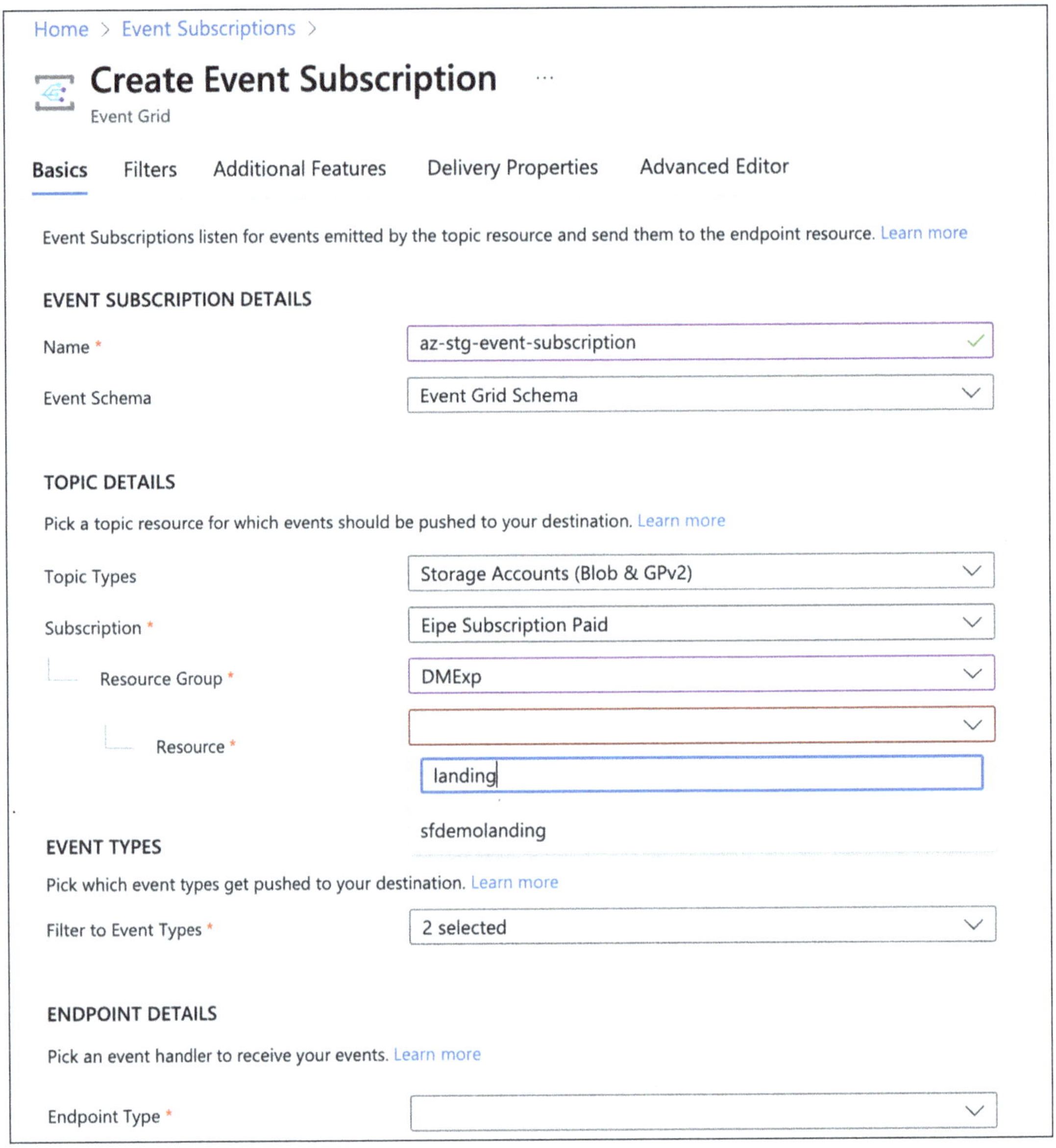

***Figure 5-13.** Azure Event subscription*

Note that selecting the storage account automatically creates a topic for it.

Choose appropriate event types and endpoint details. In this case, we select the storage queue created previously as the endpoint location.

Figure 5-14 and Figure 5-15 are the screens after successful creation.

Create Event Subscription ...
Event Grid

Basics Filters Additional Features Delivery Properties Advanced Editor

Event Subscriptions listen for events emitted by the topic resource and send them to the endpoint resource. Learn more

EVENT SUBSCRIPTION DETAILS

Name * az-stg-event-subscription

Event Schema Event Grid Schema

TOPIC DETAILS

Pick a topic resource for which events should be pushed to your destination. Learn more

Topic Type Storage account

Source Resource sfdemolanding (change)

System Topic Name * az-stg-event-topic

EVENT TYPES

Pick which event types get pushed to your destination. Learn more

Filter to Event Types * Blob Created

ENDPOINT DETAILS

Pick an event handler to receive your events. Learn more

Endpoint Type * Storage Queues (change)

Endpoint * az-storage-events-queue (change)

Figure 5-14. *Event Subscription Created*

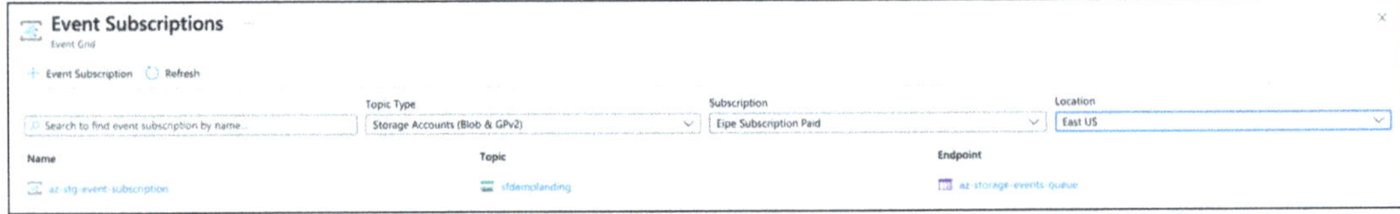

Figure 5-15. *List of Event Subscriptions*

Step 2: Creating a Notification Integration in Snowflake

A notification integration is a Snowflake object that provides an interface between Snowflake and a third-party cloud message queuing service such as Azure Event Grid.

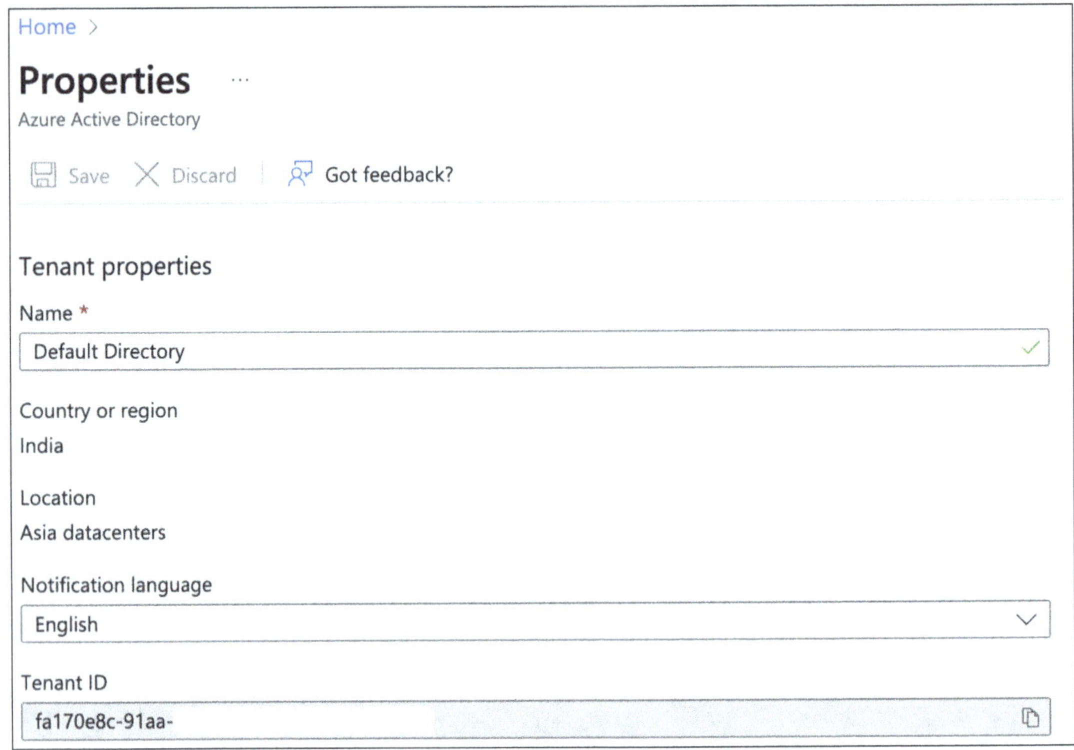

Figure 5-16. *Azure Active Directory Properties*

An **Azure Storage queue supports a single notification integration**. Referencing a single storage queue in multiple notification integrations **can result in missing data** in target tables because event notifications are split between notification integrations.

1. Create notification integration in Snowflake.

```
create notification integration sfdemo_az_notification_intg
enabled = true
type = queue
notification_provider = azure_storage_queue
azure_storage_queue_primary_uri = '<queue uri>'
azure_tenant_id = '<tenant id>';

show notification integrations;

desc notification integration az_notn_integration;
```

   Note the URL in the AZURE_CONSENT_URL column, which has the following format.

   `https://login.microsoftonline.com/<tenant_id>/oauth2/authorize?client_id=<snowflake_application_id>`

   Also, note the value in the AZURE_MULTI_TENANT_APP_NAME column.

   This is the name of the Snowflake client application created for your account.

   a) In a web browser, navigate to the URL in the AZURE_CONSENT_URL URL column. The page displays a Microsoft permissions request page.

   b) Click the **Accept** button. This allows the Azure service principal created for your Snowflake account to obtain an access token on any resource inside your tenant. Obtaining an access token succeeds only if you grant the service principal the appropriate permissions on the container (see the next step).

2. Grant access to the storage queue.

   a. Log into the Microsoft Azure portal.

   b. Navigate to **Azure Active Directory ➤ Enterprise applications**. Verify the Snowflake application is listed.

   c. Navigate to the storage account which has the queue. (In our case, it's the same storage account as before)

   d. Click Access Control (IAM). Add the role assignment.

   e. In the next screen, search for the Snowflake service principal. You may search for the keyword "Snowflake" or use the exact identity value from the AZURE_MULTI_TENANT_APP_NAME property in the `desc notification integration` output in the previous step.

   f. Grant the Snowflake app the Storage Queue Data Contributor permission.

Note that this could be provided at the individual blob container or queue level instead of providing IAM access on the storage account level.

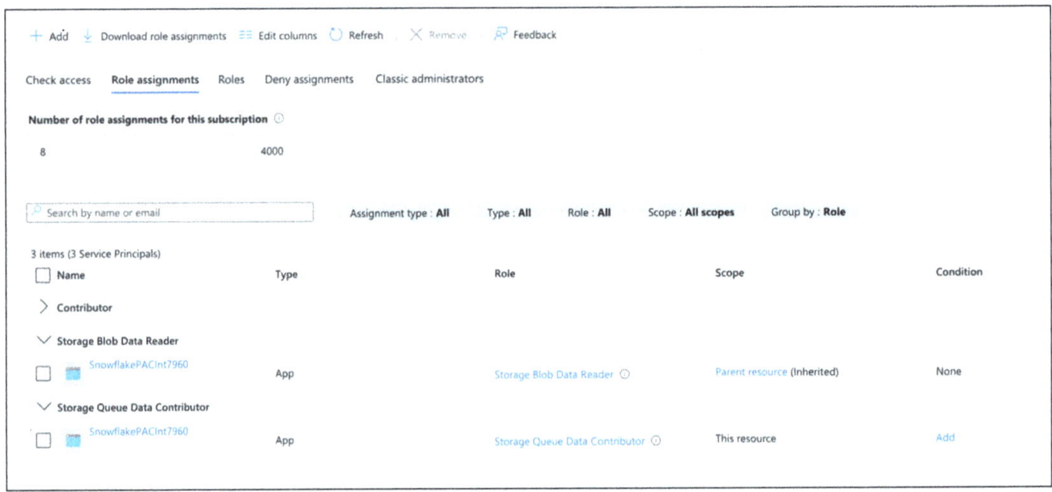

Figure 5-17. *Role Assignments*

Create a Stage and a Pipe

1. Create stage if it has not been created already.

```
create or replace stage sfdemo_landing_minibatch_az_stg
    storage_integration = sfdemo_az_storage_intg
    url = 'azure://sfdemolanding.blob.core.windows.net/customers';
```

2. Next, create a pipe.

```
create pipe sfdemo_minibatch_2
auto_ingest=true
integration = sfdemo_az_notification_intg
as
    copy into RAW_CUSTOMER(
        C_CUSTKEY
        , C_NAME
        , C_ADDRESS
        , C_NATIONKEY
        , C_PHONE
        , C_ACCTBAL
        , C_MKTSEGMENT
        , C_COMMENT
    )
    from (
    select
        $1::number
        , $2::varchar
        , $3:: varchar
        , $4::number
        , $5:: varchar
        , $6:: float
        , $7:: varchar
        , $8:: varchar
    from
        @sfdemo_landing_minibatch_az_stg
    )
   file_format =    (
        type = CSV
        field_delimiter =  ','
        record_delimiter = '\n'
        field_optionally_enclosed_by= '"'
        skip_header = 1
);
```

Verify and Test the Ingestion

After the pipe is created, you can check the status of the pipe, which should show as running.

```
select  SYSTEM$PIPE_STATUS( 'SFDEMO_MINIBATCH_2' );
```

You could also go back to Azure Event Grid to monitor the events.

Now upload the customer_sample.csv file into the storage location (azure://sfdemolanding.blob.core.windows.net/customers), wait a few seconds, and observe the copy history by running the following command.

```
select * from table(information_schema.copy_history(table_name=>'RAW_
CUSTOMER', start_time=> dateadd(hours, -1, current_timestamp())));
```

This command should show the files that were recently loaded into the table along with the stage location, last load time, row count, and so forth.

Query the RAW_CUSTOMER table to view the data.

Problem

As a data engineer for an organization, you are asked to document and gather various best practices and caveats to be considered for Snowpipe and the COPY INTO command.

Solution

Let's go over some things to consider when using COPY-INTO and Snowpipe.

- Snowpipe and the direct COPY INTO ingestion method support most of the common file formats out of the box.
    - CSV
    - JSON
    - PARQUET
    - AVRO
    - ORC
    - XML

- Files could be compressed, and Snowflake decompresses them during the ingestion process. Snowflake supports the following formats.
    - GZIP
    - BZ2
    - BROTLI
    - ZSTD
    - SNAPPY
    - DEFLATE
    - RAW_DEFLATE
- CSV (GZIP compressed) is the best format for loading to Snowflake. It is sometimes 2x to 3x faster than Parquet or ORC.
- It is advised to use a separate warehouse/cluster for data ingestion. This helps optimize and manage the cost of data ingestions.
- You may keep short-lived COPY commands together so that the cluster doesn't suspend and resume unneccssarily.
- The processing status of COPY INTO needs to be observed explicitly because it is a synchronous process, whereas Snowpipe is asynchronous.
- For Snowpipe and COPY INTO, the COPY_HISTORY view or the lower latency COPY_HISTORY function is available. Generally, files that fail to load need to be looked at by someone, so it is often good to check COPY_HISTORY less frequently than the typical file arrival rate.
- COPY commands can be checked from query history using the query_type='COPY' filter.
- Before ingesting, if you want the data to be ordered (default is natural ingestion order), you should sort an S3 bucket (using something like syncsort) before bulk load via copy. This is way faster than inserting with an ORDER BY clause.

- Use COPY INTO instead of INSERT because it utilizes the more efficient bulk loading processes.
- Use VALIDATION_MODE to validate before loading the files using COPY INTO. This acts like a pre-load test.
- Use the VALIDATE table function to validate files loaded in the COPY INTO command. This acts like a post-load test.
- COPY INTO doesn't load files already loaded into the table but you can force it by using the FORCE load parameter.
- Snowflake maintains file metadata tracked as part of the COPY INTO operations. The following are some of the details that are tracked as part of file metadata.
    - Name of each file from which data was loaded
    - File size
    - ETag for the file
    - Number of rows parsed in the file
    - Timestamp of the last load for the file
    - Information about any errors encountered in the file during loading
- This load/file metadata expires after 64 days. If the LAST_MODIFIED date for a staged data file is less than or equal to 64 days, the COPY command can determine its load status for a given table and prevent reloading (and data duplication). The LAST_MODIFIED date is the timestamp when the file was initially staged or when it was last modified, whichever is later.

  If the LAST_MODIFIED date is older than 64 days, the load status is still known if the following events occurred less than or equal to 64 days before the current date.

    - The file was loaded successfully.
    - The initial set of data for the table (i.e., the first batch after the table was created) was loaded.

However, the COPY command cannot definitively determine whether a file has been loaded already if the LAST_MODIFIED date is older than 64 days. In this case, the command skips the file by default to prevent accidental reload.

The parameter LOAD_UNCERTAIN_FILES can be changed to TRUE to load such files.

- You can set the FORCE option to load all files, ignoring load metadata if it exists. Note that this option reloads files, potentially duplicating data in a table.
- COPY INTO... PURGE = TRUE can be used to automatically delete the loaded files from the stage, or they can be deleted later using the REMOVE command. When Snowpipe is used you do not have the option to purge during the copy into operation but can only use REMOVE later to delete the files.
- File size considerations
    - To optimize the number of parallel operations for a load it is recommended that the files are 100 MB to 250 MB in size compressed.
    - Aggregate smaller files to minimize the processing overhead for each file.
    - Split larger files into a greater number of smaller files to distribute the load among the servers in an active warehouse. (Linux split command).
    - The number of data files processed in parallel is determined by the number and capacity of servers in a warehouse. XS = 1 server = can process eight concurrent files, so S would be 16 files and likewise.
    - Split large files by line to avoid records that span chunks.
    - If you are loading a Parquet file, it should be 1 GB or less.

    - Organize files in some logical path, such as subject area or create date; for example, /User/local/market/daily/2018/09/05. Many locations with fewer files are better for scanning during the (re)load operation.

- Snowflake supports transforming data while loading it into a table using the COPY command.
    - What's allowed: Column reordering, column omission, casts, truncating text strings that exceed the target column length
    - What's not allowed: FLATTEN, JOIN, GROUP BY
- An overhead to manage files in the internal load queue is included in the utilization costs charged for Snowpipe. This overhead increases in relation to the number of files queued for loading. Snowpipe charges 0.06 credits per 1000 files queued. If it takes longer than one minute to accumulate MBs of data in your source application, consider creating a new (potentially smaller) data file once per minute. This approach typically leads to a good balance between cost (i.e., resources spent on Snowpipe queue management and the actual load) and performance (i.e., load latency).
- Pipe refresh works for internal and external stages but auto_ingest='true' only for external stages.
- Snowflake recommends that you only send supported events for Snowpipe to reduce costs, event noise, and latency.
- Initial data loading
    - The Snowpipe refresh command (alter pipe <pipe_name> refresh PREFIX = '<path>' MODIFIED_AFTER = <start_time>) begins ingesting files that were recently modified up to the last seven days.
    - COPY provides greater control with the responsibility to manage the warehouse and the job execution on the user but with the advantage of picking and choosing any number of files.

- In the case of data skew, COPY jobs that don't have enough files will not utilize the warehouse efficiently because there wouldn't be enough parallel jobs to run.
- A single COPY job cannot run beyond the default job timeout of 24 hours.

Problem

Consider a scenario when the data team of an organization already has an existing data infrastructure set up in AWS and wants to trigger Snowpipe's loading process, not asynchronously as files land in the S3 landing area but only when certain conditions are met.

In other words, the organization does not want to auto ingest but to invoke Snowpipe's loading process only when needed.

To demonstrate, we utilize the Snowpipe API, which is a REST endpoint Snowflake provides for all the Snowpipe objects you create.

Prerequisites

This demonstration creates a user account with key-based authentication enabled. This is needed because calls to the public Snowpipe REST endpoints use key-based authentication rather than the typical username/password authentication, as the ingestion service does not maintain client sessions.

First, we create a public/private key pair that can be used to generate a valid JSON Web Token (JWT). These tokens are generally valid for about 60 minutes and must be regenerated. And once that is complete, we alter the user account to add the generated public key.

We use a Snowflake-provided Python library for performing the API call, and the library takes care of JWT authorization.

1. Generate RSA public/private key.

   More information is at `https://docs.snowflake.com/en/user-guide/key-pair-auth`.

We generated an encrypted private key and public key by running the following commands.

```
# generate private key
openssl genrsa 2048 | openssl pkcs8 -topk8 -v2 des3 -inform
PEM -out rsa_key_snow.p8
# generate public key
openssl rsa -in rsa_key_snow.p8 -pubout -out rsa_key_snow.pub
```

Since we opted for an encrypted private key, it asks for a passcode. Please note down the passcode or save it as an environment variable for later use.

Test if it is working fine by using the SnowSQL client.

```
# test using snowsql
snowsql -a vx15608.us-east-2.aws -u jeipe --private-key-path
rsa_key_snow.p8
```

2. Create a new user or add a public key to an existing user.

    a. Copy the public key without the headers.

    b. Remove the new line if it is present.

    c. Run the alter user command.

```
ALTER USER jeipe SET RSA_PUBLIC_KEY="MIIBIjANBgkqhki
*********************";
```

Solution

You can construct the following URL to get details about your Snowpipe API endpoint: https://{account}.snowflakecomputing.com/v1/data/pipes/{pipeName}/insertFiles?requestId={requestId}

Documentation for the Snowpipe API is at `https://docs.snowflake.com/en/user-guide/data-load-snowpipe-rest-apis`.

Let's use the same schema (RAW_CUSTOMER) as before, but we use the internal stage for the demonstration.

1. Create a stage and a pipe.

```
create stage test_customer_internal_stage;

create pipe sfdemo_pipe_api_trigger
auto_ingest=false
as
    copy into RAW_CUSTOMER(
        C_CUSTKEY
        , C_NAME
        , C_ADDRESS
        , C_NATIONKEY
        , C_PHONE
        , C_ACCTBAL
        , C_MKTSEGMENT
        , C_COMMENT
    )
    from (
    select
        $1::number
        , $2::varchar
        , $3:: varchar
        , $4::number
        , $5:: varchar
        , $6:: float
        , $7:: varchar
        , $8:: varchar
    from
        @test_customer_internal_stage
    )
   file_format =    (
        type = CSV
        field_delimiter =  ','
        record_delimiter = '\n'
        field_optionally_enclosed_by= '"'
        skip_header = 1
);
```

Note that the pipe is pointing to the internal stage, and the auto_ingest is set to false.

You may use the snowsql command line or the web interface to add the files to the stage. We added the customer_sample.csv that was used before via the web interface.

2. Write Python code using the snowflake-ingest-python open source library.

   The following code is available in the snowpipe_api_client.py file.

```
from logging import getLogger
import snowflake.connector
from snowflake.ingest import SimpleIngestManager
from snowflake.ingest import StagedFile
from snowflake.ingest.utils.uris import DEFAULT_SCHEME
from datetime import timedelta
from requests import HTTPError
from cryptography.hazmat.primitives import serialization
from cryptography.hazmat.primitives.serialization import load_pem_
private_key
from cryptography.hazmat.backends import default_backend
from cryptography.hazmat.primitives.serialization import Encoding
from cryptography.hazmat.primitives.serialization import
PrivateFormat
from cryptography.hazmat.primitives.serialization import
NoEncryption
import time
import datetime
import os
import logging

logging.basicConfig(
        filename='/tmp/ingest.log',
        level=logging.DEBUG)
logger = getLogger(__name__)
```

```
# If you generated an encrypted private key, implement this method
  to return
# the passphrase for decrypting your private key.
# instead of hardcoding get the value from environment variable or
  config files: os.environ['PRIVATE_KEY_PASSPHRASE']
def get_private_key_passphrase():
  return 'eipe'

with open("/Users/johneipe/rsa_key_snow.p8", 'rb') as pem_in:
  pemlines = pem_in.read()
  private_key_obj = load_pem_private_key(pemlines,
  get_private_key_passphrase().encode(),
  default_backend())

private_key_text = private_key_obj.private_bytes(
  Encoding.PEM, PrivateFormat.PKCS8, NoEncryption()).
  decode('utf-8')
# Assume the public key has been registered in Snowflake:
# private key in PEM format

ingest_manager = SimpleIngestManager(account='bwbyxua-xxxxx',
                                     host='bwbyxua-xxxxx.
                                     snowflakecomputing.com',
                                     user='jeipe',
                                     pipe='raw.retail.sfdemo_pipe_
                                     api_trigger',
                                     private_key=private_key_text)

# List of files, but wrapped into a class
staged_file_list = [
  StagedFile('customer_sample.csv', None),  # the second parameter
  is file size but it is optional but recommended, pass None if
  not available
]

try:
    resp = ingest_manager.ingest_files(staged_file_list)
```

```
    except HTTPError as e:
        logger.error(e)# HTTP error, may need to retry
        exit(1)

    # This means Snowflake has received file and will start loading
    assert(resp['responseCode'] == 'SUCCESS')

    # Needs to wait for a while to get result in history
    while True:
        history_resp = ingest_manager.get_history()

        if len(history_resp['files']) > 0:
            print('Ingest Report:\n')
            print(history_resp)
            break
        else:
            # wait for 20 seconds
            time.sleep(20)

        hour = timedelta(hours=1)
        date = datetime.datetime.utcnow() - hour
        history_range_resp = ingest_manager.get_history_range(date.
        isoformat() + 'Z')

        print('\nHistory scan report: \n')
        print(history_range_resp)
```

Note that we used two API calls.

- ingest_manager.ingest_files triggers Snowpipe.
- ingest_manager.get_history retrieves an ingestion report on what files were ingested and details about Snowpipe's current state.

You could also test the API using tools like POSTMAN, but then you would have to generate the JWT token using a few lines of Python code.

```
from datetime import timedelta, datetime
from cryptography.hazmat.primitives import serialization
from cryptography.hazmat.primitives.serialization import load_pem_
```

```
private_key
from cryptography.hazmat.backends import default_backend
from cryptography.hazmat.primitives.asymmetric import dsa, rsa
import os
import jwt

with open("<path to private key file>", 'rb') as pem_in:
    pemlines = pem_in.read()
    private_key = load_pem_private_key(pemlines, <private_key_passphrase>
    default_backend())
encoded_jwt = jwt.encode({"iss": "vx15608.us-east-2.jeipe", "iat":
int(datetime.utcnow().timestamp()), "exp": <provide expiry time>}, private_
key, algorithm='RS256')
```

Additional Snowflake Documentation

https://docs.snowflake.com/en/user-guide/data-load-snowpipe-intro

https://docs.snowflake.com/user-guide/data-load-snowpipe-auto

---

Recipe 5-2. Data Loading Using Streams and Tasks

Streams are also known as *stream tables, change tables*, and *offset tables* because they track changes made to tables, including inserts, updates, and deletes, as well as metadata about each change.

This process is referred to as change data capture (CDC).

Once a stream has been created for a source table, it tracks the changes made to each row in a source table as well as any new rows written into the source table,

The stream table then shows what changed, at the row level, between two transactional points of time in a table. This allows querying and consuming a sequence of change records in a transactional fashion.

Streams can be created to query change data on the following objects.

- Standard tables, including shared tables
- Views, including secure views

- Directory tables
- Event tables
- External tables

A stream stores an offset for the source object and not the actual table columns or data.

The following are the metadata columns.

METADATA$ACTION

Indicates the DML operation (INSERT, DELETE) recorded.

METADATA$ISUPDATE

Indicates whether the operation was part of an UPDATE statement. Updates to rows in the source object are represented as a pair of DELETE and INSERT records in the stream with a metadata column METADATA$ISUPDATE values set to TRUE.

Note that streams record the differences between two offsets. If a row is added and updated in the current offset, the delta change is a new one. The METADATA$ISUPDATE row records a FALSE value.

METADATA$ROW_ID

Specifies the unique and immutable ID for the row, which can be used to track changes to specific rows over time.

Types of Streams

The following stream types are available based on the metadata recorded by each.

- Standard
    - It supports streams on tables, directory tables, or views.
    - A standard stream tracks all DML changes to the source object, including inserts, updates, and deletes (including table truncates).
    - Note that net effect matters. For example, a row inserted and then deleted between two transactional points of time in a table is removed in the delta stream.

- Append-only
    - It supports streams on standard tables, directory tables, or views.
    - An append-only stream tracks row inserts only.
    - Update and delete operations (including table truncates) are not recorded.
- Insert-only
    - It supports streams on external tables only.
    - An insert-only stream tracks row inserts only; they do not record delete operations.
    - Since there is no concept of updates for external tables (Overwritten or appended files are essentially handled as new files), they appear as new inserts.

Problem

Let's say an organization has a *landing layer*, also known as a *raw layer*, with customer information from external systems like an OLTP system.

The organization's data team must build a customer table with the latest and greatest information based on this raw table.

Solution

To demonstrate this scenario, let's use the sample TPCH data provided by Snowflake (SNOWFLAKE_SAMPLE_DATA.TPCH_SF1.CUSTOMER).

1. Create a table to represent the landing or raw table.

```
-- create a raw landing table
create or replace table RAW_CUSTOMER_SRC (
  C_CUSTKEY number
  , C_NAME varchar
  , C_ADDRESS varchar
);
```

2. Insert sample data from the TPCH schema and create a stream to demonstrate its feature

```
-- insert test data
insert into RAW_CUSTOMER_SRC (c_custkey, c_name, c_address)
select c_custkey, c_name, c_address from SNOWFLAKE_SAMPLE_DATA.
TPCH_SF1.CUSTOMER
where c_custkey= 60001;
```

3. Next, create a Snowflake stream that captures the latest records from the raw table.

```
create or replace stream customer_stream on table RAW_
CUSTOMER_SRC;
```

4. Check the stream and notice that the row inserted earlier doesn't show up. This is because the stream only captures what is recorded after the stream creation.

```
select * from customer_stream;
-- returns 0 rows
```

5. Insert test data again.

```
insert into RAW_CUSTOMER_SRC (c_custkey, c_name, c_address)
select c_custkey, c_name, c_address from SNOWFLAKE_SAMPLE_DATA.
TPCH_SF1.CUSTOMER
where c_custkey= 60002;
Check the stream.select * from customer_stream;
-- returns:

C_CUSTKEY C_NAME  C_ADDRESS METADATA$ACTION METADATA$ISUPDATE
METADATA$ROW_ID
60,002  Customer#000060002  ThGBMjDwKzkoOxhz  INSERT  FALSE
2e8c5861d27......
```

6. Observe the effect of transaction locking on streams.

```
BEGIN;

-- insert test data
insert into RAW_CUSTOMER_SRC (c_custkey, c_name, c_address)
select c_custkey, c_name, c_address from SNOWFLAKE_SAMPLE_DATA.
TPCH_SF1.CUSTOMER
where c_custkey= 60003;

-- check stream
select * from customer_stream;
```

These changes are not visible because the change interval of the stream object starts at the current offset and ends at the current transactional time point, which is the beginning time of the transaction.

```
COMMIT;

-- check stream
select * from customer_stream;

C_CUSTKEY C_NAME  C_ADDRESS METADATA$ACTION METADATA$ISUPDATE
METADATA$ROW_ID
60,002  Customer#000060002  ThGBMjDwKzkoOxhz  INSERT  FALSE
2e8c5861.........
60,003  Customer#000060003  Ed hbPtTXMTAsgGhCr4HuTzK,Md2  INSERT  FALSE
ec196fc......
```

The changes surface now because the stream object uses the current transactional time as the end point of the change interval that now includes the changes in the source table.

7. Write from the stream into the target table.

8. Observe how inserts and updates affect the flow of data.

9. Create a destination table for writing from the stream.

```
-- RAW_CUSTOMER_SRC -> customer_stream -> CUSTOMER_SRC
create table CUSTOMER_SRC (
  C_CUSTKEY number
```

```
    , C_NAME varchar
    , C_ADDRESS varchar
)
```

10. Read from the stream and write into the target.

```
insert into CUSTOMER_SRC (c_custkey, c_name, c_address)
select c_custkey, c_name, c_address from customer_stream where
METADATA$ACTION='INSERT';
Check the stream.select * from customer_stream;
-- notice that the stream is empty
```

11. Insert the test data.

```
insert into RAW_CUSTOMER_SRC (c_custkey, c_name, c_address)
select c_custkey, c_name, c_address from SNOWFLAKE_SAMPLE_DATA.
TPCH_SF1.CUSTOMER
where c_custkey= 60004;
```

12. Update the test data.

```
update RAW_CUSTOMER_SRC set c_address = 'xxxxx' where
c_custkey = 60004;
```

13. Check the stream.

```
select * from customer_stream;
```

```
C_CUSTKEY C_NAME  C_ADDRESS METADATA$ACTION METADATA$ISUPDATE
METADATA$ROW_ID
60,004  Customer#000060004  xxxxx INSERT  FALSE
26456b4e...........
```

Notice that only the latest data is shown.

14. Read from the stream and write into the target.

```
insert into CUSTOMER_SRC (c_custkey, c_name, c_address)
select c_custkey, c_name, c_address from customer_stream where
METADATA$ACTION='INSERT';
```

15. Check the stream.

```
select * from customer_stream;
```

16. Insert the test data.

```
insert into RAW_CUSTOMER_SRC (c_custkey, c_name, c_address)
select c_custkey, c_name, c_address from SNOWFLAKE_SAMPLE_DATA.
TPCH_SF1.CUSTOMER
where c_custkey= 60005;
Check the stream.select * from customer_stream;
-- This shows the one row for c_custkey= 60005
```

17. Read from stream and write into target.

```
insert into CUSTOMER_SRC (c_custkey, c_name, c_address)
select c_custkey, c_name, c_address from customer_stream where
METADATA$ACTION='INSERT';
Check the stream.select * from customer_stream;
-- shows empty stream
```

18. Update the test data.

```
update RAW_CUSTOMER_SRC set c_address  = 'xxxxxx' where c_custkey= 60005;
Check the stream.select * from customer_stream;

C_CUSTKEY C_NAME  C_ADDRESS METADATA$ACTION METADATA$ISUPDATE
METADATA$ROW_ID
60,005  Customer#000060005  xxxxxx            INSERT  TRUE
af434a5d81...............
60,005  Customer#000060005  1F3KM3ccEXEtI,    DELETE  TRUE
af434a5d81.........
```

19. Read from the stream and write into the target by using merge.

```
merge into CUSTOMER_SRC c
using ( select * from customer_stream where
METADATA$ACTION='INSERT') s
on s.c_custkey = s.c_custkey
when MATCHED then update set c.c_address = s.c_address;
Check the stream.select * from customer_stream;
-- empty
```

Note that the whole stream gets cleared though only one change record was read and merged.

Problem

When should you use streams, and what are some of the caveats in using them for data teams?

Solution

The following are some of the features and considerations around using Streams.

- To view the current staleness status of a stream, execute the DESCRIBE STREAM or SHOW STREAMS command.
- A stream becomes stale when its offset is outside the data retention period for its source table (or the underlying tables for a source view).
- When a stream becomes stale, the historical data for the source table is no longer accessible, including any unconsumed change records.
- This restriction does not apply to streams on directory tables or external tables with no data retention period.
- The order in which the stream represents updates as INSERT and DELETE could appear in any random order.
- Any DML statement advances the offset for the stream. You should consume all changes in the stream at once within a single transaction OR use the CHANGES feature (more on this later).
- Use streams when you truly need CDC.
- Building SCD2 tables on top of streams is slightly more complex.
- Using streams could lead to additional Snowflake object overhead. If you have 100 tables and you plan to employ streams on all of them, you end up with 100 stream tables.
- For general ETL scenarios, if you have a landing table with a reliable timestamp column (for tracking the order of inserts and updates) and primary key columns, it is far simpler to read directly from the table rather than from the stream.

Problem

Consider a data team that wants to automate the process of reading from streams and writing to target tables so that the target tables are always updated with an acceptable lag of 10 minutes without bringing any additional technological and maintenance overhead.

Solution

There are various ways you could achieve this. You could create a cron task that runs an SQL script on Snowflake or use orchestration tools like Airflow to run the SQL code.

But all these approaches bring additional overhead for managing machines running the cron or the scheduler. In contrast, a Snowflake task is a native orchestration feature that acts like a simple task scheduler. It provides a quick and easy way to run SQL code (or Python/Java stored procs) on schedule.

A task can execute any one of the following types of SQL code.

- Single SQL statement
- Call to a stored procedure
- Procedural logic using Snowflake scripting

Like modern orchestration tools, Snowflake's tasks provide the ability to create DAGs but do not provide the features of a full-fledged orchestration tool like Airflow or Prefect.

The following is an example of a task.

```
create task task_demo
  warehouse = XSMALL_WH
  schedule = '5 MINUTE'
as
insert into my_table(ts) values(CURRENT_TIMESTAMP);
```

This task is scheduled to run every 5 minutes using an XSMALL warehouse.

A task can be scheduled using the cron expression. The following is an example.

```
schedule = 'USING CRON 0 2 * * * UTC'
```

This sets it to run every night at 2 a.m. UTC.

You could skip mentioning a user-managed warehouse and let Snowflake pick an appropriate warehouse instead. This is known as a *serverless compute model.*

For example, mention the following parameter instead of the warehouse parameter.

```
USER_TASK_MANAGED_INITIAL_WAREHOUSE_SIZE = XSMALL
```

This model enables you to rely on compute resources managed by Snowflake instead of user-managed virtual warehouses. The compute resources are automatically resized and scaled up or down by Snowflake as required for each workload. Snowflake determines the ideal size of the compute resources for a given run based on a dynamic analysis of statistics from the previous runs.

Do note that you need to specify the size of the compute resources to provision for the first run of the task before a task history is available for Snowflake to determine an ideal size.

```
create task task_demo
  user_task_managed_initial_warehouse_size = XSMALL_WH
  schedule = '5 MINUTE'
as
insert into my_table(ts) values(CURRENT_TIMESTAMP);
```

Tasks can be used to run DML operations off a stream.

The following example shows how to read from a stream created in the previous section.

```
create task load_customer_task
  warehouse = XSMALL_WH
  schedule = '5 MINUTE'
  comment = 'Task to load customer table from stream'
as
insert into CUSTOMER_SRC (c_custkey, c_name, c_address)
select c_custkey, c_name, c_address from customer_stream where
METADATA$ACTION='INSERT';
```

You can describe a task like how you describe a table.

```
desc task load_customer_task;
```

Insert test records and test the task.

```
insert into RAW_CUSTOMER_SRC (c_custkey, c_name, c_address)
select c_custkey, c_name, c_address from SNOWFLAKE_SAMPLE_DATA.TPCH_SF1.
CUSTOMER  where c_custkey= 60009;

select * from customer_stream;
```

Additional Snowflake Documentation

https://docs.snowflake.com/en/user-guide/streams-intro

https://docs.snowflake.com/en/user-guide/tasks-intro

Recipe 5-3. Data Loading Using Kafka

If you're unfamiliar with Kafka, it's a scalable, fault-tolerant, publish-subscribe messaging system that enables you to build distributed applications. It provides a data collection and distribution infrastructure to write and read streams of records.

You may use Kafka if it's already acting as a source of data stream, if event records need to be distributed to multiple sinks, or if real-time analytics needs to be performed as fast as possible and closer to the source. But if Kafka is not part of your infrastructure, there is no additional benefit for using Kafka to load data into Snowflake.

Problem

Consider a scenario where your organization already uses Kafka to collect and distribute consumer records. You are tasked with bringing the data into Snowflake for analytics and historical record keeping.

Prerequisites

To demonstrate a Kafka to Snowflake integration, we use the cloud services provided by Confluent Cloud (`https://confluent.cloud/`) and the Snowflake Sink provided by the Snowflake team.

Create a Cluster and a Topic in Confluent Cloud

We created a cluster called cluster_primary and a topic called topic_tpch_orders with three partitions. Figure 5-18 is a screenshot of the Confluent Cloud Platform with the topics created.

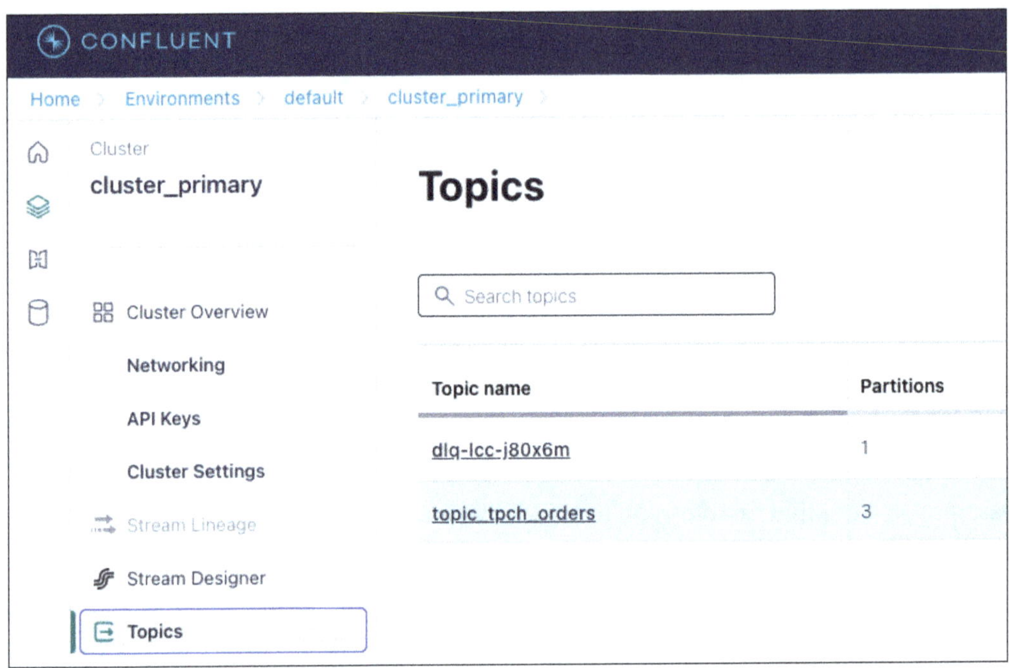

Figure 5-18. *Confluent Cloud*

Make sure you have access to the TPCH sample database in your Snowflake account.

Take a CSV extract of 1000 rows of data by running this query in the Snowflake interface and clicking export.

```
select * from SNOWFLAKE_SAMPLE_DATA.TPCH_SF1.ORDERS order by
O_ORDERKEY limit 1000;
```

Solution

Loading data from Kafka topics into Snowflake is simple and can be performed in a few easy steps, especially if you are using Confluent Cloud.

The following are the steps in setting up the connection using the publicly available Snowflake Sink for Kafka.

Step 1

Create a python client that acts as the producer for events. The code is available in the kafka_orders_ingest.py file.

There are three sections in the code. First, we read the CSV and convert all records into JSON objects. Next, we create a Kafka producer using the confluent_kafka Python library. Finally, we read each JSON object and send it to the topic configured in the code.

To make the code work, you must provide client configuration by registering the Python code as a client in Confluent Cloud. It is fairly simple: navigate to the Clients section, select the appropriate type of client, and copy the generated API key and secret. The interface also suggests and provides details for the schema registry, but we're skipping it for this demo.

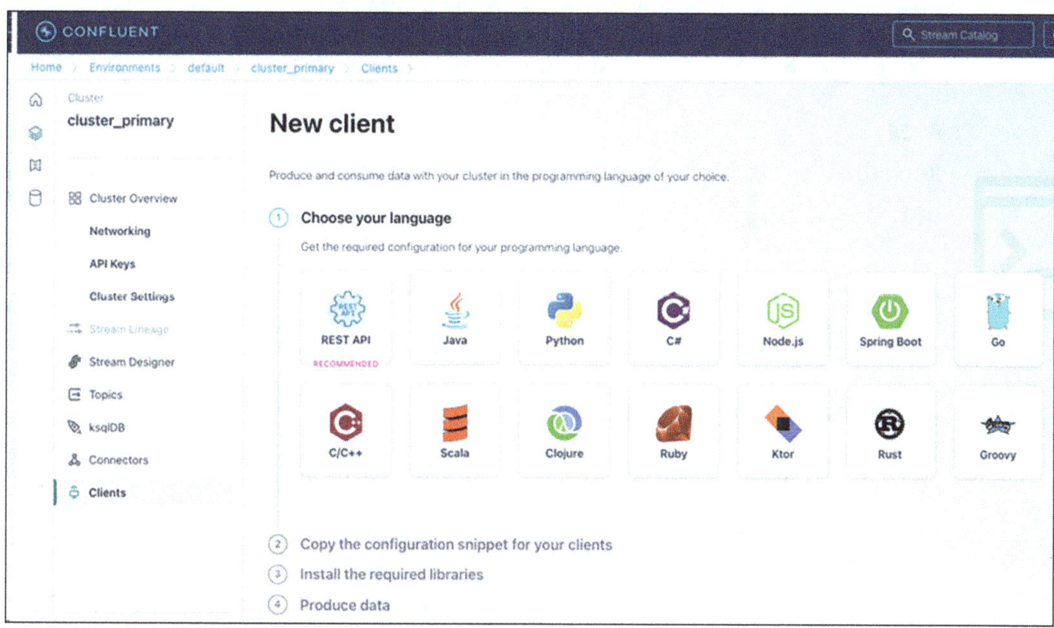

Figure 5-19. *Confluent - Generating Client code*

Step 2

Generate a Snowflake key pair.

Before configuring a Kafka connector that can sink data to Snowflake, you must generate a key pair. Snowflake authentication requires 2048-bit (minimum) RSA.

1. Generate RSA public/private key.

   More information is at https://docs.snowflake.com/en/user-guide/key-pair-auth.

   We generated an unencrypted private and public keys by running the following commands.

   ```
   # generate private key
   openssl genrsa -out rsa_key_snow.pem 2048
   # generate public key
   openssl rsa -in rsa_key_snow.pem -pubout -out rsa_key_snow.pub
   ```

   If you have opted for an encrypted private key, it asks for a passcode. Please note down the passcode for later use.

   Test if it is working fine by using the SnowSQL client.

   ```
   # test using snowsql
   snowsql -a vx15608.us-east-2.aws -u jeipe --private-key-path
   rsa_key_snow.p8
   ```

2. Create a new user or add a public key to an existing user.

   - Copy the public key without the headers.
   - Remove the new line if it is present.
   - Run the alter user command.

   ```
   ALTER USER jeipe SET RSA_PUBLIC_KEY="MIIBIjANBgkqhki
   ***********************************************************
   **************************************";
   ```

Step 3

Configure Snowflake Sink connector.

The Snowflake Sink connector provides the following features.

- **Database authentication**: Uses private key authentication.

- **Snowflake ingestion methods**: Kafka connector now supports two different ingestion methods configurable by the **snowflake.ingestion.method** property.
    - SNOWPIPE (default)
    - SNOWPIPE_STREAMING

The default ingestion method for Kafka connector uses SNOWPIPE, which internally uses Snowflake's Snowpipe REST API behind the scenes for buffered record-to-file ingestion.

How is SNOWPIPE_STREAMIN different?

- Snowpipe Streaming is the latest data ingestion method offering high-throughput, low-latency streaming data ingestion at a low cost.
- Rows are directly streamed from data sources into Snowflake tables, allowing faster, more efficient data pipelines.
- There is no intermediary step in cloud storage before data is available in Snowflake, reducing end-to-end latency.
- It provides exactly-once delivery, ordered ingestion, and error handling with dead-letter queue (DLQ) support.
- Though we haven't done formal benchmarking, online sources support the claim that Snowpipe Streaming can increase speed by up to 10x while credit usage only increased by roughly 5%.
- For new implementations, you should go with SNOWPIPE_STREAMING.

Both methods support the following input data formats.

- Avro
- JSON Schema
- Protobuf
- JSON (schemaless) input data formats

You may use the schema registry to enforce a schema, but this example does not configure a schema and uses JSON as the data format.

1. As the first step, search for Snowflake Connect in the Confluent dashboard and select it as shown in Figure 5-20.

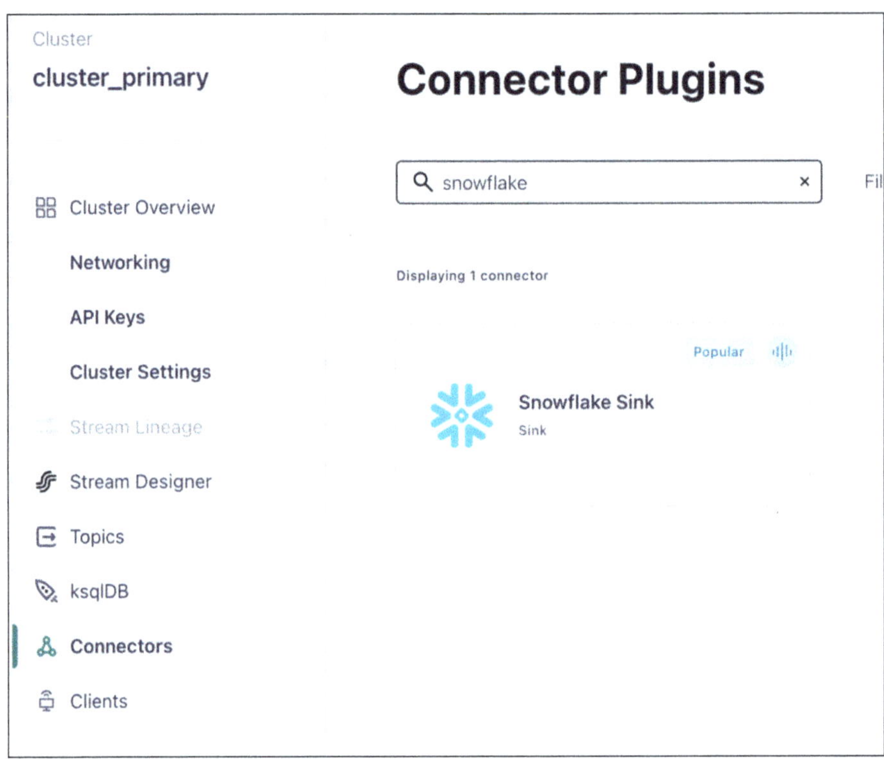

***Figure 5-20.** Confluent - Searching Connectors*

2. Choose the topic in the next screen.

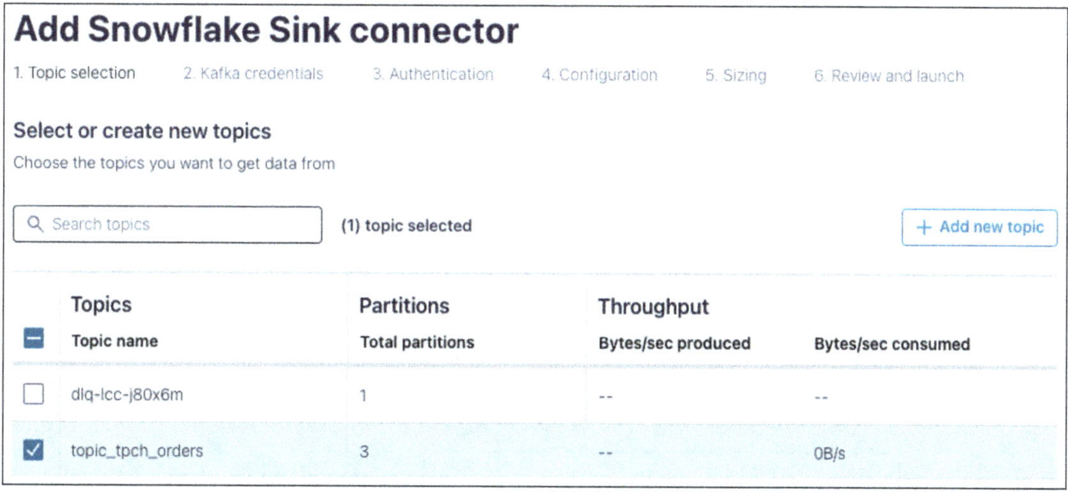

Figure 5-21. Topics in your cluster

3. Set credentials and access.

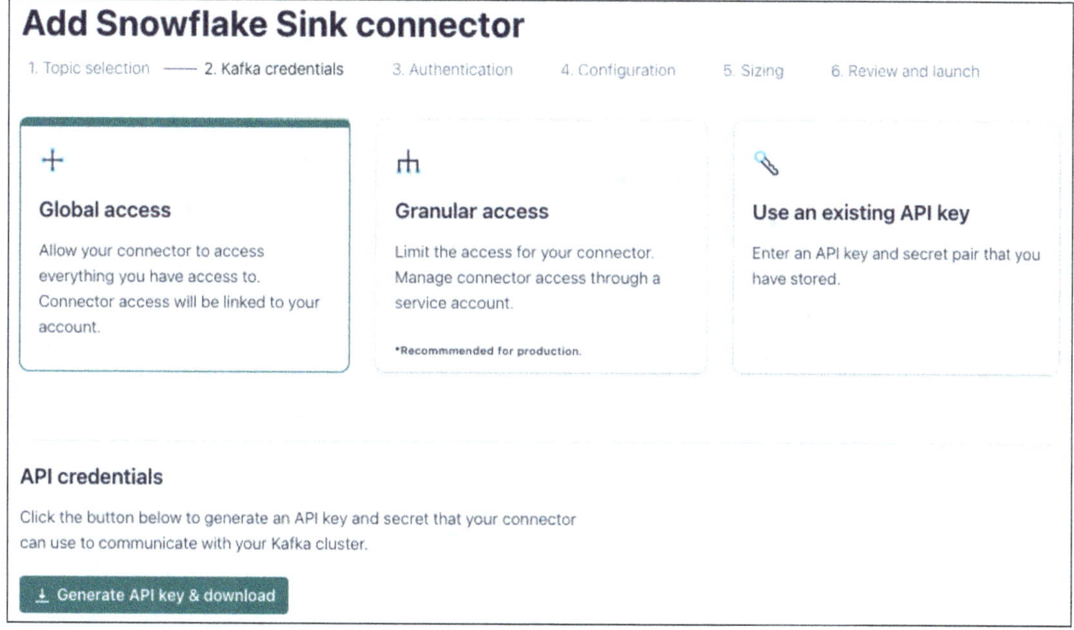

Figure 5-22. Configuring Sink Connector

4. Configure Snowflake details for the connector on the next page. Fill in the private key details. If you used an encrypted key, paste the passcode in the decryption key of the private key field.

Figure 5-23. *Configuring Sink Connector - 1*

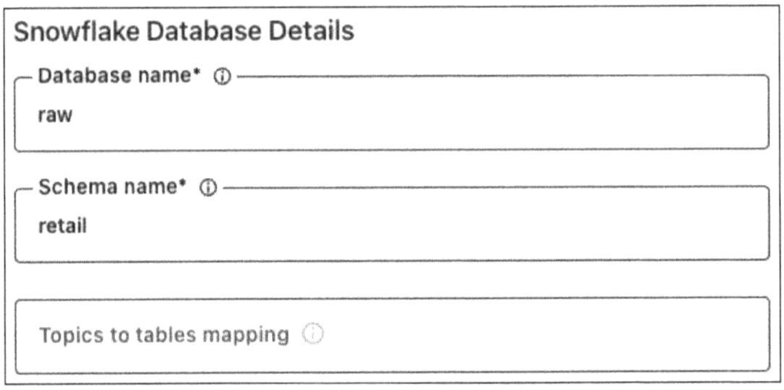

Figure 5-24. *Configuring Sink Connector - 2*

5. Expand the section to find the IP address in the allowed list. This is needed because the connector running in Confluent Cloud is ingesting data into Snowflake.

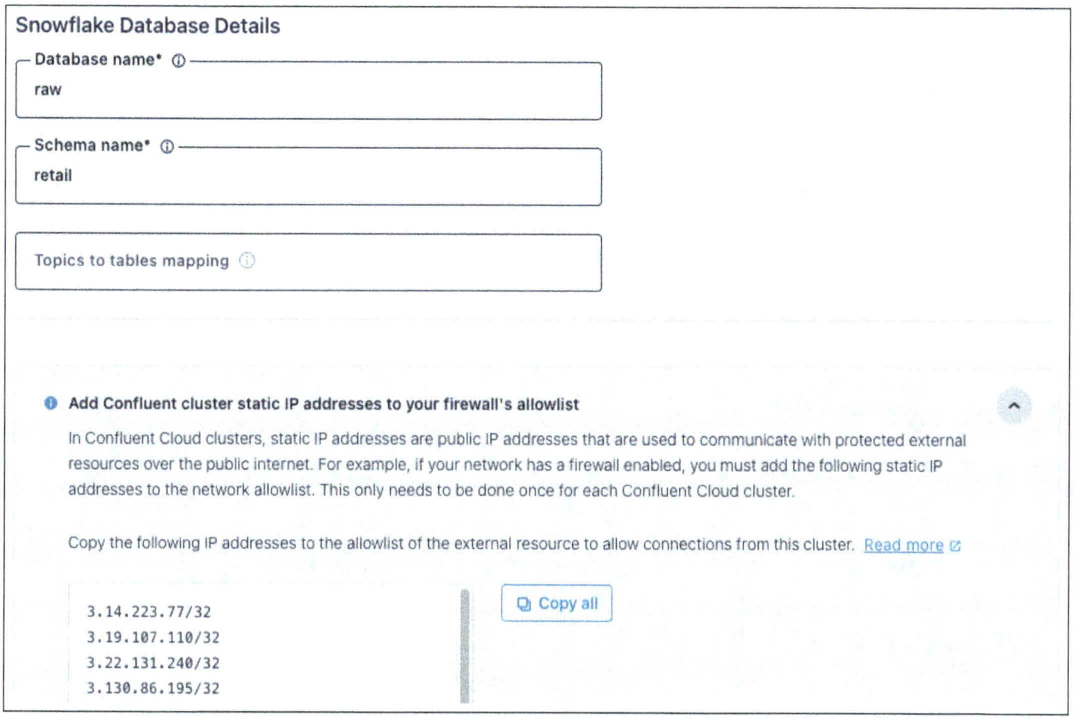

Snowflake Database Details

Database name*

raw

Schema name*

retail

Topics to tables mapping

Add Confluent cluster static IP addresses to your firewall's allowlist

In Confluent Cloud clusters, static IP addresses are public IP addresses that are used to communicate with protected external resources over the public internet. For example, if your network has a firewall enabled, you must add the following static IP addresses to the network allowlist. This only needs to be done once for each Confluent Cloud cluster.

Copy the following IP addresses to the allowlist of the external resource to allow connections from this cluster. Read more

```
3.14.223.77/32
3.19.107.110/32
3.22.131.240/32
3.130.86.195/32
```

Copy all

Figure 5-25. *Configuring Sink Connector - 3*

You could use the Snowflake admin interface or SQL to create and apply network policies.

The following shows the general syntax.

```
use role accountadmin;
create or replace policy default_policy allowed_ip_list =
('3.14.223.77/32',...);
alter account set network_policy = 'default_policy'
```

6. In the next screen, select the preferred ingestion method and format.

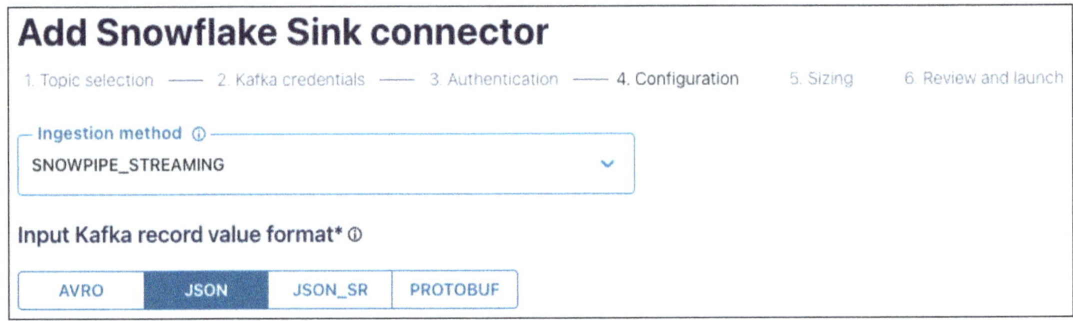

Figure 5-26. *Configuring Sink Connector - 4*

7. In the next screen, select sizing for the connector instance.

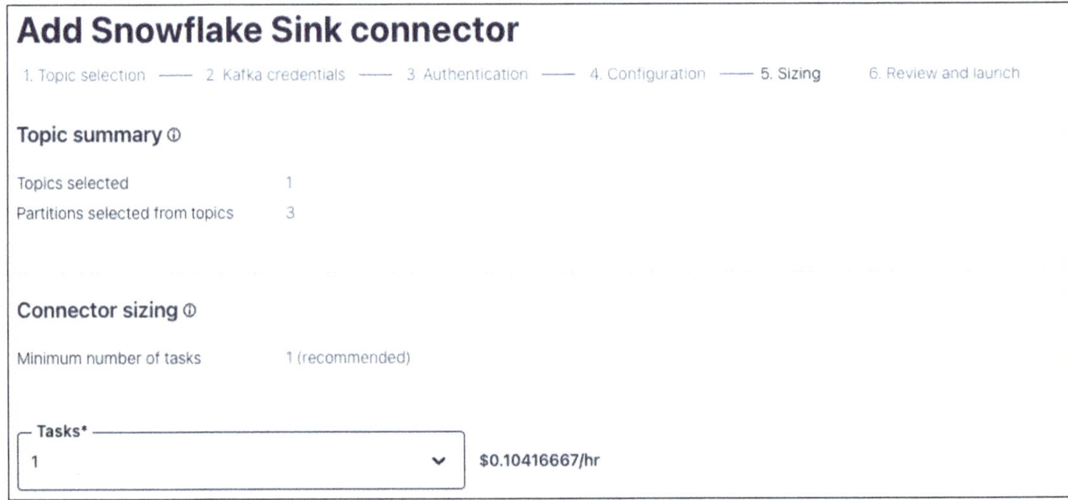

Figure 5-27. *Configuring Sink Connector - 5*

8. Set a name for the new connector.

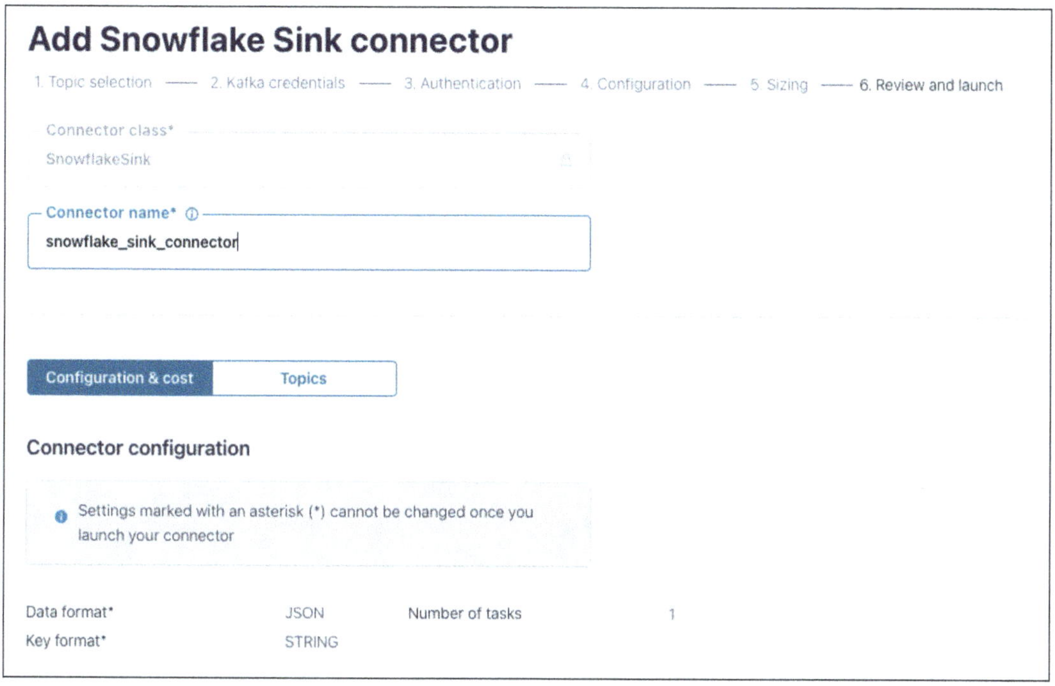

Figure 5-28. *Configuring Sink Connector - setting name*

9. Run the Python client and test the connector.

Run the Python event producer (kafka_orders_ingest.py) test the ingestion from client to kafka to Snowflake.

If the events are received, you will see the bytes consumed have increased.

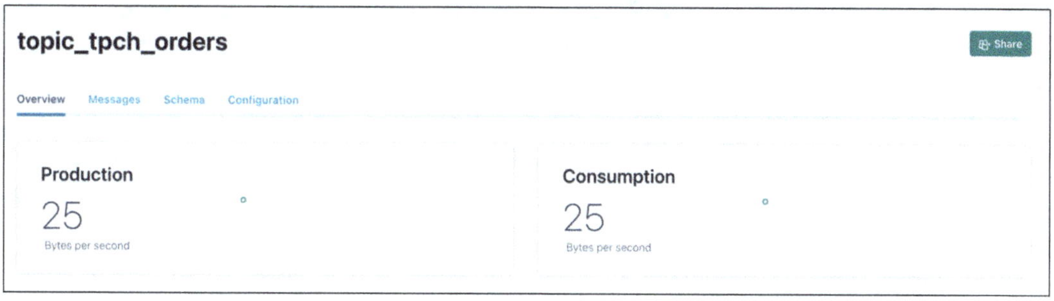

Figure 5-29. *Snowflake Sink*

If the events are routed via the connector, you should see a non-zero value for the messages processed.

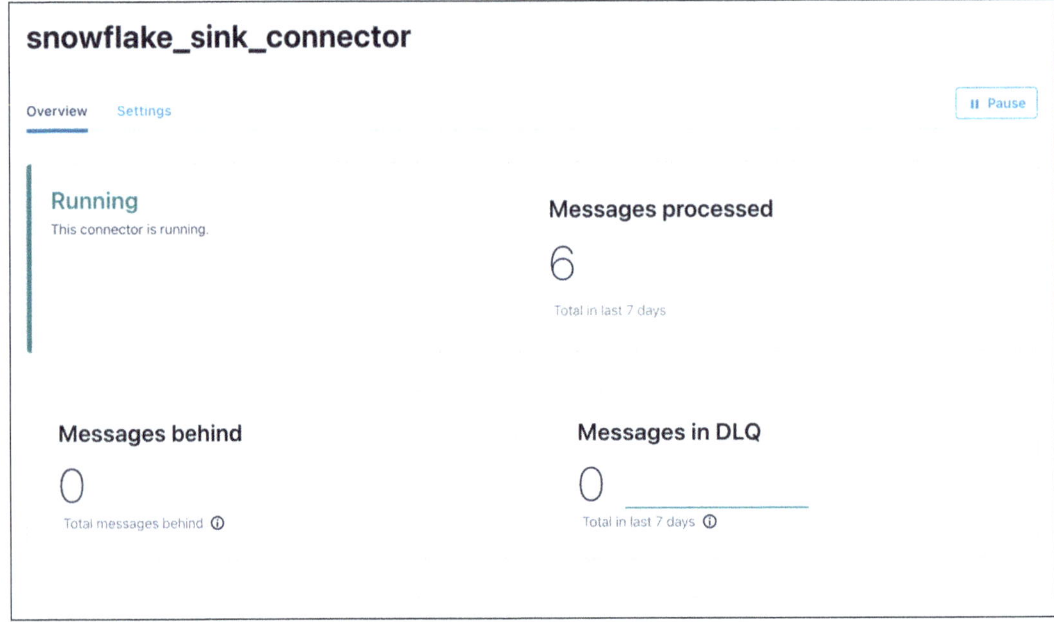

Figure 5-30. *Monitoring Sink*

Finally, if it's in Snowflake, you should see the table created and populated with two columns: record_metadata and record_content.

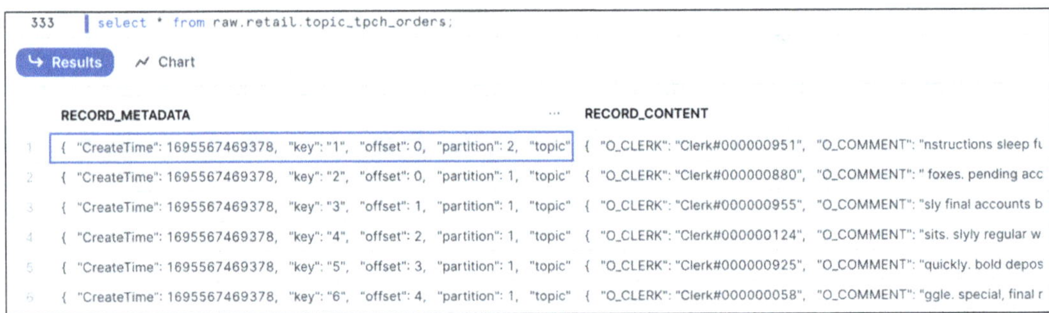

Figure 5-31. *Query Results*

The following is a sample RECORD_METADATA value.

```
{
  "CreateTime": 1695567469378,
  "key": "1",
  "offset": 0,
  "partition": 2,
  "topic": "topic_tpch_orders"
}
```

What we receive as part of RECORD_METADATA could be controlled via the following properties.

- **snowflake.metadata.createtime**: If this value is set to false, the CreateTime property value is omitted from the metadata column. The default is true.
- **snowflake.metadata.topic**: If this value is set to false, the topic property value is omitted from the metadata column. The default is true.
- **snowflake.metadata.offset.and.partition**: If the value is set to false, the Offset and Partition property values are omitted from the metadata column. The default is true.
- **snowflake.metadata.all**: If the value is set to false, the metadata in the metadata column is empty. The default value is true.

Note that the Snowflake connector sends files based on these connector configuration properties.

```
Buffer.count.records = 10000
Buffer.flush.time = 120 seconds
Buffer.flush.size = 5 MB
```

Additional Snowflake Documentation

`https://docs.snowflake.com/en/user-guide/kafka-connector-overview`

`https://docs.confluent.io/cloud/current/connectors/cc-snowflake-sink/cc-snowflake-sink.html`

Recipe 5-4. Change Tracking

Problem

Let's say you are faced with a situation when you need control over reading the changes on a table instead of reading off from a stream object. How can we enable this change tracking, and how is it different from streams?

Solution

As you saw in the previous section, a stream object records DML changes made to tables and metadata about each change so that actions can be taken using the changed data.

Now, using the CHANGES clause on a table or view enables you to query the change tracking metadata for that table or view within a specified interval without creating a stream with an explicit transactional offset.

How does it differ for streams?

- In the CHANGES clause, the user or the client code controls the interval of scan (i.e., the window for which you need the metadata, whereas when you create a STREAM), the stream automatically records the changes for you in the stream, and you are to query the stream to read the latest DML changes.
- Using the CHANGES clause, you can run multiple queries multiple times. You can still retrieve the change tracking metadata between different transactional start and endpoints, but in STREAMS, it is removed from the stream once you consume the data.

- The change tracking is enabled for a table when either is true.
    - Change tracking is enabled on the table (using ALTER TABLE ... CHANGE_TRACKING = TRUE)
    - A stream is created for the table (using CREATE STREAM).
- Using the CHANGES clause gives you more flexibility to choose and consume the changes between any time window of your choice and lets you read the metadata any number of times. But it comes with the complexity of managing the time intervals and the last consumed offset.
- The AT | BEFORE clause is required for querying the changes metadata. It is used to set the current offset for the change tracking metadata.
- The optional END clause sets the end timestamp for the change interval. If no END value is specified, the current timestamp is used as the end of the change interval.
- The value for TIMESTAMP or OFFSET must be a constant expression.
- The smallest time resolution for TIMESTAMP is milliseconds.
- Both streams and changes metadata cannot go beyond the set Time Travel on the table.

The following is an example of using change tracking but enabled by creating a stream

```
create or replace table RAW_CUSTOMER_SRC (
  C_CUSTKEY number
  , C_NAME varchar
  , C_ADDRESS varchar
);

-- we create a stream
create or replace stream customer_stream on table RAW_CUSTOMER_SRC;

-- insert data into the raw table
insert into RAW_CUSTOMER_SRC (c_custkey, c_name, c_address)
```

```
select c_custkey, c_name, c_address from SNOWFLAKE_SAMPLE_DATA.TPCH_SF1.
CUSTOMER  where c_custkey= 60001;

-- check the stream
select * from customer_stream;

-- insert data again into the raw table
insert into RAW_CUSTOMER_SRC (c_custkey, c_name, c_address)
select c_custkey, c_name, c_address from SNOWFLAKE_SAMPLE_DATA.TPCH_SF1.
CUSTOMER  where c_custkey= 60002;

-- mark the timestamp
set t1 = (DATEADD(minute, -2, current_timestamp));

 -- Query the change tracking metadata in the table during the interval
from $t1 to the current time.
 -- Return the full delta of the changes.
 select *
 from raw_customer_src
   changes(information => default)
   at(timestamp => $t1 );

-- Consume the stream

create table CUSTOMER_SRC (
  C_CUSTKEY number
  , C_NAME varchar
  , C_ADDRESS varchar
)

merge into CUSTOMER_SRC c
using ( select * from customer_stream where METADATA$ACTION='INSERT') s
on s.c_custkey = s.c_custkey
when MATCHED then update set c.c_address = s.c_address;

-- check the stream
select * from customer_stream;
-- it should show empty stream

-- change the timestamp
```

```
set t1 = (DATEADD(minute, -3, current_timestamp));
```

```
-- Check the metadata again and it should still work
-- You might have to change how many minutes (-3 worked for me but you
could place a static timestamp value too)
-- you need to set the variable t1
select *
 from raw_customer_src
   changes(information => default)
   at(timestamp => $t1 );
```

You could enable change tracking also using the CHANGES clause.

```
alter table raw_customer_src set change_tracking = TRUE;
```

When querying the changes you could use the DEFAULT or APPEND_ONLY option.

```
select *
 from raw_customer_src
   changes(information => append_only)
   at(timestamp => $t1 );
```

This returns appended rows only. Querying append-only changes can be more performant than querying standard (default) changes.

Additional Snowflake Documentation

https://docs.snowflake.com/en/sql-reference/constructs/changes

---

Recipe 5-5. Dynamic Tables

Dynamic tables are a type of table in Snowflake that automatically materialize the results of a query as a table.

Dynamic Tables can join and aggregate across multiple source objects and incrementally update results as sources change. You could also chain dynamic tables and thereby create a simple ETL pipeline.

As of this writing, it is in private preview and does not support certain non-deterministic built-in functions nor any custom UDF/UDTF function defined as VOLATILE (which marks it as non-deterministic).

Dynamic tables only support SQL and not Snowpark code but you can call your Snowpark UDFs as part of the SQL code. You cannot create dynamic tables on top of external tables, streams, or materialized views.

Problem

Consider a situation when your data team wants you to build a summary model of order information that is always current. The solution should be simple with no additional complexity.

Solution

There are several ways to transform your source data to create summary models. For example, you could use streams and tasks, CTAS, custom stored procedures, or snowpark code.

Snowflake introduced a feature called dynamic tables which is an excellent option when the following conditions are true.

- You don't want to write code to track data dependencies and manage data refresh.
- You don't need or want to avoid the complexity of streams and tasks.
- You need to materialize the results of a query of multiple base tables.
- You don't need fine-grained refresh schedule control.
- Your requirement is simple and does not facilitate the use of stored procedures or snowpark code.

To demonstrate the situation, consider a source table called ORDERS and CUSTOMERS. The requirement is to build a summary table that shows the total order price by order status and customer segment.

We utilize the TPCH data provided within the Snowflake sample database to feed our tables with information.

```
create or replace TABLE RAW.RETAIL.ORDERS (
  O_ORDERKEY NUMBER(38,0),
  O_CUSTKEY NUMBER(38,0),
  O_ORDERSTATUS VARCHAR(1),
  O_TOTALPRICE NUMBER(12,2),
  O_ORDERDATE DATE,
  O_ORDERPRIORITY VARCHAR(15)
);

create or replace TABLE RAW.RETAIL.CUSTOMERS (
  C_CUSTKEY NUMBER(38,0),
  C_NAME VARCHAR(25),
  C_ADDRESS VARCHAR(40),
  C_NATIONKEY NUMBER(38,0),
  C_MKTSEGMENT VARCHAR(10)
);

insert into RAW.RETAIL.CUSTOMERS
select C_CUSTKEY, C_NAME, C_ADDRESS, C_NATIONKEY, C_MKTSEGMENT from
SNOWFLAKE_SAMPLE_DATA.TPCH_SF100.CUSTOMER;

insert into RAW.RETAIL.ORDERS
select O_ORDERKEY, O_CUSTKEY, O_ORDERSTATUS, O_TOTALPRICE, '2024-01-01',
O_ORDERPRIORITY
from SNOWFLAKE_SAMPLE_DATA.TPCH_SF100.ORDERS where date(O_
ORDERDATE)='1992-01-01';
```

The syntax for creating dynamic tables is as follows.

```
create or replace dynamic table <table name>
  lag = ' 1 hour'
  warehouse = COMPUTE_WH
  as <select... query>;
```

This example creates a summary table called ORDER_SUMMARY.

```
create or replace dynamic table RAW.RETAIL.ORDER_SUMMARY
 TARGET_LAG = '5 minutes'
  WAREHOUSE = COMPUTE_WH
  AS
    select o_orderstatus, c_mktsegment, sum(o_totalprice) as TOTAL_PRICE
    from RAW.RETAIL.orders o
inner join RAW.RETAIL.CUSTOMERS c on c.c_custkey=o.o_custkey group by all;
```

Check whether the table is loaded.

```
select * from RAW.RETAIL.ORDER_SUMMARY;
```

You can look for all dynamic tables within a schema using the show command.

```
show dynamic tables in schema raw.retail;
```

Try inserting more records into orders to test if the dynamic table is loaded fine.

```
insert into RAW.RETAIL.ORDERS
select O_ORDERKEY, O_CUSTKEY, O_ORDERSTATUS, O_TOTALPRICE, '2024-01-02',
O_ORDERPRIORITY
from SNOWFLAKE_SAMPLE_DATA.TPCH_SF100.ORDERS where date(O_ORDERDATE)=
'1992-01-02';
```

```
select * from RAW.RETAIL.ORDER_SUMMARY;
```

Another option to view the dynamic table information is from the asset explorer in Snowflake UI (Snowsight), as shown in Figure 5-32.

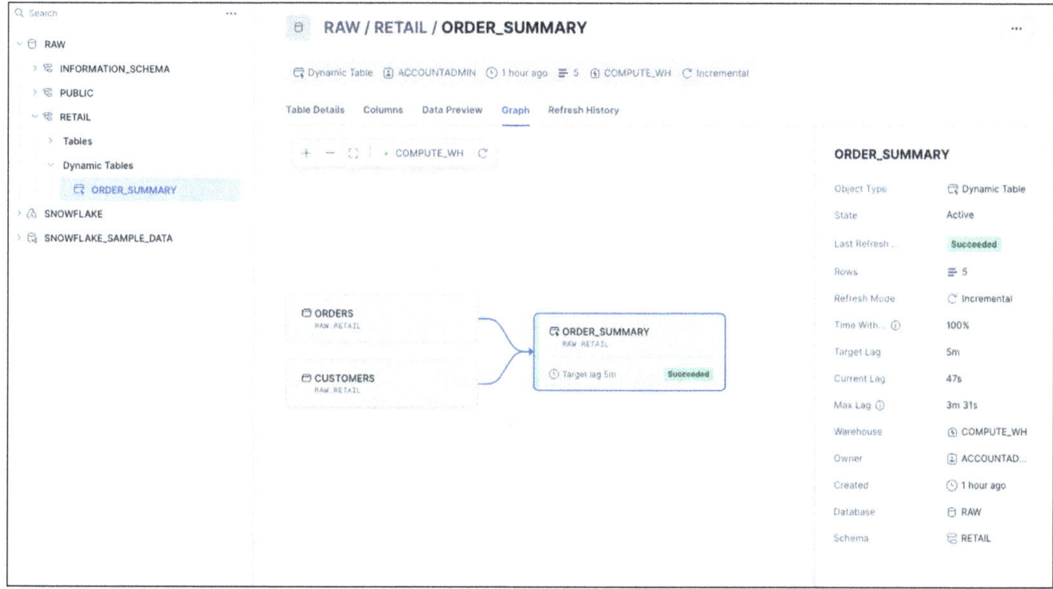

Figure 5-32. *Chained Dynamic Tables*

Note the following features and challenges of dynamic tables.

- You can create a chain of dynamic tables.
- Dynamic tables work by enabling change tracking on the underlying source tables.
- The dynamic table refresh process operates in one of two ways.
    - Incremental refresh: This automated process analyzes the dynamic table's query and calculates changes since the last refresh. It then merges these changes into the table.
    - Full refresh: When the automated process can't perform an incremental refresh, it conducts a full refresh.
- A dynamic table's refresh mode is determined at creation time and immutable afterward. If not specified explicitly, the refresh mode defaults to AUTO, which selects a refresh mode based on factors such as query complexity or unsupported constructs, operators, or functions.

- As of this writing, it does not support the following.
    - Certain non-deterministic functions (but functions like SEQ, CURRENT_ROLE, CURRENT_DATE, etc., are supported)
    - Any UDF/UDTF function defined as VOLATILE (which marks it as non-deterministic)
    - External functions and Snowpark transformations written in Python, Java, or Scala
    - Sources that are external tables, streams, and materialized views
    - Sources that are views on dynamic tables
- Using ORDER BY in a dynamic table's definition might produce results sorted in an unexpected order. You can use ORDER BY when querying the dynamic table.
- You could create streams over dynamic tables but not the other way around.

Additional Snowflake Documentation

`https://docs.snowflake.com/en/user-guide/dynamic-tables-refresh`

Recipe 5-6. Iceberg Tables

Iceberg is an open source table format that Netflix originally developed to address various challenges encountered within Apache's Hive Hadoop project. After its initial development in 2018, Netflix donated Iceberg to the Apache Software Foundation as a completely open source, openly managed project. It remedies many of the shortcomings of its predecessors like Hive. It has quickly become one of the most popular open source table formats among others like Hudi and Delta Lake.

Apache Iceberg decouples the processing engine from the table format, so you can choose any processing engine that works for you. Many other processing engines like Spark, Dermio, and Presto now support the Iceberg format.

Snowflake supports Iceberg tables using the Apache Parquet file format, but Apache Iceberg works on top of formats like Parquet, ORC, and Avro.

In Snowflake, you could use the new Apache Iceberg table format if you want to build your data platform on open standards to benefit from this interoperability and the ACID compliance, schema evolution, and Time Travel. This also means your data is always stored in an external cloud storage like AWS S3 or Azure Blob Storage that the user manages.

Snowflake can have different types of Iceberg tables depending on where the catalog and metadata are managed for Iceberg tables.

- Snowflake-managed Iceberg tables
    - Snowflake manages the metadata and catalog for these tables.
    - These tables can support all Snowflake features with read and write access.
    - This is performance optimized and has performance close to native Snowflake tables.
- Externally-managed Iceberg tables
    - An external system such as AWS Glue or Spark manages the metadata and catalog.
    - These tables can support read-only access in Snowflake and do not support automatic metadata refresh within Snowflake.
- Dynamic Iceberg tables
    - These are dynamic tables that read from Snowflake-managed Iceberg tables. You can think of this like regular dynamic tables but read off from an Iceberg table instead of a regular table.
    - Dynamic Iceberg tables that integrate with data lakes and hence let you read and write data in external cloud storage such as AWS S3 or Azure Blob Storage. This essentially brings together the benefits of dynamic tables and Snowflake-managed Iceberg tables.
    - These tables support ACID transactions, schema evolution, hidden partitioning, and table snapshots.

- Since these are dynamic tables, it uses declarative SQL to define the desired end state and lets Snowflake handle scheduling and refreshing the data.
- Performance is optimized through incremental processing, which processes only changed data to improve performance and reduce costs compared to full data refreshes.

Problem

Consider a scenario where your organization moves some of your workflows to Snowflake from your AWS platform. Most of the data is already in Apache Iceberg format defined as AWS Glue tables.

You are required to utilize this data in Snowflake without copying over the data but utilize the computing capabilities of Snowflake.

Solution

The good news is you have data in Apache Iceberg format in AWS and, hence, can leverage the existing infrastructure to use the data in Snowflake.

As the scope of this recipe is not to delve into AWS Glue in detail, we assume that some orders of mock data are already generated and loaded as Parquet files in S3 and an Iceberg table in AWS Glue.

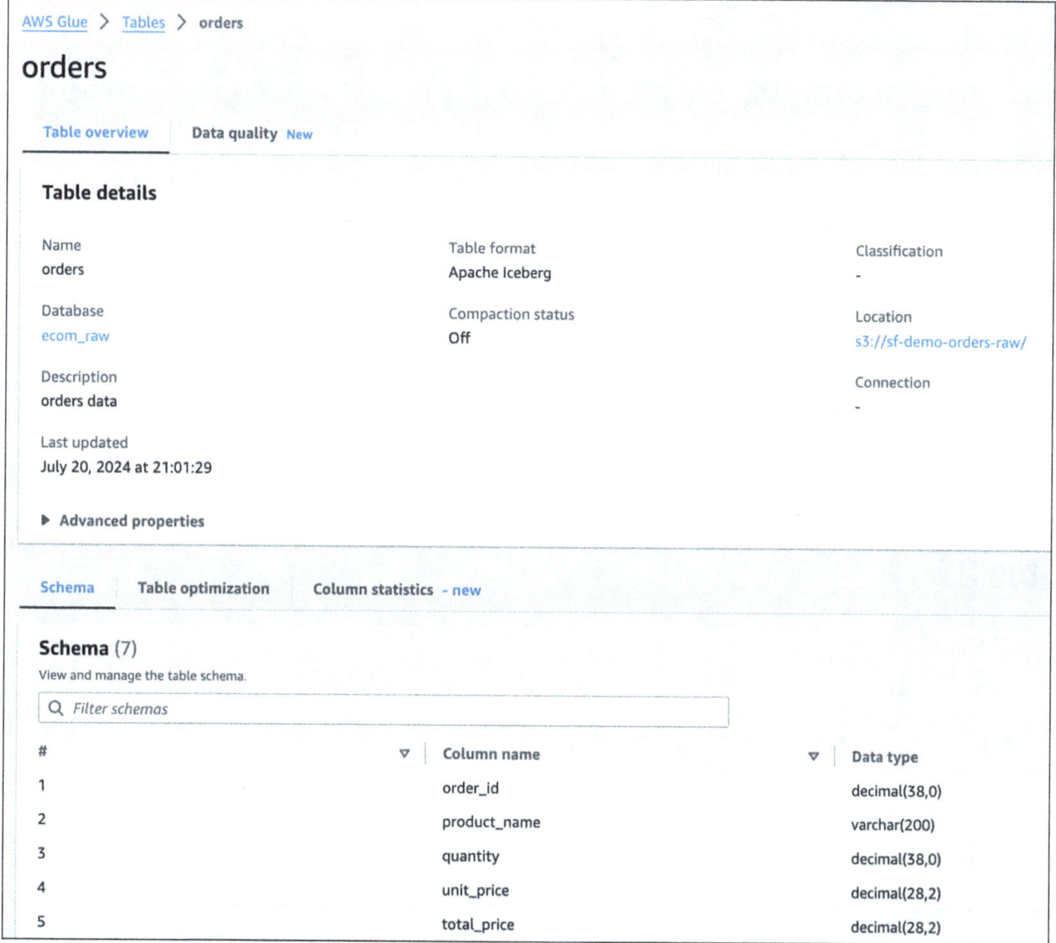

Figure 5-33. *Glue table*

Now that you have some Parquet files stored as Iceberg tables in AWS Glue, you need to perform the following steps to integrate and use the AWS data from Snowflake.

Step 1

Configure the Snowflake external volume integration with S3.

Create an external volume and configure it to work with your Snowflake account.

```
CREATE OR REPLACE EXTERNAL VOLUME EXT_VOL_S3_DATA
  STORAGE_LOCATIONS =
     (
        (
```

```
            NAME = '<name>'
            STORAGE_PROVIDER = 'S3'
            STORAGE_BASE_URL = 's3://<enter your S3 bucket name>/<iceberg
            data path>/'
            STORAGE_AWS_ROLE_ARN = 'arn:aws:iam::<enter your AWS account
            ID>:role/<your AWS Role that was created>'
         )
);
```

- STORAGE_BASE_URL: enter your S3 bucket name
- STORAGE_AWS_ROLE_ARN: enter your AWS account ID and the role

Step 2

Grant the IAM user permissions to access bucket objects.

Run the describe external command to get the Snowflake values of the object.

```
DESC EXTERNAL VOLUME EXT_VOL_S3_DATA;
```

The STORAGE_LOCATIONS output row has a property value in JSON that looks like the following.

```
{"NAME":"<name>",
"STORAGE_PROVIDER":"S3",
"STORAGE_BASE_URL":"s3://<s3 path>/",
"STORAGE_ALLOWED_LOCATIONS":["s3://<s3 path>,"],
"STORAGE_AWS_ROLE_ARN":"arn:aws:iam::< AWS account ID>:role/<AWS Role>",
"STORAGE_AWS_IAM_USER_ARN":"arn:aws:iam::********:user/*******",
"STORAGE_AWS_EXTERNAL_ID":"*********************",
"ENCRYPTION_TYPE":"NONE",
"ENCRYPTION_KMS_KEY_ID":""
}
```

Note the STORAGE_AWS_IAM_USER_ARN and STORAGE_AWS_EXTERNAL_ID values.

Next, paste the preceding values into the AWS IAM role's trust relationships.

The following is a sample trust relationship JSON value.

```
{
    "Version": "2012-10-17",
    "Statement": [
      {
        "Effect": "Allow",
        "Principal": {
            "Service": "glue.amazonaws.com"
        },
        "Action": "sts:AssumeRole"
        },
        {
        "Effect": "Allow",
        "Principal": {
        "AWS": "<snowflake storage arn>"
        },
        "Action": "sts:AssumeRole",
        "Condition": {
        "StringEquals": {
            "sts:ExternalId": "<snowflake external id ext volume>"
            }
        }
      },
      {
      "Effect": "Allow",
      "Principal": {
      "AWS": "<snowflake glue arn>"
      },
      "Action": "sts:AssumeRole",
      "Condition": {
      "StringEquals": {
            "sts:ExternalId": "<snowflake external id glue catalog>"
```

```
                }
            }
        }
    ]
}
```

From the Snowflake output, copy the value from STORAGE_AWS_IAM_USER_ARN and paste it into the IAM policy by replacing <snowflake storage arn> with that value.

Next, copy the Snowflake STORAGE_AWS_EXTERNAL_ID and paste it into the IAM policy by replacing the <snowflake external id ext volume>.

We will update the external ID for the AWS Glue Data Catalog later.

Step 3

Now, we create the catalog integration, which is a native Snowflake object that helps to integrate with the AWS Glue Data Catalog.

```
CREATE or REPLACE CATALOG INTEGRATION GLUE_CATALOG_INT
  CATALOG_SOURCE=GLUE
  CATALOG_NAMESPACE='iceberg'
  TABLE_FORMAT=ICEBERG
  GLUE_AWS_ROLE_ARN='arn:aws:iam::<enter your AWS account ID>:role/<your
  AWS Role that was created>'
  GLUE_CATALOG_ID='<enter your AWS account ID>'
  ENABLED=TRUE;
```

- GLUE_AWS_ROLE_ARN: enter your AWS account ID and AWS Role that was created
- GLUE_CATALOG_ID: enter your AWS account ID

Step 4

Grant the IAM user permissions to access the Glue Catalog.

Run the describe external command to get the Snowflake values of the object.

```
DESC CATALOG INTEGRATION GLUE_CATALOG_INT;
```

Capture the values shown for GLUE_AWS_IAM_USER_ARN and GLUE_AWS_EXTERNAL_ID.

Go back to the IAM trust policy and update the following.

- <snowflake glue arn> with the value from the Snowflake output GLUE_AWS_IAM_USER_ARN
- <snowflake external id glue catalog> with the value from the Snowflake output GLUE_AWS_EXTERNAL_ID

Step 5

We can now create an Iceberg table by referencing the AWS Glue Data Catalog.

```
CREATE OR REPLACE ICEBERG TABLE ORDERS_TBL
EXTERNAL_VOLUME='EXT_VOL_S3_DATA'
CATALOG='GLUE_CATALOG_INT'
CATALOG_TABLE_NAME='ORDERS';
```

The AWS Glue Data Catalog should recognize CATALOG_TABLE_NAME.

Once the Iceberg table is defined, you can query it like any other table.

```
select * from ORDERS_TBL
```

Problem

Consider another scenario where your organization requires you to build its data lakehouse in Snowflake but on open standards like Apache Iceberg. You are also required to use AWS S3 for storage considering the majority of its current and previous operations are deployed and managed in AWS.

Solution

Let's assume that, as the first step, the data team is trying to define and store some of its customer data. We use the Snowflake-provided TPCH sample data to demonstrate how it can be stored and managed using Snowflake-managed Iceberg tables.

There are two Snowflake objects that you create to define this managed table.

First, you create an external volume that connects Snowflake to any of the external storage locations for the Iceberg data files. Next, you create the iceberg table using the standard create table syntax or CTAS (if a default volume and catalog parameters are set for the database in use).

Step 1: Create an External Volume

An external volume is a native Snowflake object that stores information about your cloud storage locations and identity and access management (IAM) entities. The syntax and steps differ based on the cloud object storage that you choose.

External volumes are account-level objects; a single external volume can support one or more Iceberg tables.

Considerations for creating external volumes.

- Snowflake only supports external volumes to S3 storage buckets in the same region that hosts your Snowflake account.
- Snowflake doesn't support external volumes with bucket names that contain dots (for example, my.s3.bucket).
- Snowflake uses virtual-hosted-style paths and HTTPS to access data in S3. However, S3 does not support SSL for virtual-hosted-style buckets with dots in the name.
- Permissions in AWS to create and manage IAM policies and roles.

In this case, we need to create an external volume for AWS S3 US East 2 since the Snowflake account is hosted there.

Step 1-1: Create an IAM Policy to Grant Access to S3

1. Log in to the AWS IAM dashboard.
2. Select Policies and Create Policy.
3. Use JSON to specify your policy and paste or type in the contents of your policy.

   You can use a policy definition like the following. Be sure to appropriately replace it with your S3 bucket name.

```
{
   "Version": "2012-10-17",
   "Statement": [
         {
            "Effect": "Allow",
            "Action": [
               "s3:PutObject",
               "s3:GetObject",
               "s3:GetObjectVersion",
               "s3:DeleteObject",
               "s3:DeleteObjectVersion"
            ],
            "Resource": "arn:aws:s3:::<my_bucket>/*"
         },
         {
            "Effect": "Allow",
            "Action": [
               "s3:ListBucket",
               "s3:GetBucketLocation"
            ],
            "Resource": "arn:aws:s3:::<my_bucket>",
            "Condition": {
               "StringLike": {
                     "s3:prefix": [
                        "*"
                     ]
               }
            }
         }
   ]
}
```

4. Save the policy with a name.

Step 1-2: Create an IAM Role

1. Create a role in the AWS IAM console.
2. Click "Create role" and choose AWS Account as the Trusted entity type.
3. Under the "An AWS account" section, select the "This account" option. Select the check box.
4. Select the "Require external ID" option. Enter an external ID of your choice; for example, iceberg_external_volume_external_id. In a later step, you modify the trust relationship and grant access to Snowflake.
5. Click Next to go to the screen where you select the policy that needs to be attached to the role. Select the policy that you created earlier.
6. Click Next to enter the role name and finish the role creation process.

Step 1-3: Create an External Volume in Snowflake

Now, you create an external volume in Snowflake using the following syntax.

```
CREATE OR REPLACE EXTERNAL VOLUME iceberg_external_volume
  STORAGE_LOCATIONS =
     (
        (
           NAME = 'my-s3-us-west-2'
           STORAGE_PROVIDER = 'S3'
           STORAGE_BASE_URL = 's3://<my_bucket>/'
           STORAGE_AWS_ROLE_ARN = '<arn value of your AWS role>'
           STORAGE_AWS_EXTERNAL_ID = '<external id created earlier>'
        )
     );
```

The following is an example based on this scenario.

```
CREATE OR REPLACE EXTERNAL VOLUME iceberg_external_volume
STORAGE_LOCATIONS =
 (
    (
       NAME = 's3-us-east-2'
       STORAGE_PROVIDER = 'S3'
       STORAGE_BASE_URL = 's3://sf-demo-customers-iceberg/'
       STORAGE_AWS_ROLE_ARN = 'arn:aws:iam::296080767349:role/s3-role-
       snowflake-iceberg'
       STORAGE_AWS_EXTERNAL_ID = 'iceberg_external_volume_external_id'
    )
 );
```

This definition is based on the fact that the storage bucket used is sf-demo-customers-iceberg, and the external ID specified during the AWS role creation is iceberg_table_external_id.

Step 1-4: Grant IAM User Permission to Access Bucket Objects

1. Describe the external volume that was previously created.
2. Since our external volume only uses one storage location, look for the value of the property STORAGE_LOCATION_1.

	parent_property	property	property_type	property_value	property_default
1		ALLOW_WRITES	Boolean	true	true
2	STORAGE_LOCATIONS	STORAGE_LOCATION_1	String	{"NAME":"s3-us-east-2","STORAGE_PROVIDER":"S3","STORAGE_BASE_URL":"s3://sf-	
3	STORAGE_LOCATIONS	ACTIVE	String		

Figure 5-34. *Results of Describe External Volume*

3. Capture the STORAGE_AWS_IAM_USER_ARN value from the property value. This is the AWS IAM user created for your Snowflake account. You may capture the STORAGE_AWS_EXTERNAL_ID if you skipped it during the volume creation process. In this case, we explicitly mentioned it, which is used as the external ID.

4. Log in to the AWS IAM dashboard and select the previously created role for the external volume.

5. Select the Trust Relationships tab to edit the trust policy.

6. Modify the principal with the STORAGE_AWS_IAM_USER_ARN value that was recorded earlier. As stated earlier, if you hadn't supplied an external ID, you must replace the externalId value with what was captured for STORAGE_AWS_EXTERNAL_ID.

IAM > Roles > s3-role-snowflake-iceberg > Edit trust policy

Edit trust policy

```
{
    "Version": "2012-10-17",
    "Statement": [
        {
            "Effect": "Allow",
            "Principal": {
                "AWS": "arn:aws:iam::119873109848:user/oz630000-s"
            },
            "Action": "sts:AssumeRole",
            "Condition": {
                "StringEquals": {
                    "sts:ExternalId": "iceberg_external_volume_external_id"
                }
            }
        }
    ]
}
```

Figure 5-35. *Editing IAM Trust Policy*

7. Select "Update policy" to save your changes.

Step 2: Create an Iceberg Table

Create an Iceberg table using the standard CREATE ICEBERG TABLE syntax.

We have specified CATALOG = 'SNOWFLAKE' so that the table uses Snowflake as the Iceberg catalog.

Though the BASE_LOCATION parameter is optional, we have used it to instruct Snowflake where to write table data and metadata. Leaving it empty causes Snowflake to write data and metadata to the location specified in the external volume definition.

Since we used the table name itself as the base location, it helps to organize the data within the volume as Snowflake writes data and metadata under a directory with the same name as this table name.

Soon after creating the table, you notice that Snowflake has created a folder in the bucket.

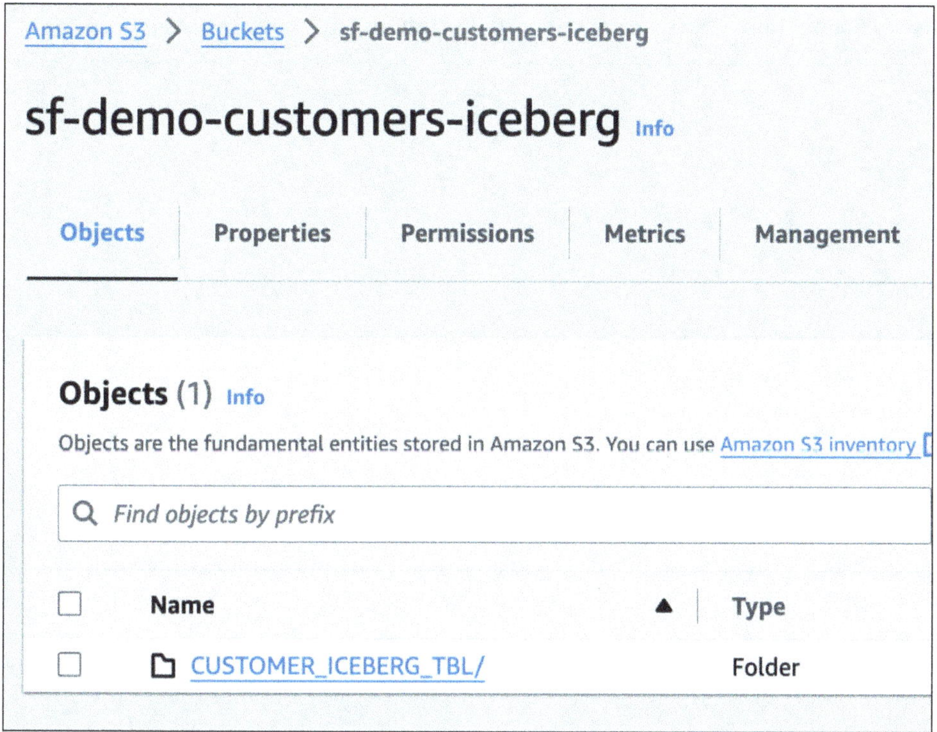

Figure 5-36. *Iceberg Table*

Let's add some sample data and see how it looks.

```
insert into CUSTOMER_ICEBERG_TBL
 select c_custkey, c_name, c_nationkey, c_phone, c_acctbal FROM snowflake_
sample_data.tpch_sf1.customer;
```

If you check in AWS S3, you notice that the folder now contains data and metadata folders, the data folder containing the Parquet files, and the metadata folder containing JSON and Avro files.

Additional Snowflake Documentation

`https://docs.snowflake.com/en/user-guide/tables-iceberg`

`https://docs.snowflake.com/user-guide/tables-iceberg-create`

`https://docs.snowflake.com/user-guide/tables-iceberg-manage`

CHAPTER 6

Programmable Data Pipelines

Though SQL is a fundamental skill for data engineers, one might need to go beyond plain SQL often in today's world. Using programming languages (like Python) and other frameworks like Spark provides the flexibility to write complex programmatic constructs within their data pipeline.

Language constructs and data structures like DataFrames have gained significant popularity over SQL and traditional drag-and-drop ETL (extract, transform, load) tools in the world of data for several reasons.

- **Flexibility and scalability**: DataFrames, which are essentially two-dimensional tabular data structures with functionalities and features built around them, offer a high degree of flexibility and scalability. They can handle large volumes of data efficiently and support a wide range of data manipulation operations, making them well-suited for complex data transformations and processing pipelines. In contrast, drag-and-drop ETL tools can be limited in their ability to handle highly complex or dynamic data scenarios.
- **Code-based approach**: Programmable data pipelines using DataFrames leverage programming languages like Python, allowing data engineers and analysts to write code to define and execute their data pipelines. This code-based approach promotes transparency, version control, and reproducibility, which are crucial in data-driven environments. Additionally, it enables seamless integration with other data processing tools and libraries within the same programming ecosystem.

D. Dayton and J. Eipe, *Snowflake Recipes*, https://doi.org/10.1007/979-8-8688-0938-5_6

- **Integration with machine learning and data science**: By using programmable data pipelines with DataFrames, organizations can seamlessly integrate data preprocessing, feature engineering, and model training stages, enabling a more streamlined and cohesive data science lifecycle.
- **Community and ecosystem**: Programming languages and data processing frameworks like Python, R, and Apache Spark have active and vibrant communities that contribute to developing libraries, tools, and documentation. This rich ecosystem provides access to a wide range of resources, including prebuilt functions, utilities, and best practices, enabling faster development and easier maintenance of data pipelines.
- **Customization and extension**: With a code-based approach, data engineers and analysts can easily customize and extend their data pipelines to meet specific business requirements or adapt to changing data sources or formats. This level of customization is often more challenging or limited in drag-and-drop ETL tools, which may have rigid or proprietary architectures.

This chapter looks at how to use the Snowflake connectors, stored procs, Snowpark API, and an in-depth exploration of the Python implementation of Snowpark.

Recipe 6-1. Using Client APIs

This recipe looks at different methods to connect to Snowflake from your tooling or framework.

Most ETL tools and frameworks now support Snowflake out of the box. That means most GUI-based tools would either have the Snowflake drivers installed or provide an easy way to add the Snowflake driver or connector to the user.

You choose the appropriate driver or connector if you are required to install/add it by yourself. Assuming you have pipelines written and run from custom Java frameworks, you would use the JDBC driver; if you use Python-based frameworks (like Airflow), you would select an appropriate Python connector from the PyPI package repository.

The following sections demonstrate how to use the Python client API and Java JDBC API to run Snowflake SQL.

Note that we already used the Python connector in Chapter 5 to work with the Snowpipe API example.

The scope of the following sections is to provide only an introduction and not an in-depth analysis of every connector/driver available for Snowflake.

Problem

Your organization has already incorporated Snowflake as its primary cloud data warehouse solution. You are now tasked with repointing your SQL used in your data orchestration tools like Airflow to connect to Snowflake with minimal code or infrastructure changes.

Solution

Since Apache Airflow is a framework built and run on Python runtime, you would need to look for an appropriate Snowflake Python Connector.

Snowflake Python Connector

The Snowflake documentation states that the connector is a native Python package with no dependencies on JDBC or ODBC. It can be installed using pip on Linux, macOS, and Windows platforms where Python is installed.

The connector supports developing applications using the Python Database API v2 specification (PEP-249), including using the following standard API objects.

- Connection objects for connecting to Snowflake
- Cursor objects for executing DDL/DML statements and queries

Step 1: Install the Connector

You can get the Python connector from the official Snowflake drivers documentation page [`https://docs.snowflake.com/en/developer-guide/python-connector/python-connector`], which eventually directs you to download it using the Python pip package installer.

Open your terminal and install the latest connector using the command,

```
$ pip install snowflake-connector-python
```

Note that the latest version requires Python version 3.8 or later.

Step 2: Test the Connector

```
import snowflake.connector

conn = snowflake.connector.connect(
    user='****',
    password='**********',
    account='vx15608.us-east-2.aws',
    warehouse='XSMALL_WH',
    database='RAW',
    schema='RETAIL'
    )
```

Replace all the preceding parameters with what is appropriate to your Snowflake account.

```
cur = conn.cursor()
try:
    cur.execute("SELECT current_date, current_timestamp")
    for (col1, col2) in cur:
        print('{0}, {1}'.format(col1, col2))
finally:
    cur.close()
```

Snowflake provides utility functions like fetchone, fetchmany, and fetchall.

You may use fetchone or fetchmany if the result set is too large to fit into Python's memory.

Do note that on Snowflake, it runs the query as it is provided, and using functions like fetchone has no effect.

If you need to reduce the payload size (network payload) and compute the Snowflake warehouse cost, you need to change the query appropriately (e.g., by using a limit clause at the end).

```
col1, col2 = conn.cursor().execute("select C_NAME, C_ACCTBAL from
SNOWFLAKE_SAMPLE_DATA.TPCH_SF1.CUSTOMER").fetchone()
print('{0}, {1}'.format(col1, col2))

cur = conn.cursor().execute("select C_NAME, C_ACCTBAL from SNOWFLAKE_
SAMPLE_DATA.TPCH_SF1.CUSTOMER")
ret = cur.fetchmany(3)

results = conn.cursor().execute("select C_NAME, C_ACCTBAL from SNOWFLAKE_
SAMPLE_DATA.TPCH_SF1.CUSTOMER").fetchall()
for rec in results:
    print('%s, %s' % (rec[0], rec[1]))
```

Snowflake Connector also provides a dictcursor that returns the resultset as dictionary pairs.

```
from snowflake.connector import DictCursor
cur = conn.cursor(DictCursor)
try:
    cur.execute("select C_NAME, C_ACCTBAL from SNOWFLAKE_SAMPLE_DATA.TPCH_
    SF1.CUSTOMER")
    for rec in cur:
        print('{0}, {1}'.format(rec['C_NAME'], rec['C_ACCTBAL']))
finally:
    cur.close()
```

This code is available in the repository [file: snowflake_py_conn_client].

The SQL code would not require many changes since Snowflake supports standard SQL, including a subset of ANSI SQL:1999 and the SQL:2003 analytic extensions. Snowflake also supports common variations for many commands where those variations do not conflict with each other. For example, Snowflake only has one true fixed point data type called NUMBER. It also supports a variety of variations like NUMERIC, INT, INTEGER, BIGINT, SMALLINT, TINYINT, and BYTEINT so that the same definition and code works without issues when it is ported from another database.

Problem

Your organization has already incorporated Snowflake as its primary cloud data warehouse solution, and it has certain Java applications that are used to run SQL against the prior data warehouse. These Java functions now need to run SQL against Snowflake.

Solution

Assuming that the application is a stand-alone Java application, changing the code and connecting to Snowflake using the Snowflake JDBC driver is easy.

Snowflake JDBC Driver

Snowflake provides a JDBC type 4 driver that supports core JDBC functionality. The recommended version of the JDBC driver is >= 3.15.0 and must be installed in a 64-bit environment and requires Java LTS (long-term support) versions 1.8 or higher.

The driver can be used with any Java application using the JDBC programming interface.

Step 1: Install the JDBC driver

Download the driver [`https://repo1.maven.org/maven2/net/snowflake/snowflake-jdbc`] manually or using your build platform like Maven.

Add it to the compilation path of your Java program.

Step 2: Test the Connector

```
// build connection properties
Properties properties = new Properties();
properties.put("user", "****");
properties.put("password", "**********");
properties.put("warehouse", "XSMALL_WH");
properties.put("db", "SNOWFLAKE_SAMPLE_DATA");
properties.put("schema", "TPCH_SF1");
// properties.put("tracing", "all");

String connectStr = "jdbc:snowflake://vx15608.us-east-2.aws.
snowflakecomputing.com";
```

Replace all these parameters with what is appropriate to your snowflake account.

Since this is a JDBC connector, it follows the JDBC spec and interface. That means you could use the JDBC methods and interfaces to create the connection and statement objects.

Once the statement object is created, you could call the execute(), executeQuery(), or the executeUpdate() methods on the object.

```
Connection connection = DriverManager.getConnection(connectStr,
properties);
Statement statement = connection.createStatement();
ResultSet resultSet = statement.executeQuery("select C_NAME, C_ACCTBAL from
SNOWFLAKE_SAMPLE_DATA.TPCH_SF1.CUSTOMER limit 10");
```

This code is a runnable project in the repository [file: sfdemo-jdbc/SnowflakeConnectionDemo.java].

Do note that based on which version of Java you are using it might necessitate adding certain JVM arguments during runtime. We had to use the --add-opens=java.base/java.nio=ALL-UNNAMED argument.

The SQL code would not require many changes since Snowflake supports standard SQL, including a subset of ANSI SQL:1999 and the SQL:2003 analytic extensions. Snowflake also supports common variations for many commands where those variations do not conflict with each other. For example, Snowflake only has one true fixed point data type called NUMBER. It also supports a variety of variations like NUMERIC, INT, INTEGER, BIGINT, SMALLINT, TINYINT, and BYTEINT so that the same definition and code works without issues when it is ported from another database.

Problem

You are tasked to programmatically manage Snowflake via its API from your enterprise application. You need the ability to remotely control aspects of your Snowflake like managing users, warehouses, and tasks.

Solution

There are two ways you could manage Snowflake-specific objects like "warehouses".

- Snowflake Python API (do note that this is different from what you have seen)
- Snowflake REST API

Snowflake Python API

With the Snowflake Python API, you can manage the following objects.

- Tasks
- -Databases, schemas, and tables
- Virtual warehouses
- Resources in Snowpark Container Services, including compute pools, image repositories, and services

Step 1: Install the API Client

You can get the Python API library from the official Snowflake drivers documentation page [`https://docs.snowflake.com/en/developer-guide/snowflake-python-api/snowflake-python-installing`].

Open your terminal and install the latest connector using the following command.

```
$ pip install snowflake
```

Note that the latest version requires Python version 3.8 or later. It also installs dependency snowflake-connector-python, snowflake-snowpark-python, and snowflake.core packages.

Step 2: Test the Connector

To test the connection, you would need to create a connection object using the snowflake-connector-python package or a session object using the snowflake-snowpark-python package. You could do both without issues since both packages are installed as a dependency.

```
import snowflake.snowpark
session = snowflake.snowpark.Session.builder.configs(connection_
parameters).create()

//or

import snowflake.connector
connection = snowflake.connector.connect(connection_parameters)
```

Once the connection or session object is created, you then create the Root object (this is again from a dependent package called snowflake.core)

```
from snowflake.core import Root
```

With a Root object created from your connection to Snowflake, you can access objects and methods of the Snowflake Python API.

The Root object is the root of the resource tree modeled by the Snowflake Python API.

```
databases = root.databases.iter(like="r%")
for database in databases:
  print(database.to_dict())
```

Complete example with both approaches are available in the repository [file: snowflake_py_library_client.py].

Snowflake REST API

The REST API provides a mechanism to run SQL statements via the REpresentational State Transfer (REST) interface. Since all Snowflake objects can be managed using SQL, the REST API could be used to run something like "alter warehouse xyz..."

The official documentation is at Snowflake SQL REST API (`https://docs.snowflake.com/en/developer-guide/sql-api/index`).

Additional Snowflake Documentation

`https://docs.snowflake.com/en/developer-guide/python-connector/python-connector-connect`

`https://docs.snowflake.com/en/user-guide/snowflake-client-repository`

Recipe 6-2. Using Snowpark API

Snowflake rolled out Snowpark API sometime around mid-2021. It is a set of libraries and runtimes in Snowflake that securely deploy and process non-SQL code, including Python, Java, and Scala.

This feature opened a new paradigm shift in data programmability. Until then, one could only use snowflake connectors and drivers to execute SQL and load the data natively into their preferred runtime (Python, Java, or Scala). For example, if your data application is written in Python, you would use the Python connectors to execute SQL and then capture the results in a pandas DataFrame or Python array to perform further operations.

Snowpark offers language extensibility and better developer experience by providing a DataFrame-based API with lazy execution mode and the ability to build complex pipelines that look similar to common DataFrame APIs like Spark but stay performant since the operations are pushed down to the Snowflake engine.

Snowpark API was first made available in Scala language and is now available for Java and Python.

Snowpark: Client Side and Server Side

Snowpark code can be run on the client side, which could be your data applications running on your server or Snowflake worksheets (from the Snowflake web UI).

You can use the Snowpark consists of libraries, including the DataFrame API and native Snowpark machine learning (ML) APIs for model development and deployment from the client side.

When Snowpark code is bundled via packages, UDFs/UDTFs/VUDFs, or Stored Procedures, it gets completely executed on the server side. On the server side, your Python, Java, and Scala code could run in the Compute Warehouse or Snowpark Container Services.

This chapter focuses on Snowpark Python API, and Chapter 10 delves more into Snowpark and also on using third-party Python packages.

Snowpark Python

Snowpark Python was a game changer primarily because of the following reasons.

- It enables data engineering and data science teams to use the familiar Python DataFrame syntax.
- Snowflake also opens access to Anaconda libraries through a highly secure sandbox environment, providing consistent governance and security policies.
- Python code can be made 100% push down by creating Snowpark UDFs.
- Availability of Snowpark-optimized compute warehouses. These are recommended for workloads with large memory requirements, such as ML training. Snowpark-optimized warehouses provide 16x memory per node compared to a standard Snowflake virtual warehouse.

Snowpark Python includes the following exciting capabilities.

- Python (DataFrame) API
- Python (pandas) API
- Python Scalar user-defined functions (UDFs)
- Python UDF batch API (vectorized UDFs)
- Python table functions (UDTFs)
- Python stored procedures
- Integration with Anaconda

Problem

You have been tasked to explore the Snowpark Python API so that some of the prior data pipelines written in PySpark could be ported to Snowflake with minimal changes.

Solution

Let's first look at what a DataFrame is in Snowflake.

This is the definition from the Snowflake docs: "Represents a lazily-evaluated relational dataset that contains a collection of Row objects with columns defined by a schema (column name and type)."

A DataFrame is considered lazy because it encapsulates the computation or query required to produce a relational dataset. The computation is not performed until you call a method that performs an action."

If you know the Apache Spark framework, this should sound familiar to the DataFrames in Spark. Different programming implementations of Snowpark have the same underlying DataFrame data structure and DataFrame API.

In comparison to Spark, Snowpark includes the following benefits.

- It supports interacting with data within Snowflake using libraries and patterns built for Snowflake.
- It provides better performance than Spark.
- It supports authoring Snowpark code using local tools such as Jupyter, Visual Studio Code, or IntelliJ IDEA.
- It supports pushdown for all operations, including Snowflake UDFs. Snowpark pushes all data transformation and heavy lifting to the Snowflake data cloud.
- It is a single platform. There is no requirement for a separate cluster outside of Snowflake for computations.

Now, let's look at common operations that could be performed on a DataFrame.

Broadly, the operations on DataFrame can be divided into two types.

- Transformations can produce a new DataFrame from one or more existing DataFrames. Note that transformations are lazy and don't cause the DataFrame to be evaluated. If the API does not provide a method to express the SQL that you want to use, you can use functions.sqlExpr() as a workaround.
- Actions cause the DataFrame to be evaluated. When you call a method that performs an action, Snowpark sends the SQL query for the DataFrame to the server for evaluation.

To demonstrate how Snowpark Python code is created for a basic requirement, let's look at a trivial scenario using the SNOWFLAKE_SAMPLE_DATA dataset.

Let's say you want to find the top three products by order quantity for each year from the orders table.

You would write the following in SQL.

```
select
    receipt_year, type, quantity::number(19,0) as quantity
from (
    select
        sum(li.l_quantity) quantity
        , year(li.l_receiptdate) receipt_year
        , p.P_TYPE type
        , row_number() OVER (PARTITION BY receipt_year ORDER BY quantity
          DESC) rnum
    from snowflake_sample_data.tpch_sf1.lineitem li
    left outer join snowflake_sample_data.tpch_sf1.part p
    on p.P_PARTKEY = li.L_PARTKEY
    group by receipt_year, type
)
where rnum<=3
order by receipt_year, quantity desc;
```

The expected result is shown in Figure 6-1.

	RECEIPT_YEAR	PRODUCT_TYPE	QUANTITY
1	1992	LARGE PLATED BRASS	132731
2	1992	LARGE PLATED STEEL	131793
3	1992	ECONOMY ANODIZED STEEL	130787
4	1993	ECONOMY ANODIZED STEEL	169405
5	1993	PROMO BRUSHED BRASS	167152
6	1993	LARGE PLATED STEEL	165244

Figure 6-1. *Query Results*

Let's write the same query as a set of steps using CTEs.

```
with df_items as (
    select l_quantity, year(l_receiptdate) as receipt_year, l_partkey from
    snowflake_sample_data.tpch_sf1.LINEITEM
),
df_parts as (
    select p_partkey, p_type from snowflake_sample_data.tpch_sf1.PART
)
, df_joined as (
    select receipt_year, l_quantity, p_type as product_type from df_items
    li inner join df_parts p on p.p_partkey = li.l_partkey
)
, df_grpd as (
    select sum(l_quantity) as quantity, receipt_year, product_type from
    df_joined group by 2,3
)
, df_ordered as (
    select quantity,receipt_year, product_type, row_number() over
    (partition by receipt_year ORDER BY quantity desc) rnum from df_grpd
)
select receipt_year, product_type, quantity::number(19,0) as quantity from
df_ordered where rnum<=3 order by receipt_year, quantity desc;
```

Looking at this query, it is easy to transform this to the equivalent Snowpark Python code.

```
# The Snowpark package is required for Python Worksheets.
# You can add more packages by selecting them using the Packages control
and then importing them.

import snowflake.snowpark as snowpark
from snowflake.snowpark.functions import col
from snowflake.snowpark.dataframe_reader import *
from snowflake.snowpark.functions import *
from snowflake.snowpark.window import *
from snowflake.snowpark.types import IntegerType

def main(session: snowpark.Session):

    # session object provides table() function to read from a mentioned
    table and select() function to select specific columns
    # below line of code creates a dataframe from tpch_sf1.lineitem table
    by selecting 2 columns and returns it in a DataFrame
    df_items = session.table("snowflake_sample_data.tpch_sf1.lineitem").
    select("l_quantity", year(col("l_receiptdate")).as_("receipt_year"),
    "l_partkey")

    # session object provides sql() function using which you could pass in
    any SQL command
    # below line of code runs sql command to select 2 columns from tpch_
    sf1.part table and returns it in a DataFrame
    df_parts = session.sql("select p_partkey, p_type from snowflake_sample_
    data.tpch_sf1.part")

    # join() function lets you perform a join (defaults to inner join
    unless specified) with the current DataFrame with another DataFrame
    (right) on a list of columns (on).
    # below line of code joins the 2 Dataframes created above on the
    "partkey" and renames "p_type" column as "product_type"
```

```
df = df_items.join(df_parts,df_items.col("l_partkey") == df_parts.
col("p_partkey")).select("receipt_year", "l_quantity", df_parts["p_
type"].as_("product_type"))

# group_by() function groups rows by the columns specified by
expressions (similar to GROUP BY in SQL)
# below line of code groups the dataset by "receipt_year" and "product_
type" to aggregate on "l_quantity"
# additionally "l_quantity" column is renamed as "quantity"
df = df.group_by("receipt_year","product_type").agg(sum("l_quantity").
as_("quantity"))

# with_column() funtion lets you engineer and construct new columns
# below line of code adds a new column rownum by creating a unique row
number for each row within the window partition.
df = df.with_column("rownum", row_number().over(Window.partition_
by(col("receipt_year")).order_by(col("quantity").desc())))

# below code selects specific columns from the Dataframes, does a
casting for quantity and filters the top 3 rows using the rownum column
df = df.select( "receipt_year", "product_type", cast(df["quantity"],
IntegerType()).as_("quantity"), "rownum").filter(col("rownum")<=3).
order_by("receipt_year", col("quantity").desc())

# finally we remove the rownum column as it offers no business value
df = df.drop("rownum")

return df
```

This code is also available in the source code repository [file: snowpark_python.py].

Problem

You have been tasked to rewrite some of the pandas code used in your project to be compatible with running it at scale in Snowflake using Snowpark API.

Solution

You have two options.

- Rewrite the code using Snowpark Python API. This involves major rework but benefits from all the capabilities of Native Snowpark API.
- Use the newly added Snowpark pandas API feature. This is the winner here since it offers the following advantages.
  - Familiar interface to Python developers by providing a pandas-compatible layer that can run natively in Snowflake.
  - This API bridges the convenience of pandas with the scalability of mature data infrastructure. pandas code can now run and scale by leveraging pre-existing query optimization techniques within Snowflake.
  - Code changes are minimal compared to the first approach.

Let's use the Snowpark pandas API as our choice of solution.

The Snowpark pandas API lets you work with much larger datasets, so you can avoid the time and expense of porting your pandas pipelines to other frameworks or using larger and more expensive machines. It runs workloads natively in Snowflake through transpilation to SQL, enabling it to take advantage of parallelization and the data governance and security benefits of Snowflake.

Step 1: Install the Connector

You need to install the following Python package to use this private preview feature as of this writing.

```
$ pip install "snowflake-snowpark-python[modin]"
```

Step 2: Change Import Statements in Your Code

```
import modin.pandas as pd

# Import the Snowpark pandas plugin for modin.
import snowflake.snowpark.modin.plugin
```

Snowpark pandas AP provides the same API signature as native pandas (pandas 2.2.1) but provides scalable computation with Snowflake. Native pandas library executes operations immediately and materializes results fully in memory after each operation. This eager evaluation of operations might lead to increased memory pressure as data needs to be moved extensively within a machine. But Snowpark pandas API uses a similar eager evaluation model of pandas, but internally builds a lazily-evaluated query graph to enable optimization across operations.

You can use Snowpark pandas to read and transform data in Snowflake, but you can also convert it to a native Python pandas DataFrame and from a Snowpark DataFrame to a pandas DataFrame. This is explained in the below Table 6-1.

Table 6-1. *Snowpark Dataframe Operations*

Operation	Input	Output	Notes
to_pandas	Snowpark pandas DataFrame	native pandas DataFrame	Materialize all data to the local environment. If the dataset is large, this may result in an out-of-memory error.
pd.DataFrame(...)	DataFrame Input	Snowpark pandas DataFrame	This should be reserved for small DataFrames. Creating a DataFrame with large amounts of local data might incur performance issues due to data uploading.
session.write_pandas	pandas DataFrame	Snowflake table	The result can be subsequently loaded into Snowpark pandas with `pd.read_snowflake` using the table name specified in the `write_pandas` call.

Additional Snowflake Documentation

`https://docs.snowflake.com/en/developer-guide/snowpark/index`

`https://docs.snowflake.com/en/developer-guide/snowpark/python/snowpark-pandas`

Recipe 6-3. Using Snowflake Functions and Stored Procs

Like other databases, Snowflake also provides user-defined functions and procedures.

The reasoning behind why one should use functions and procedures can be encapsulated in four points.

- **Modularity and reusability**: Allow developers to encapsulate logic into reusable units
- **Transaction control**: Ease of wrapping code in transactions, allowing developers to ensure data consistency and integrity
- **Abstraction of complexity**: Helps to abstract complex database operations into simpler interfaces
- **Reduced network traffic**: By executing logic directly within the database server, functions and procedures can reduce network traffic between the application and the database

This recipe looks at UDFs and stored procs in different language implementations and how to define and use them from your client or Snowflake worksheets.

Stored Procedures

Stored procedures are written to construct reusable and modular code within a database.

It contains logic that you write in any of the supported languages, and once the procedure is constructed, you can call it from any SQL code.

Snowflake procedures and functions support the languages listed in Table 6-2.

Table 6-2. Languages Supported by Snowflake

Language	Capability As Part Of	Handler Location	Runtime Version Supported
Javascript	Default feature	In-line	NA
SQL	Default feature (SnowSQL scripting)	In-line	NA
Python	Snowpark	In-line or staged	Python runtimes: 3.8, 3.9, 3.10, 3.11
Java	Snowpark	In-line or staged	JVM runtimes: 11.x, 17.x
Scala	Snowpark	In-line or staged	Scala 2.12.9 and 2.12.x only JVM runtimes: 11.x, 17.x

The handler location states if it is,

- **In-line**: the stored procedure logic or body should be defined where the stored procedure is defined.
- **Staged**: the stored procedure logic can be defined in an external file or package and can be imported when the stored procedure is defined.

When should you use a staged location?

- When you already have precompiled reusable code
- When the handler code is too large or complex to go in-line and/or if it needs to be managed by a different person/team
- When you need handler code from multiple functions or procedures (Staged code can contain multiple handler functions in which each can be used by a different UDF or procedure.)

This book looks at stored procs and functions in Java, Python, and SQL languages and provides an example of Python staged invocation.

User-Defined Functions

Snowflake user-defined functions (UDFs) are custom functions that you create to perform specific tasks within your Snowflake environment. They allow you to encapsulate complex or repetitive logic into a single function, making your code more organized and easier to manage.

There are two types of Snowflake UDFs.

- Scalar returns one output row for each input row.
- Tabular returns a tabular value for each input row.

Problem

What are the limitations to consider when developing stored procs and UDFs in Snowflake?

Solution

- Stored Procedures are not atomic, you need to specifically wrap the code in the procedure in a transaction.
- Stored Procedure may or may not return a value, but a UDF should always return a value.
- Stored procs and functions utilize two types of privileges.
    - Privileges directly on the object itself. Currently, there are only two types of privileges: usage and ownership (defaults to the creator)
    - Privileges on the database objects (e.g., tables) that the stored procedure or function accesses or uses.
- A stored procedure runs with either the *caller's* or *owner's* rights. It cannot run with both at the same time.
    - A procedure marked with caller's rights operates using the privileges of the caller, offering the key benefit of accessing information related to the caller or the caller's ongoing session. On the other hand, an owner's rights stored procedure generally operates with the privileges of the procedure's owner. Choose this

if you want to delegate tasks to another role, allowing them to run with the owner's privileges. It is useful to provide limited access without granting broader privileges.

  - A function always runs on the caller's rights.

- Although your handler can use functionality in external libraries, Snowflake security restrictions disable some capabilities, such as writing to files.
- Snowflake security restriction also prevents accessing Snowflake objects in the UDFs. The only exception is SQL UDFs. The SQL UDFs only support "select" statements, not DML or DDL.
- Multiple Snowflake UDFs can be called in a single SQL statement but you could invoke only one stored procedure in a single executable statement.
- Stored procedure calls cannot be part of an expression or combined with other calls in a single statement, but a UDF can be.
- You can use subqueries as arguments to procedures and functions.

  For example, call calculate_square (select vol from tbl limit 1).
- Avoid too many nested calls or recursive calls.
- Avoid storing large binary or text values in variables.
- Functions are memoizable.
- Data type-specific considerations for stored procedures are based on the language runtime you choose. Please refer to the documentation for details.

Problem

You have been tasked with creating and exploring UDFs and stored procedures in Java, Python, SQL, and JavaScript and what works best for different use cases.

Solution

In Snowflake, the code structure remains uniform for creating and using UDFs and Procedures, irrespective of the implementation language.

The following is the code template for functions.

```
create or replace function <name>( [<arg_name> <arg_data_type> ] [ , ... ])
returns <result_data_type>
language <language>
as
   $$
      <actual_code>
   $$
;
```

The following is the code template for procedures.

```
create or replace procedure <name>( [<arg_name> <arg_data_type> ] [
, ... ])
returns <result_data_type>
language <language>
execute as [CALLER | OWNER]
as
   $$
      <actual_code>
   $$
;
```

- Note that the argument data types (arg_data_type) and the return types (result_data_type) must follow the SQL data types.
- The same number of arguments gets carried over into the <actual code>, as shown in the following examples.
- The data type specified as part of the stored procedure or user-defined function definition is an SQL data type, as the code you write is called from SQL. However, its underlying handler uses data types from the handler's language, such as Java, Python, or Scala. At runtime, Snowflake converts between the SQL types and handler

types for arguments and return values based on these mapping rules as defined in `https://docs.snowflake.com/en/developer-guide/udf-stored-procedure-data-type-mapping`.

UDTFs and VUDFs are discussed in the next chapter.

The following examples utilize the SNOWFLAKE_SAMPLE_DATA.TPCH_SF1.ORDERS dataset to demonstrate a simple example of finding the order volume increase between two years. The first year is called the base year (which defaults to a constant if not provided), and the second year, which is chosen for comparison, is a mandatory input.

These samples are also added to the source code repository for easy access [folder: udfs-procs/].

SQL UDF and SQL Stored Procedure

To demonstrate SQL UDF and procedure, we construct two functions—one being an overloaded version of the other.

The UDFs are then called from within the SQL procedure.

Both the following functions utilize SQL code directly, which is only possible within SQL UDFs.

```
SET default_year = 1995;

create or replace function get_order_volume(chosen_year NUMBER)
returns NUMBER
AS
$$
    select count(*) from SNOWFLAKE_SAMPLE_DATA.TPCH_SF1.ORDERS where
    year(o_orderdate) = chosen_year
$$
;
create or replace function get_order_volume()
returns NUMBER
AS
$$
    select count(*) from SNOWFLAKE_SAMPLE_DATA.TPCH_SF1.ORDERS where
    year(o_orderdate) = $default_year
$$
```

This approach uses a session variable and the overloading feature of functions in Snowflake.

You could also use the default argument approach and create just one function.

```
create or replace function get_order_volume(chosen_year NUMBER
default 1995)
returns NUMBER
AS
$$
    select count(*) from SNOWFLAKE_SAMPLE_DATA.TPCH_SF1.ORDERS where
    year(o_orderdate) = chosen_year
$$
;
```

Test the function by calling it from the Snowflake worksheets.

```
select get_order_volume();
```

The following procedure calculates the increase.

```
create or replace procedure get_order_increase(chosen_year NUMBER)
returns FLOAT NOT NULL
language sql
execute as caller
as
$$
declare
    order_vol_base_year NUMBER DEFAULT 0;
    order_vol_chosen_year NUMBER DEFAULT 0;
    res1 RESULTSET DEFAULT (select get_order_volume(:chosen_year)
    as count);
    res2 RESULTSET DEFAULT (select get_order_volume() as count);
    c1 CURSOR for res1;
    c2 CURSOR for res2;
begin
    for row_variable in c1 do
        order_vol_base_year := row_variable.count;
    end for;
```

```
    for row_variable in c2 do
        order_vol_chosen_year := row_variable.count;
    end for;
return (order_vol_base_year-order_vol_chosen_year)*100/order_vol_base_year;
END;
$$
;
```

Test the procedure by running the following.

```
call get_order_increase(1997);
```

Javascript UDF and Javascript Stored Procedure

To demonstrate JavaScript UDF and procedure, we construct a function that calculates an increase in the order volume compared to a base order value, and the procedure utilizes this UDF to calculate the % order increase.

```
create or replace function calc_order_increase(order_y1 VARCHAR, order_y2
VARCHAR)
returns VARCHAR
language javascript
as
$$
    incr = (parseInt(ORDER_Y2)-parseInt(ORDER_Y1))*100/parseInt(ORDER_Y1);
    return incr.toString();
$$
;
```

The Javascript procedure is as follows.

```
create or replace procedure get_order_increase(chosen_year VARCHAR)
returns VARCHAR
language javascript
execute as caller
as
$$
    const DEFAULT_YEAR = 1995;
```

```
/** Javascript stored procedures provides an implicit object called
"snowflake" which can be used to execute SQL commands */
res = snowflake
        .createStatement({
            sqlText: `select count(*) from SNOWFLAKE_SAMPLE_DATA.
            TPCH_SF1.ORDERS where year(o_orderdate) = ${chosen_
            year}::number`
        }).execute();
/**
 * The res variable is a ResultSet object. You only see one row at
 a time in a ResultSet, just as you can see one row at a time in a
 SQL cursor.
 * Typically, after you retrieve a ResultSet, you iterate through it by
 repeating the following operations:
 **  Call next() to get the next row.
 **  Retrieve data from the current row by calling methods such as
 getColumnValue().
 */
res.next();
order_vol_chosen_year = res.getColumnValue(1);

res = snowflake
        .createStatement({
            sqlText: `select count(*) from SNOWFLAKE_SAMPLE_DATA.
            TPCH_SF1.ORDERS where year(o_orderdate) = ${DEFAULT_
            YEAR}::number`
        }).execute();
res.next();
order_vol_base_year = res.getColumnValue(1);
res = snowflake
        .createStatement({
            sqlText: `select calc_order_increase('${order_vol_base_
            year}', '${order_vol_chosen_year}')`
        }).execute();
res.next();
```

```
    return res.getColumnValue(1);
$$
;
```

Test the procedure by running the following.

```
call get_order_increase(1997);
```

Do note the following considerations while using JavaScript procedures.

- You cannot call the JavaScript eval() function.
- JavaScript stored procedures support access to the standard JavaScript library. Note that this excludes many objects and methods typically provided by browsers.
- You cannot import or use additional modules.
- JavaScript code is executed within a restricted engine, preventing system calls from the JavaScript context (e.g., no network and disk access) and constraining the system resources available to the engine, specifically memory.
- Argument names are case-insensitive in the SQL portion of the stored procedure code but are case-sensitive in the JavaScript portion.

Python UDF and Python Stored Procedure

To demonstrate Python UDF and procedure, we construct a function similar to the JavaScript example that calculates an increase in the order volume compared to a base order value, and the procedure utilizes this UDF to calculate the % order increase.

```
create or replace function calc_order_increase(order_y1 INTEGER, order_y2
INTEGER)
returns DOUBLE
language python
runtime_version = '3.8'
handler = 'main'
as
$$
```

```
def main(order_y1, order_y2):
  return (order_y2-order_y1)*100/order_y1
$$;
```

Python procedure is as follows,

```
create or replace procedure get_order_increase(chosen_year int)
returns VARCHAR
language python
runtime_version = '3.8'
packages=('snowflake-snowpark-python')
handler = 'main'
as
$$

DEFAULT_YEAR = 1995;

def main(session, chosen_year):

    df_order_vol_chosen_year = session.sql("select count(*) as count from
    SNOWFLAKE_SAMPLE_DATA.TPCH_SF1.ORDERS where year(o_orderdate) = ?",
    params=[chosen_year])

    df_order_vol_base_year = session.sql("select count(*) as count from
    SNOWFLAKE_SAMPLE_DATA.TPCH_SF1.ORDERS where year(o_orderdate) = ?",
    params=[DEFAULT_YEAR])

    #this is when the query executes; lazy loading; collect returns an
    immutable named tuple
    order_vol_chosen_year = df_order_vol_chosen_year.collect()[0]['COUNT']
    order_vol_base_year = df_order_vol_base_year.collect()[0]['COUNT']

    res = session.sql("select calc_order_increase(?,?) as result",[order_
    vol_base_year, order_vol_chosen_year])

    return res.collect()[0]['RESULT']
$$;
```

Test the procedure by running,

```
call get_order_increase(1997);
```

Java UDF and Java Stored Procedure

To demonstrate Java UDF and procedure, we construct a function similar to the Python example, that calculates increase in the order volume compared to a base order value. The procedure utilizes this UDF to calculate the % order increase.

```
create or replace function calc_order_increase(order_y1 INTEGER, order_y2
INTEGER)
returns DOUBLE
language java
handler='CalcUtil.main'
target_path='@~/CalcUtil.jar'
as
$$
class CalcUtil {
    public static double main(int order_y1, int order_y2) {
        return (double)(order_y2-order_y1)*100/order_y1;
    }
}
$$;

create or replace procedure get_order_increase(chosen_year NUMBER)
returns DOUBLE
language java
runtime_version = 11
handler = 'OrderVolumeCheck.main'
packages=('com.snowflake:snowpark:latest')
as
$$
import com.snowflake.snowpark_java.*;
import java.text.MessageFormat;
class OrderVolumeCheck {

    static int DEFAULT_YEAR=1995;

    public double main(Session session, int chosen_year) {
```

```
        DataFrame df_order_vol_chosen_year = session.sql(String.
        format("select count(*) as count from SNOWFLAKE_SAMPLE_DATA.TPCH_
        SF1.ORDERS where year(o_orderdate) = %d",chosen_year));
        Row[] result_1 = df_order_vol_chosen_year.collect();
        int order_vol_chosen_year =  result_1[0].getInt(0);

        DataFrame df_order_vol_base_year = session.sql(String.
        format("select count(*) as count from SNOWFLAKE_SAMPLE_DATA.TPCH_
        SF1.ORDERS where year(o_orderdate) = %d",DEFAULT_YEAR));
        Row[] result_2 = df_order_vol_base_year.collect();
        int order_vol_base_year =  result_2[0].getInt(0);

        DataFrame df = session.sql(String.format("select calc_order_
        increase(%d,%d) as result",order_vol_base_year, order_vol_
        chosen_year));
        return (df.collect())[0].getDouble(0);
    }
}
$$
;
```

Test the procedure by running,

```
call get_order_increase(1997);
```

Recipe 6-4. What and How of Snowpark Python API

This recipe shows how to run Snowpark code, different approaches, and how Snowpark code is executed by Snowflake.

Problem

Consider a scenario where you, as a data engineer, are tasked with writing some data pipelines in Snowpark Python. You are required to consider and explore different ways to run Snowpark code in Snowflake.

Solution

Snowflake lets you run Snowpark code in various ways.

- Local/client code, which is pushed down as SQL
- Wrapping local code (from your local machine or snowflake worksheets) as functions and registering them as UDFs or stored procs
- Create Snowflake UDFs and stored procs directly

Let's look at the same scenario used in the prior recipes in Snowpark Python for both UDF and stored procs that cover all three methodologies.

Approach 1: Local Code Using the Snowpark API

The following code uses the same scenario as in the previous examples to calculate order volume increase from between two years.

```
from snowflake.snowpark import Session

connection_parameters = {
    "account": "",
    "user": "",
    "password": "",
    "warehouse": ""
}

session = Session.builder.configs(connection_parameters).create()

DEFAULT_YEAR = 1995

def local_order_increase(order_y1: int, order_y2: int) -> float:
  return (order_y2-order_y1)*100/order_y1

def get_order_increase(chosen_year: int) -> float:

    df_order_vol_chosen_year = session.sql("select count(*) as count from
    SNOWFLAKE_SAMPLE_DATA.TPCH_SF1.ORDERS where year(o_orderdate) = ?",
    params=[chosen_year])
```

```
    df_order_vol_base_year = session.sql("select count(*) as count from
    SNOWFLAKE_SAMPLE_DATA.TPCH_SF1.ORDERS where year(o_orderdate) = ?",
    params=[DEFAULT_YEAR])

    #this is when the query executes; lazy loading; collect returns an
    immutable named tuple
    order_vol_chosen_year = df_order_vol_chosen_year.collect()[0]['COUNT']
    order_vol_base_year = df_order_vol_base_year.collect()[0]['COUNT']

    return local_order_increase(order_vol_base_year,order_vol_chosen_year)

res = get_order_increase(1996)
print(res)
session.close()
```

This code is added to the code repository [snowpark-approaches/snowpark_local.py].

The following are things to keep in mind while using this approach.

- All your code runs on your local machine.
- The transformations are compiled and translated to SQL queries, which are observable from the query history. This eases debugging on Snowflake.

Figure 6-2 is from the query history. Snowflake populates the query tag with a trace log on to your local Python code base to the line of code that triggered this operation.

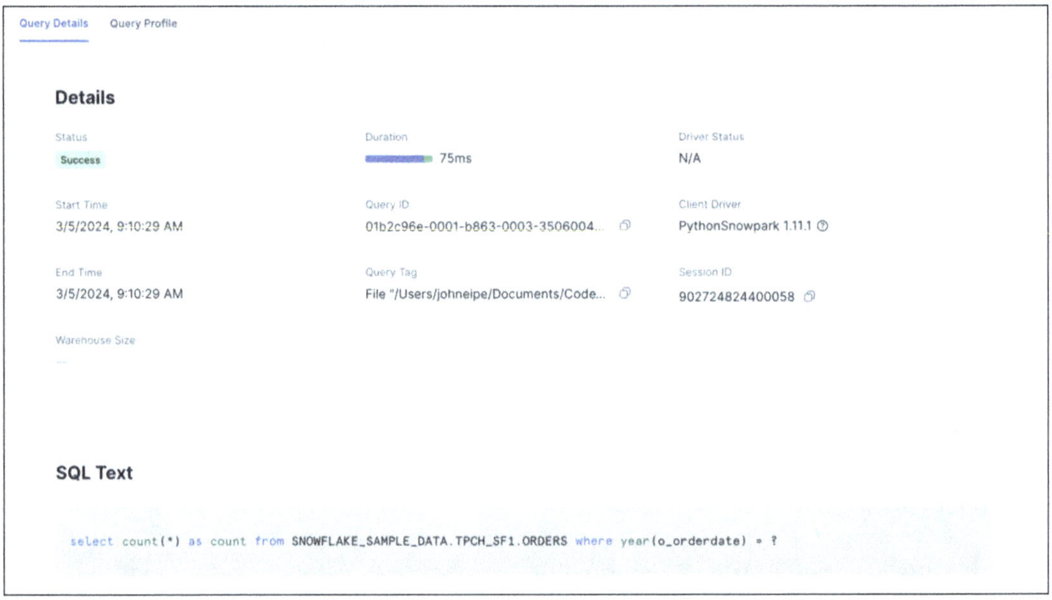

Figure 6-2. *Query Details*

You can easily utilize any library you want without any hassle.

The only caveat you should care about is not to bring huge datasets into local memory and should try to push down as far as possible all heavy compute to Snowflake.

Approach 2: Local Functions as UDFs and Stored Procs

In this approach you could register your local functions as a UDF or stored procs.

The following code assumes that you have created a COMMONS database and a schema called UTILS under it.

```
from snowflake.snowpark.functions import udf,sproc
from snowflake.snowpark import Session
from snowflake.snowpark.dataframe_reader import *
from snowflake.snowpark.functions import *

connection_parameters = {
    "account": "",
    "user": "",
    "password": "",
    "warehouse": "",
```

```
    "database":"COMMONS",
    "schema":"UTILS"
}

session = Session.builder.configs(connection_parameters).create()

@udf(name="calc_order_increase", is_permanent=True, stage_location="@
client_code", replace=True)
def local_order_increase(order_y1: int, order_y2: int) -> float:
  return (order_y2-order_y1)*100/order_y1

DEFAULT_YEAR = 1995

@sproc(name="get_order_increase", is_permanent=True, stage_location="@
client_code", replace=True, packages=["snowflake-snowpark-python"])
def local_get_order_increase(session: Session, chosen_year: int) -> str:

    df_order_vol_chosen_year = session.sql("select count(*) as count from
    SNOWFLAKE_SAMPLE_DATA.TPCH_SF1.ORDERS where year(o_orderdate) = ?",
    params=[chosen_year])

    df_order_vol_base_year = session.sql("select count(*) as count from
    SNOWFLAKE_SAMPLE_DATA.TPCH_SF1.ORDERS where year(o_orderdate) = ?",
    params=[DEFAULT_YEAR])

    #this is when the query executes; lazy loading; collect returns an
    immutable named tuple
    order_vol_chosen_year = df_order_vol_chosen_year.collect()[0]['COUNT']
    order_vol_base_year = df_order_vol_base_year.collect()[0]['COUNT']

    res = session.sql("select calc_order_increase(?,?) as result",[order_
    vol_base_year, order_vol_chosen_year])

    return res.collect()[0]['RESULT']

session.close()
```

This code is added to the repository [snowpark-approaches/snowpark_client_udf_proc.py].

The following are things to keep in mind while using this approach.

- All your code runs on your local machine, but the registered functions don't.
- Though you use constants in the registered functions, they are substituted without issues.
- The return types for these functions are necessary because Snowflake needs this information to generate the stored proc or UDF.
- Debugging using Snowflake query history is not easy.

Figure 6-3 shows query history. As you see, the code is wrapped in a Snowflake generated UDF/procedure.

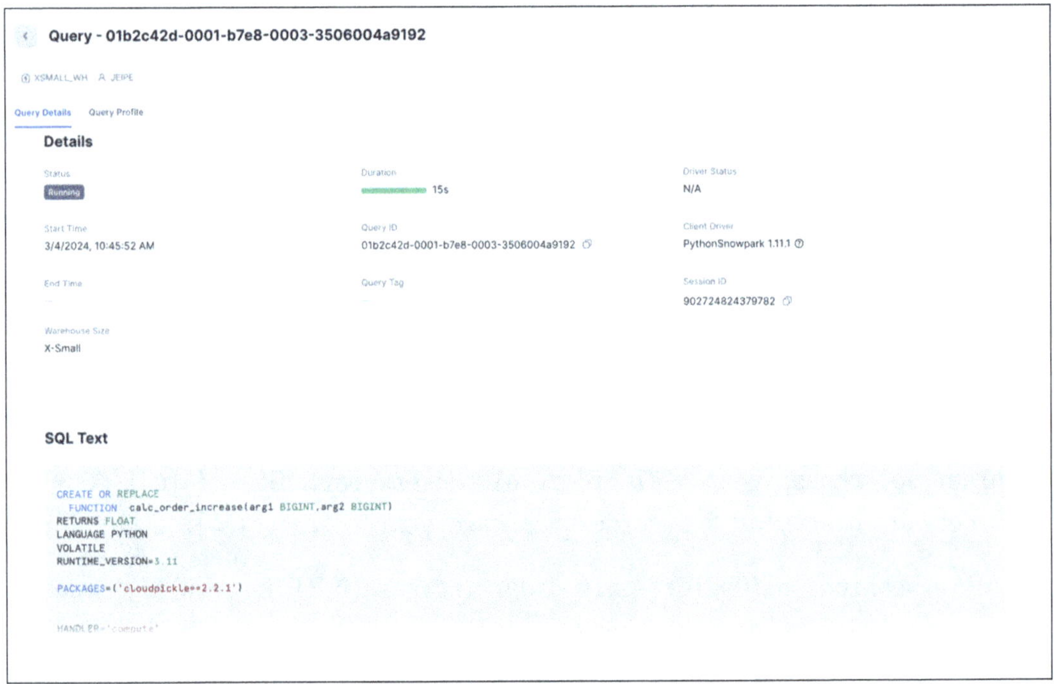

Figure 6-3. *Query Result Details*

You see the preceding SQL text because the procedure is likewise created (Figure 6-4).

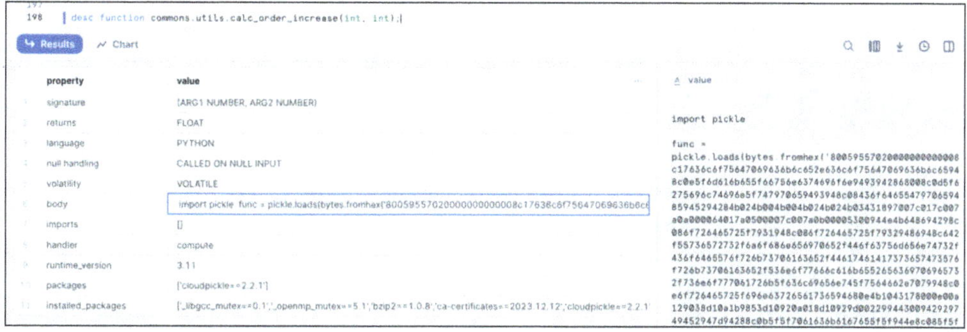

***Figure 6-4.** Query Results*

The actual code is available in this body as commented out text.

The following is what the generated code looks like.

```
CREATE OR REPLACE PROCEDURE COMMONS.UTILS.GET_ORDER_INCREASE("ARG1"
NUMBER(38,0))
RETURNS VARCHAR(16777216)
LANGUAGE PYTHON
RUNTIME_VERSION = '3.11'
PACKAGES = ('snowflake-snowpark-python','cloudpickle==2.2.1')
HANDLER = 'compute'
EXECUTE AS OWNER
AS '
import pickle

func = pickle.loads(bytes.fromhex(''800595970........6948652302e''))
# The following comment contains the source code generated by snowpark-
python for explanatory purposes.
# DEFAULT_YEAR  # variable of type <class ''int''>
# @sproc(name="get_order_increase", is_permanent=True, stage_location="@
client_udf", replace=True, packages=["snowflake-snowpark-python"])
# def local_get_order_increase(session: Session, chosen_year: int) -> str:
#
#     df_order_vol_chosen_year = session.sql("select count(*) as count
from SNOWFLAKE_SAMPLE_DATA.TPCH_SF1.ORDERS where year(o_orderdate) = ?",
params=[chosen_year])
#
```

```
#      df_order_vol_base_year = session.sql("select count(*) as count from
       SNOWFLAKE_SAMPLE_DATA.TPCH_SF1.ORDERS where year(o_orderdate) = ?",
       params=[DEFAULT_YEAR])
#
#      #this is when the query executes; lazy loading; collect returns an
       immutable named tuple
#      order_vol_chosen_year = df_order_vol_chosen_year.collect()[0]
       [''COUNT'']
#      order_vol_base_year = df_order_vol_base_year.collect()[0][''COUNT'']
#
#      res = session.sql("select calc_order_increase(?,?) as result",[order_
       vol_base_year, order_vol_chosen_year])
#
#      return res.collect()[0][''RESULT'']
#
# func = local_get_order_increase

def compute(session,arg1):
    return func(session,arg1)
';
```

- All packages should be mentioned as part of the registration including snowflake-snowpark-python.
- You still have the option to utilize any library but if it must be packaged along with the code then it has to follow the security constraints specified by Snowflake [covered in next chapter].
- You need a global Session object to register the UDFs and procs. This was weird because we don't use the session directly for anything as you see in the code.
- You need the Snowflake connection set to a default database and schema on which the current user has appropriate grants. This is why we used a different database and schema instead of going with SNOWFLAKE_SAMPLE_DATA as the default database. (Note that SNOWFLAKE_SAMPLE_DATA is a shared database provided by Snowflake on which we cannot have write privileges.)

- If you need to unpickle the hex code generated by snowflake, you can use the following code

```
import pickle
import inspect
bytes = bytes.fromhex('<copy the hex code here')
func = pickle.loads(bytes)

# this will print the function's code
print(inspect.getsource(func))

# this will print all global variables used/available within the
function
print(func.__globals__)
```

Approach 3: UDFs and Stored Procs Directly

This approach is detailed as part of the"Python UDF and Python Stored Procedures" subsection in the previous recipe.

The following is the Python UDF.

```
create or replace function calc_order_increase(order_y1 INTEGER, order_y2
INTEGER)
returns DOUBLE
language python
runtime_version = '3.8'
handler = 'main'
as
$$
def main(order_y1, order_y2):
  return (order_y2-order_y1)*100/order_y1
$$;
```

Python procedure is as follows,

```
create or replace procedure get_order_increase(chosen_year int)
returns VARCHAR
language python
runtime_version = '3.8'
```

```
packages=('snowflake-snowpark-python')
handler = 'main'
as
$$

DEFAULT_YEAR = 1995;

def main(session, chosen_year):

    df_order_vol_chosen_year = session.sql("select count(*) as count from
    SNOWFLAKE_SAMPLE_DATA.TPCH_SF1.ORDERS where year(o_orderdate) = ?",
    params=[chosen_year])

    df_order_vol_base_year = session.sql("select count(*) as count from
    SNOWFLAKE_SAMPLE_DATA.TPCH_SF1.ORDERS where year(o_orderdate) = ?",
    params=[DEFAULT_YEAR])

    #this is when the query executes; lazy loading; collect returns an
    immutable named tuple
    order_vol_chosen_year = df_order_vol_chosen_year.collect()[0]['COUNT']
    order_vol_base_year = df_order_vol_base_year.collect()[0]['COUNT']

    res = session.sql("select calc_order_increase(?,?) as result",[order_
    vol_base_year, order_vol_chosen_year])

    return res.collect()[0]['RESULT']
$$;
```

You may refer the section or see the code in the repository [snowpark-approaches/snowpark_udf_proc.py].

The following are things to keep in mind while using this approach.

- The majority of code are SQL statements, so it needs to be run tools and features like Flyway, Snowflake's GIT Integration, and Liquibase for continuous deployment and delivery, or it has to be run manually on Snowflake using SnowCLI or Worksheets.
- Your constants should be defined within the function or procedure.
- Debugging the functions and procs from within Snowflake is not easy.

- All packages used should be included in the registration, including snowflake-snowpark-python.
- You still have the option to utilize any library, but if it must be packaged and referred from a stage, it must follow the security constraints specified by Snowflake [covered in the next chapter].
- None of the code runs on your local.
- You need the Snowflake connection set to a default database and schema on which the current user has appropriate grants.

Approach 4: A Blended Approach

I sometimes prefer a blended approach of keeping the usage or call statement and the DataFrame code locally as far as possible and then creating and registering UDFs or Procedures only when the whole compute needs to be pushed down to Snowflake.

You get the added advantage of easy debugging, a good trace in Snowflake query history, and the advantage of using any library you want as long as it doesn't interfere with the pushed-down UDFs or procs.

You may refer to the code in the repository [snowpark-approaches/ snowpark_blended.py].

Problem

How does Snowpark Python code run behind the scenes in Snowflake?

Solution

As you have seen in the previous sections, Snowpark provides a data programmability framework to explore and transform your data, and you can choose to leverage Snowflake for data processing.

How Snowpark code gets executed depends on how it is called.

- If a DataFrame query runs from your client code, it gets translated via the Snowflake Connector for Python into SQL text and gets executed by the SQL engine.

- If it is a function defined at the client side but registered to run as a UDF/proc, the code gets serialized and sent to the python sandbox by the Snowflake Connector for Python. The code runs in the secure sandbox and if that code contains DataFrame expressions, it gets translated to SQL query and sent to the SQL engine.
- The communication between the SQL engine and the secure sandbox happens inside Snowflake and is invisible to the end user.

The interaction between each component is demonstrated using the Figure 6-5 provided below.

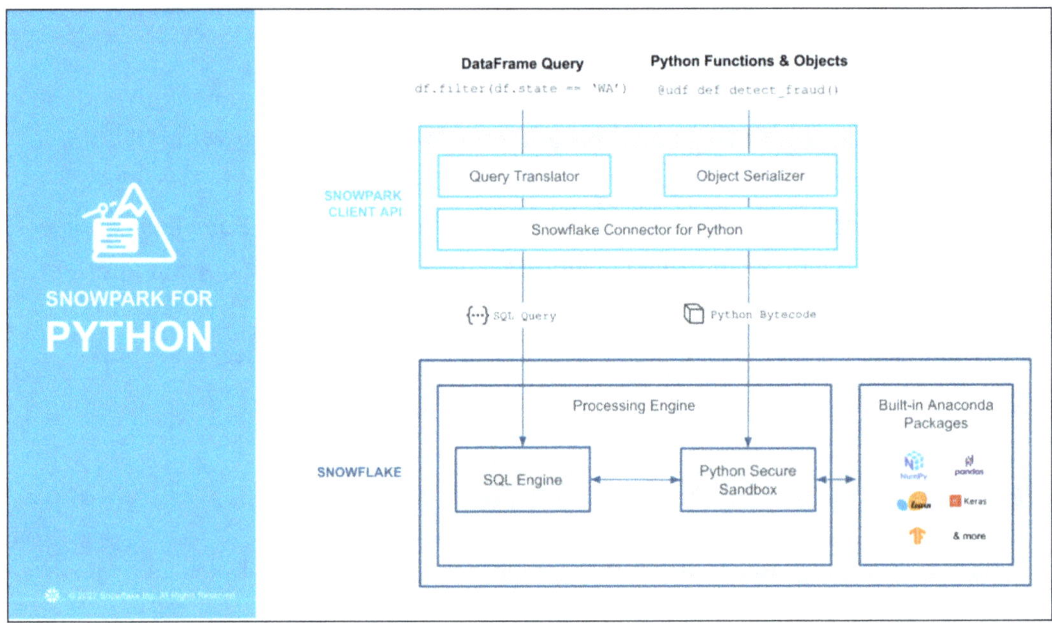

Figure 6-5. *Snowpark for Python*

Additional Snowflake Documentation

https://docs.snowflake.com/en/developer-guide/udf/udf-overview

https://docs.snowflake.com/en/developer-guide/stored-procedure/stored-procedures-overview

https://docs.snowflake.com/en/developer-guide/stored-procedures-vs-udfs

---

Recipe 6-5. Snowflake SQLAlchemy Toolkit

SQLAlchemy is a Python SQL toolkit and Object Relational Mapping (ORM). Developers use this library to work with Python objects and not write separate SQL queries. They can use Python to access and work with databases.

Problem

You must utilize some of the data assets in Snowflake from your Python web application. You are tasked to utilize SQLAlchemy to query from Snowflake because the web application already uses SQLAlchemy as part of its dependency.

Solution

The only requirement for Snowflake SQLAlchemy is the Snowflake Connector for Python; however, the connector does not need to be installed because installing Snowflake SQLAlchemy installs the connector automatically.

The following is a basic example of using the SQLAlchemy library.

```
#!/usr/bin/env python
from sqlalchemy import create_engine

engine = create_engine(
    'snowflake://{user}:{password}@{account_identifier}/'.format(
        user="",
        password="",
        account_identifier=""
    )
)
try:
    connection = engine.connect()
    results = connection.execute('select current_version()').fetchone()
    print(results[0])
finally:
    connection.close()
    engine.dispose()
```

SQLAlchemy provides a MetaData class. It acts like a container object that keeps together many different features of a database (or multiple databases) described.

To represent a table, use the Table class.

The MetaData object is a facade around a Python dictionary that stores Table objects keyed to their string name.

The following example queries the ORDERS table from the SNOWFLAKE_SAMPLE_DATA.

```
from sqlalchemy import create_engine, select, Table, MetaData

engine = create_engine(
    'snowflake://{user}:{password}@{account_identifier}/
    {database}/'.format(
        user="",
        password="",
        account_identifier="",
        database='SNOWFLAKE_SAMPLE_DATA'
    )
)

try:
    # Create a metadata object
    metadata = MetaData()
    # Create a session
    session = engine.connect()

    # Reflect the table from the database
    orders = Table('ORDERS', metadata, autoload_with=engine,
    schema='TPCH_SF1')

    # Query the table
    stmt = select(orders.c.o_orderkey, orders.c.o_shippriority).
    where(orders.c.o_orderdate == "1995-05-30").where(orders.c.o_
    orderstatus == "F")

    print(stmt)
```

```
    for row in session.execute(stmt):
        print("o_orderkey: ", row.o_orderkey)
        print("o_shippriority: ", row.o_shippriority)
        print("---------")

finally:
    session.close()
    engine.dispose()
```

This code is added to the repository [/snowflake_sqlalchemy.py].

Additional Snowflake Documentation

https://www.sqlalchemy.org/

https://docs.snowflake.com/en/developer-guide/python-connector/sqlalchemy

---

CHAPTER 7

Data Reusability and Monetization

In the landscape of contemporary business, technology, and digital transformation, the value of data transcends the traditional role as a mere repository of information. Data is now a dynamic asset capable of driving innovation, shaping strategies, and fueling economic growth. As organizations continue to accumulate vast amounts of data, the imperative is to secure and derive tangible value from it.

This chapter embarks on a journey into data reusability and monetization, two interrelated concepts that form the bedrock of progressive data management strategies. In an era where data is not just a byproduct but a strategic resource, understanding how to unleash its full potential is crucial.

We explore strategies for making data reusable across diverse contexts within an organization. From breaking down silos to fostering a culture of data democratization, we illuminate the path toward creating data sets that serve myriad business needs and foster innovation.

Moving beyond reusability, we delve into the transformative notion of data as a product. Treating data as a product involves a shift in mindset, where organizations view their data as a valuable commodity with intrinsic market value. We explore how this perspective can lead to the creation of data assets tailored for internal consumption and positioned for external monetization. Moreover, we show how the Snowflake Marketplace can help drive a successful outcome.

Whether you're seeking to foster collaboration internally, capitalize on data assets externally, or navigate the complexities of data monetization, this exploration of data reusability and monetization serves as your guide in unlocking the full potential of your organization's most valuable resource.

D. Dayton and J. Eipe, *Snowflake Recipes*, https://doi.org/10.1007/979-8-8688-0938-5_7

Recipe 1-1. Data Democratization

Data democratization is a strategic approach aimed at making data accessible and understandable to an organization's broad spectrum of users, breaking down traditional barriers to access, and fostering a culture where data is a shared resource rather than a siloed asset. This concept seeks to empower individuals across various departments and roles, ensuring that data-driven insights are not confined to a select few but are available to anyone who can benefit from them. This involves addressing the following key pillars that nurture a data-driven culture: ensuring accessibility, fostering responsible usage, and promoting continuous improvement.

- Accessibility lays the foundation.
- Governance provides the guardrails.
- Scalability ensures longevity.
- Data governance forms the ethical backbone.
- Continuous feedback completes the cycle.

By addressing these key pillars, organizations can embark on a successful data democratization journey, transforming their workforce into informed decision-makers and unlocking the true potential of their data assets.

Problem

Imagine a sprawling healthcare system encompassing multiple hospitals and clinics, each serving a diverse patient population across a broad geographical footprint. This system prides itself on delivering high-quality care, but a hidden obstacle impedes its progress: fractured data.

Each facility within the system operates independently, utilizing a patchwork of electronic health record (EHR) systems and data storage practices. While seemingly practical at the facility level, this siloed approach creates a critical roadblock on the broader system-wide scale. The consequences are far-reaching.

- **Fragmented insights**: Clinical data, the lifeblood of informed healthcare decisions, becomes imprisoned within isolated silos. Doctors, nurses, and others lack a holistic view of their patients, hindering comprehensive diagnoses, treatment plans, and preventive measures. Researchers remain blind to larger trends and patterns across the system, limiting their ability to develop groundbreaking discoveries.

- **Restricted access**: Crucial stakeholders across the system—from frontline clinicians to administrators and researchers—face limited access to essential data. This information scarcity hampers informed decision-making at all levels. Clinicians grapple with incomplete patient histories, making treatment recommendations a delicate dance in the dark. Administrators struggle with legacy systems to identify operational inefficiencies and optimize resource allocation without a comprehensive view of system-wide performance.
- **Inefficient operations**: Disconnected data analysis creates a domino effect of inefficiencies. The system's inability to readily pool and analyze data across facilities hinders the identification of operational redundancies and resource wastage. Optimizing patient flow, managing bed availability, and streamlining administrative processes become arduous tasks due to fragmented data sources.
- **Slowed innovation**: The system's potential for groundbreaking advancements in patient care and clinical research remains untapped. The inability to leverage the collective power of its data across facilities stifles the development of data-driven healthcare initiatives and personalized care models. Predictive analytics for early disease detection, targeted interventions for chronic conditions, and personalized treatment plans all remain elusive goals due to data silos.

This fragmented architecture creates a perfect storm hindering the system's ability to deliver on its core mission: providing high-quality, efficient, and innovative healthcare. To fully unlock the transformative potential of its data, the system faces a critical juncture—a paradigm shift toward data democratization, and integrated analytics is no longer an option but a necessity.

Solution

Data democratization is not just a technology initiative but a cultural shift that requires buy-in from all levels of an organization. It has the potential to transform industries like healthcare by unlocking the power of data to improve patient care, drive innovation, and optimize operations. By empowering individuals to access and leverage data, organizations can make more informed, data-driven decisions that lead to better outcomes for all stakeholders.

To address the healthcare system's fractured data landscape and unlock the potential of data-driven healthcare, using Snowflake Data Cloud offers a transformative path forward.

How It Works

Here's how Snowflake's key features address the core challenges.

Unifying Fragmented Data

- **Modern data architecture**: Consider levering Snowflake's multi-cloud data lake to build a data mesh, data fabric, distributed data cleanroom, or data lakehouse to seamlessly ingest EHR data from diverse sources (cloud, hybrid, on-prem) and formats (structured, unstructured, semi-structured) across all facilities ensuring data is available and accessible.
- **Secure data sharing**: Robust data governance policies and role-based access controls ensure data privacy and compliance with regulatory standards, fostering secure collaboration within the system.

Empowering Data Access and Analytics

- **Single platform for all users**: Snowflake provides a unified platform accessible to clinicians, administrators, researchers, and other stakeholders, democratizing data access and enabling self-service analytics.

- **User-friendly interface**: The intuitive Snowsight empowers users with varying technical expertise to query and analyze data without requiring extensive coding knowledge, fostering data literacy across the system. All the while providing more advanced interfaces like Python, Java, Go, and others.
- **Scalability**: Snowflake's cloud-based architecture effortlessly scales *up* with several warehouse sizing options and *out* with multi-cluster warehouses, all to accommodate growing data volumes and user needs, ensuring continuous performance and future-proofing the system's analytics capabilities.

Transforming Operations and Innovation

- **Real-time insights**: Snowflake's near-real-time data ingestion and query capabilities enable immediate access to insights, empowering clinicians to make informed decisions at the point of care and administrators to optimize resource allocation in real time.
- **Predictive analytics**: Advanced analytics and machine learning capabilities within Snowflake facilitate the development of predictive models for early disease detection, targeted interventions, and personalized treatment plans, fostering a proactive and patient-centric approach to healthcare.
- **Collaborative research**: Securely sharing and analyzing data across facilities accelerates research and clinical trials, fostering collaboration among researchers and leading to groundbreaking discoveries.

By embracing Snowflake Data Cloud, the healthcare system can break down data silos, democratize access, and unleash the transformative power of integrated analytics. This paradigm shift can revolutionize care delivery, streamline operations, and accelerate groundbreaking advancements in healthcare, ultimately delivering better patient outcomes, enhanced efficiency, and a more innovative, data-driven healthcare model. Though the preceding illustrates healthcare as an example, a similar discussion holds for other industries.

Additional Snowflake Documentation

https://docs.snowflake.com/en/user-guide/data-sharing-intro

https://docs.snowflake.com/en/user-guide/ui-snowsight

https://docs.snowflake.com/en/user-guide/warehouses-overview

https://docs.snowflake.com/en/developer-guide/snowpark-ml/index

Recipe 1-2. Data as a Product (DaaP)

In the contemporary landscape of data-driven decision-making, organizations increasingly recognize their data's inherent value. Data is no longer merely a byproduct of business operations; it is evolving into a strategic asset that can be packaged and delivered as a product. This concept, Data as a Product (DaaP), represents a paradigm shift in how organizations perceive, manage, and monetize their data.

At its core, DaaP involves treating data as a distinct and valuable offering akin to a traditional product or service. Rather than viewing data solely as an internal resource for operational needs, organizations adopting DaaP recognize its potential to create new revenue streams, enhance customer experiences, and drive innovation.

To be successful in implementing DaaP, several key elements must be considered.

- **Quality and accuracy**: The foundation of any successful DaaP initiative is high-quality, accurate data. Organizations must invest in robust data governance practices, ensuring that the data offered as a product is reliable and meets consumers' expectations.

- **Accessibility and usability**: DaaP involves making data accessible to a broader audience, including both internal stakeholders and external customers. User-friendly interfaces, clear documentation, and seamless integration capabilities ensure that data products are easily consumable.

- **Security and compliance**: As data is transformed into a product, organizations must prioritize security and compliance. Implementing robust security measures and adhering to relevant data protection regulations are imperative to build consumer trust and avoid legal ramifications.
- **Monetization strategy**: A well-defined monetization strategy is crucial for the success of DaaP. Organizations need to determine how they price and package their data products, considering market demand, value proposition, and competitive landscape factors.
- **Data governance and ethics**: Ethical considerations are paramount when offering DaaP. Organizations must establish clear guidelines for the ethical use of data, addressing issues such as privacy, consent, and responsible data practices.
- **Agile infrastructure**: DaaP requires an agile and scalable infrastructure that can adapt to changing market demands and evolving data landscapes. Cloud-based solutions and flexible architectures play a pivotal role in supporting the dynamic nature of data products.
- **Data literacy**: Both internal and external consumers of data products need to be data-literate. Organizations should invest in training programs to enhance the data literacy of their workforce and provide educational resources for external users.

Successful implementation of DaaP involves a holistic approach that encompasses data quality, accessibility, security, monetization, governance, infrastructure, and data literacy. Organizations that effectively navigate these considerations can unlock the full potential of their data, transforming it from a passive resource into a dynamic and valuable product driving innovation and revenue growth. Let's look at an example and how Snowflake can be leveraged.

Problem

Consider a coastal town—call it Sunhaven Cove. This gem, famed for its idyllic beaches and bustling cultural heart, navigates a double-edged dilemma. On the one hand, its tourism industry, the lifeblood of its economy, suffers from volatile revenue streams heavily impacted by seasonal fluctuations. On the other hand, uneven crime distribution across the city creates pockets of insecurity, threatening both residents' well-being and the allure of Sunhaven Cove's vibrant charm.

Traditional strategies have yielded tepid results, leaving Sunhaven Cove yearning for transformative solutions. The fragmented nature of existing data, siloed across disparate government departments and tourism agencies, impedes comprehensive analysis and hinders the formulation of effective interventions.

Solution

Sunhaven Cove is looking for a holistic approach that leverages the power of its data as a strategic asset, not just a passive byproduct.

This need for data-driven innovation presents a unique challenge. Sunhaven Cove must do the following.

- **Data democratization**: As discussed earlier, bridge the data silos separating tourism statistics, crime records, and infrastructure information, creating a consolidated data lake that fosters holistic insights.
- **Develop data-driven models**: Utilize advanced analytics and machine learning techniques to do the following.
    - **Predict tourism trends**: Identify patterns in tourist behavior and forecast seasonal fluctuations, enabling proactive revenue management strategies.
    - **Pinpoint crime hotspots**: Uncover spatial and temporal patterns in criminal activity, informing targeted deployments of law enforcement resources.
    - **Optimize resource allocation**: Allocate budget and personnel across departments based on data-driven insights, maximizing efficiency and impact.

By overcoming these technical hurdles, Sunhaven Cove can transform its data into a product-driven solution.

- **Dynamic tourism packages**: Offering targeted deals and promotions based on predicted demand, smoothing out revenue fluctuations and attracting tourists during off-peak seasons.
- **Predictive policing**: Deploying law enforcement resources before crimes occur, proactively safeguarding high-risk areas, and preventing criminal activity.
- **Data-driven infrastructure investments**: Allocating resources toward public amenities and tourism attractions in areas likely to generate the highest return on investment.

Sunhaven Cove's journey holds immense potential, not just for the city itself but for urban-scale data utilization nationwide. Its successfully crafting a DaaP approach to tackling complex challenges could pave the way for a new era of data-driven governance, transforming cities into dynamic, data-powered ecosystems.

Solution

Consider how Sunhaven Cove could utilize its diverse data assets—tourism data from website visits, lodge bookings, Wi-Fi access, and public safety data from crime reports, traffic flow, and camera footage—to develop two data products.

Sunhaven Cove Tourism Insights

- Sunhaven Cove's first step is to identify consumers (tourism industry players), their use cases (e.g., targeted marketing, seasonality mitigation), and the data contract (anonymized, aggregated datasets) to brainstorm around potential product features and a high-level design.

Tip Consider using the Data Product Canvas framework.

- The Data Product Canvas (DPC) is a visual framework, similar to a business model canvas, designed to guide the development and commercialization of data products. It facilitates collaboration between diverse stakeholders, fosters a structured approach to data monetization, and helps navigate the complexities of turning data into a thriving business asset.

The following lists the benefits of using the DPC.

- **Clarity and focus:** Provides a structured framework for defining the data product's purpose, value proposition, and target audience, ensuring clarity and alignment among stakeholders.
- **Cross-functional collaboration:** Bridges the gap between technical and business teams by creating a shared language and visual representation, fostering efficient collaboration and informed decision-making.
- **Reduced risk:** By prompting thorough consideration of key aspects like data sources, infrastructure, and consumer needs, the DPC helps mitigate potential risks associated with data product development and launch.
- **Flexibility and adaptability:** The DPC allows for iteration and adjustments throughout the development process, enabling teams to adapt to changing market dynamics and consumer needs.
- **Enhanced communication:** Facilitates effective communication with potential customers and partners by providing a clear and concise overview of the data product's capabilities and value proposition.

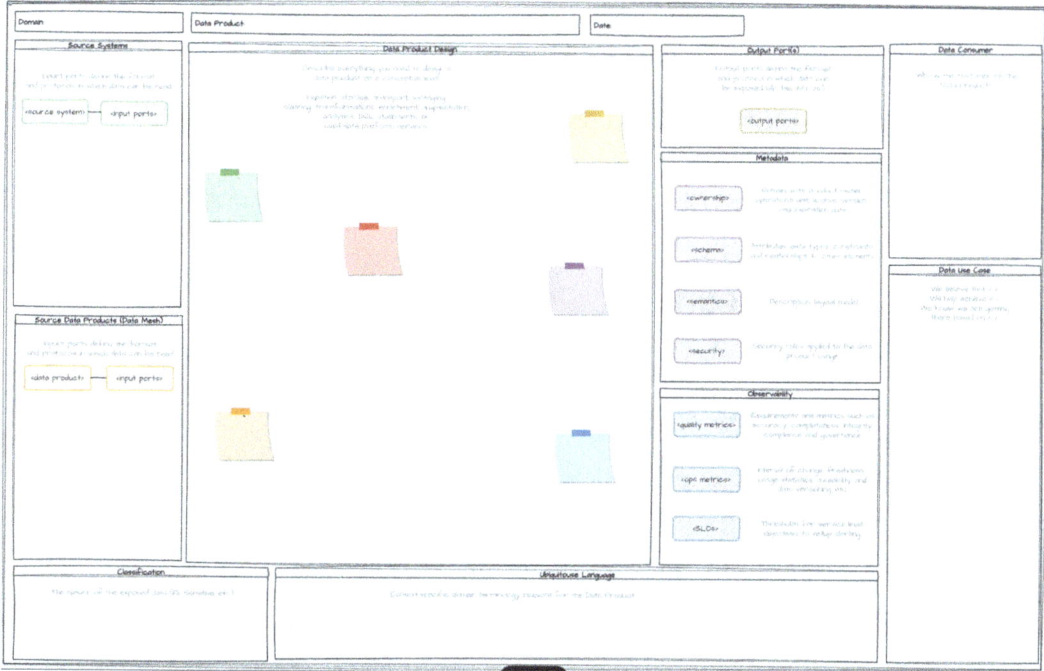

Figure 7-1. Data Product Canvas Template

While the Data Product Canvas template shown in Figure 7-1 provides a solid foundation, it's important to remember it's a tool, not a complete solution. Successful data product development requires a comprehensive strategy beyond the canvas, encompassing market research, data governance, pricing models, and effective marketing and sales initiatives. The DPC offers a valuable framework for organizations exploring the exciting world of data products. Companies can gain a competitive edge in the data-driven economy by embracing its collaborative approach and leveraging its potential to translate data into tangible business value.

Now that the canvas has been created and iterated, the Sunhaven team can identify the features needed for their tourism product.

- **Product features**
    - Curated datasets on tourist demographics, preferences, and travel patterns.
    - Predictive analytics on tourist behavior and future trends.

    - Targeted marketing recommendations for attracting new demographics and optimizing campaigns.

- **Monetization**: Subscription-based service with flexible pricing models based on data volume and user needs for local small businesses to support the community and increase tourism.

Sunhaven Cove Safe Streets

The data canvas guides the development of this API by aligning data with the needs of consumers (local law enforcement) and their use cases (crime prevention, resource allocation).

- **Product Features**
    - Real-time crime statistics and hotspot predictions.
    - Traffic flow optimization data for congestion mitigation.
- **Monetization**: Free public API to foster community engagement and data democratization.

Sunhaven Cove anticipates a tidal wave of prosperity powered by its data-driven initiatives. Tourism Insights promises to usher in a golden age of customized marketing campaigns, attracting new demographics and smoothing out seasonal dips in visitor numbers. Tourists can expect improved experiences thanks to strategically placed infrastructure and events tailored to their preferences. Meanwhile, Safe Streets acts as a guardian angel for residents and visitors alike. Real-time crime data empowers law enforcement to proactively patrol high-risk areas, creating a safer environment for everyone. Additionally, optimized traffic flow promises to ease congestion, reducing frustration for residents and ensuring faster emergency response times. In short, Sunhaven Cove is poised to become a thriving tourist destination and a model of public safety, all thanks to the transformative power of its data products and the use of Snowflake.

Snowflake proves the perfect platform for Sunhaven Cove's DaaP journey, providing a rock-solid foundation for their data products to weather any storm. Its secure and scalable cloud platform drives quality, privacy, and effortless management of its ever-growing data volume. Whether it's the surge of tourist data feeding Tourism Insights or the constant stream of crime and traffic data fueling Safe Streets, Snowflake handles it all seamlessly. But Snowflake's capabilities extend beyond mere storage. Its integrated marketplace simplifies data product distribution, offering flexible pricing models for both subscription and API-based offerings. Sunhaven Cove can effortlessly reach its target audience, whether tourism stakeholders hungry for insights or law enforcement agencies seeking real-time data. Most importantly, Snowflake's adaptable infrastructure scales effortlessly alongside demand. As Sunhaven Cove's data products mature and attract more users, Snowflake seamlessly ramps up its warehouses, ensuring its DaaP initiatives remain future-proof and perpetually impactful.

Additional Snowflake Documentation

`https://www.snowflake.com/en/data-cloud/marketplace/`

`https://docs.snowflake.com/en/user-guide/warehouses-multicluster`

`https://docs.snowflake.com/en/guides-overview-govern`

Recipe 1-3. Snowflake Marketplace

So far, the focus has been on strategy, but at some point, the strategy needs to evolve from theory to practice, queue the Snowflake Marketplace—the global data exchange within the Snowflake Data Cloud, emerges as a transformative solution, empowering organizations to transform their data into a lucrative, monetizable force.

Imagine a secure and vibrant trading floor, not for commodities or stocks, but for data. This is the essence of Snowflake Marketplace—a curated ecosystem connecting data owners with a vast network of potential buyers. Leading businesses across diverse industries, data scientists, and research institutions eagerly seek your data's unique insights. This fosters a mutually beneficial scenario: as the data owner, you access new revenue streams while safeguarding privacy and security; data consumers gain invaluable information fueling their growth and innovation.

But the Snowflake Marketplace transcends a mere data exchange. It's a meticulously curated ecosystem, guaranteeing high-quality offerings and seamless transactions. Think of it as your own data boutique, showcasing your meticulously prepared assets in diverse formats—from raw datasets to pre-analyzed insights and AI-powered models. You retain complete control over packaging and pricing, tailoring offerings to specific buyer needs and maximizing monetization potential.

Gone are the days of inaccessible data trapped in siloed databases. Snowflake Marketplace fosters transparency and ease in data and application sharing. Its intuitive interface simplifies listing assets, managing access controls, and tracking your monetization journey. Robust security and compliance features, a hallmark of Snowflake, ensure your data remains secure throughout the process, fostering trust and encouraging confident participation from buyers and sellers.

The benefits of Snowflake Marketplace extend far beyond immediate financial gains. Participating in this dynamic data exchange unlocks a wealth of new opportunities. Discover emerging market trends and analyze competitor strategies through anonymized data insights. Connect with potential partners and collaborators whose data complements your own, opening doors to groundbreaking innovations and joint ventures. The possibilities are truly limitless.

Whether you're a data-driven enterprise seeking to leverage your information wealth or a data-hungry organization fueling your next breakthrough, Snowflake Marketplace offers a revolutionary approach to data monetization. It's a chance to shed the limitations of data silos and embrace a collaborative, value-driven ecosystem where everyone flourishes. Are you ready to unleash the hidden potential of your data and transform it into a powerful engine for growth? Step into the vibrant world of Snowflake Marketplace and unlock the limitless possibilities that await.

Problem

A retail titan in the e-commerce arena navigates a cutthroat landscape where customer loyalty hangs by a thread. Though their data warehouse holds terabytes of information, their ability to glean actionable insights remains severely curtailed. The problem? Scarcity. While detailed within its niche, customer data lacks the breadth and diversity necessary for robust analytics and personalized marketing campaigns.

This data scarcity manifests itself in several critical challenges. Traditional segmentation techniques, reliant on a multitude of data points, stumble amid the narrow data pool. The resulting customer segments, coarse and imprecise, fail to capture the full spectrum of customer behavior and preferences. This leads to marketing campaigns, armed with blunt segmentation tools, misfire their messages at broad, ill-defined audiences. This cast net approach squanders marketing expenditure and fails to resonate with individual customers. The brand's dream of delivering hyper-personalized experiences tailored to individual needs and preferences remains elusive. Lacking the granularity and diversity of data, their personalization efforts fall flat, failing to trigger the emotional connections that drive customer loyalty.

The consequences are stark.

- **Customer churn**: Generic marketing campaigns fail to engage customers, leading to rising churn rates and dwindling customer lifetime value.
- **Inefficient marketing spend**: Wasted ad impressions and irrelevant promotional offers hemorrhage marketing dollars, eroding profitability and hindering further data acquisition initiatives.
- **Competitive disadvantage**: Data-driven rivals, wielding rich customer profiles and sophisticated analytics tools, personalize experiences with uncanny precision, leaving the brand in the dust.

Solution

This predicament demands a strategic shift, moving beyond simply collecting more data. They require a data enrichment strategy, one that leverages external data sources. Strategically partnering with data providers to access complementary datasets enriches customer profiles with demographic, behavioral, and contextual insights. With more quality data, the engineering teams can now embrace advanced analytics techniques. Employing machine learning models and advanced statistical methods to extract hidden patterns and correlations from their existing and new data, unlocking its latent potential.

By undertaking this data enrichment metamorphosis, our retailer can transform their shallow data pool into a reservoir of actionable insights.

- **Targeted segmentation**: Utilizing enriched customer profiles, precise segments can be created based on nuanced demographics, psychographics, and purchase behaviors. This refined targeting ensures messages resonate deeply with individual customers.
- **Personalized at scale**: Advanced analytics allow for dynamically generated product recommendations, targeted promotions, and tailored content, delivering hyper-personalized experiences that foster customer loyalty and boost engagement.

- **Data-driven decision-making**: Every marketing campaign, every product development, every customer interaction—all informed by accurate, insightful data. This data-driven approach minimizes risk, optimizes resource allocation, and fuels sustainable growth.

This journey serves as a cautionary tale for any organization grappling with the complexities of data-driven marketing in a resource-constrained environment. It's a testament to the transformative power of embracing data enrichment strategies, data exchanges, and data sharing, not just for e-commerce giants but for all businesses seeking to navigate the treacherous waters of customer loyalty and competitive advantage in the digital age.

How It Works

Let's walk through the steps to navigate the Snowflake Marketplace and unlock valuable demographic data for your marketing campaigns.

The following are the prerequisites.

- **Identified data needs**: Ensure demographic data requirements are clearly defined through research and analysis.
- **Marketplace access**: Verify you possess the necessary permissions (ORGADMIN, ACCOUNTADMIN, or a custom role with CREATE DATABASE and IMPORT SHARE) to access the marketplace.

Let's dive in.

Accessing the Marketplace

- **First-time users**: If you haven't explored the marketplace before, an ORGADMIN must accept the consumer terms of service (Figure 7-2).

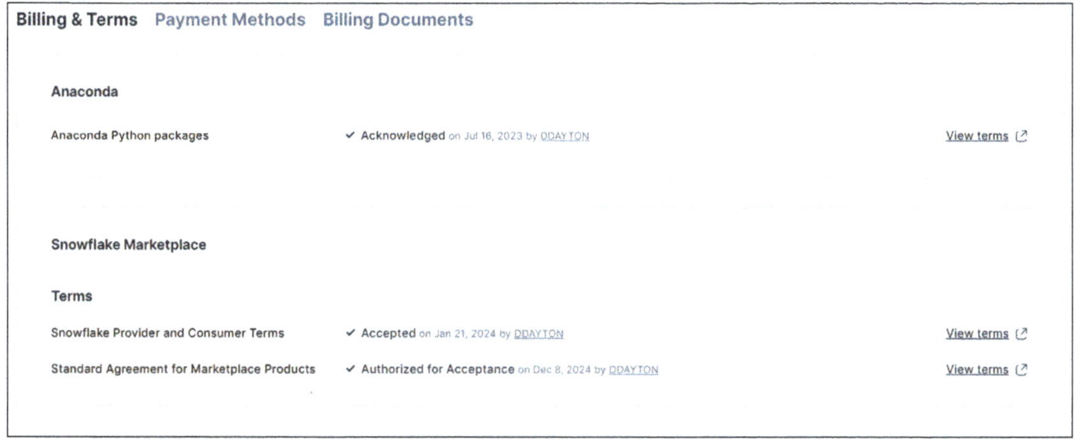

Figure 7-2. *Snowflake Marketplace consumer terms of service*

- **Payment method**: Since the marketplace offers free, trial, and paid data sources, a payment method is required. Snowflake currently uses Stripe as a third-party payment service.
- **Authorized users**: Once permissions are granted, navigate to `https://app.snowflake.com/marketplace` (Figure 7-3) and log in.

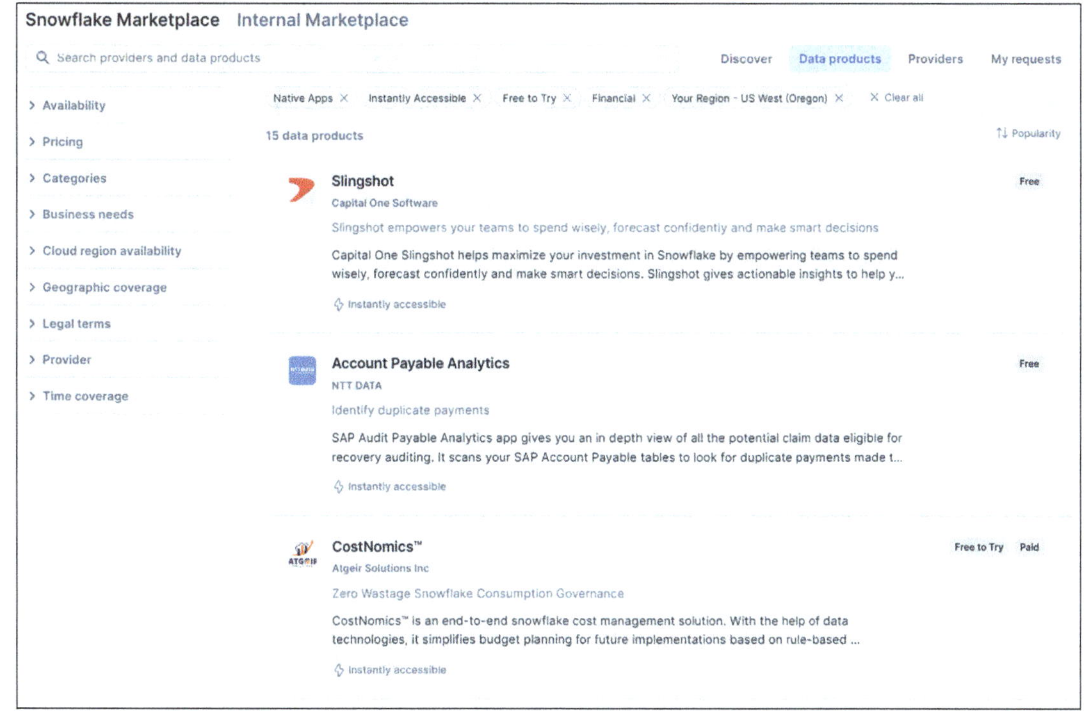

Figure 7-3. *Snowflake Native Apps*

Browsing the Treasure Trove

- **Search bar**: Utilize the search bar to enter keywords like *demographics* or *consumer insights* shown in Figure 7-4, or specific target audience characteristics.

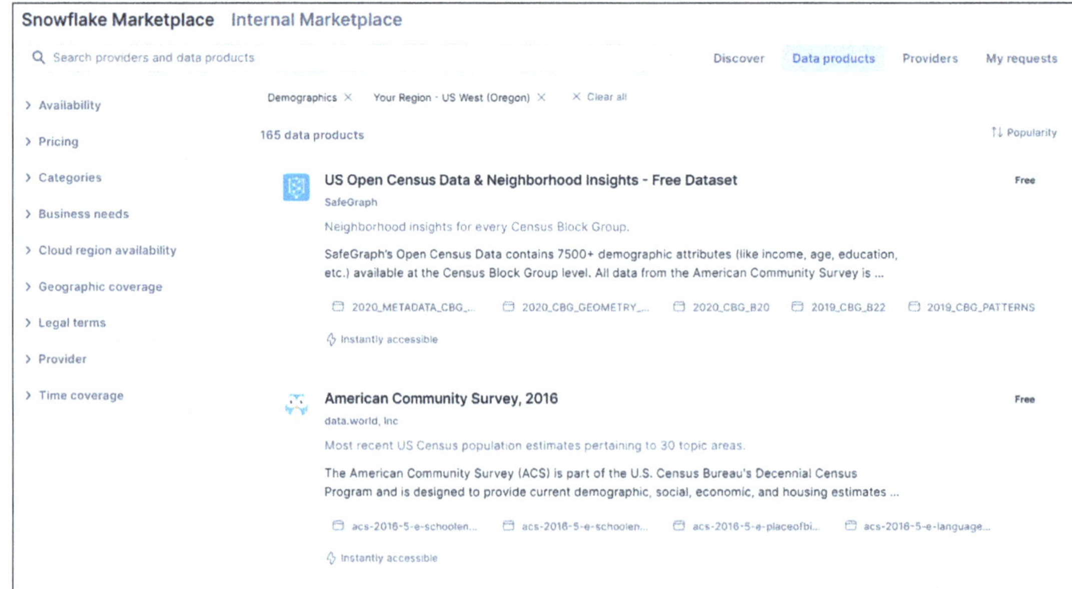

Figure 7-4. *Snowflake Marketplace search for Demographics data*

- **Filters**: More than 60 results were returned. Let's refine the search further using filters like categories and geolocation as illustrated in Figure 7-5.

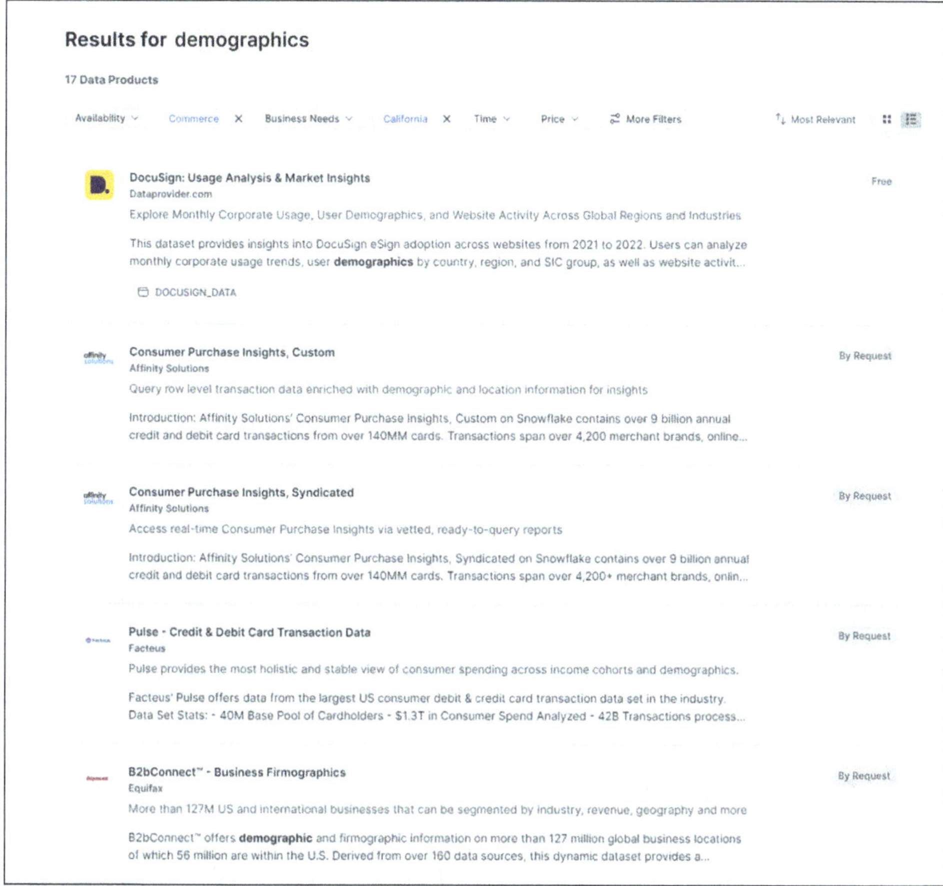

Figure 7-5. *Snowflake Marketplace*

Finding Your Data Gem

- **Data listings**: Each listing provides details on the data provider, data description, sample data, pricing, and applicable terms of service.
- **Reviews and ratings**: Read user reviews and ratings to gain insights into data quality and provider reputation.
- **Ask the provider**: Don't hesitate to contact the data provider with specific questions or requests for clarification.

Purchase insights are crucial to the analysis so let's look at Affinity Solutions offering Consumer Purchase Insights, Custom in Figure 7-6.

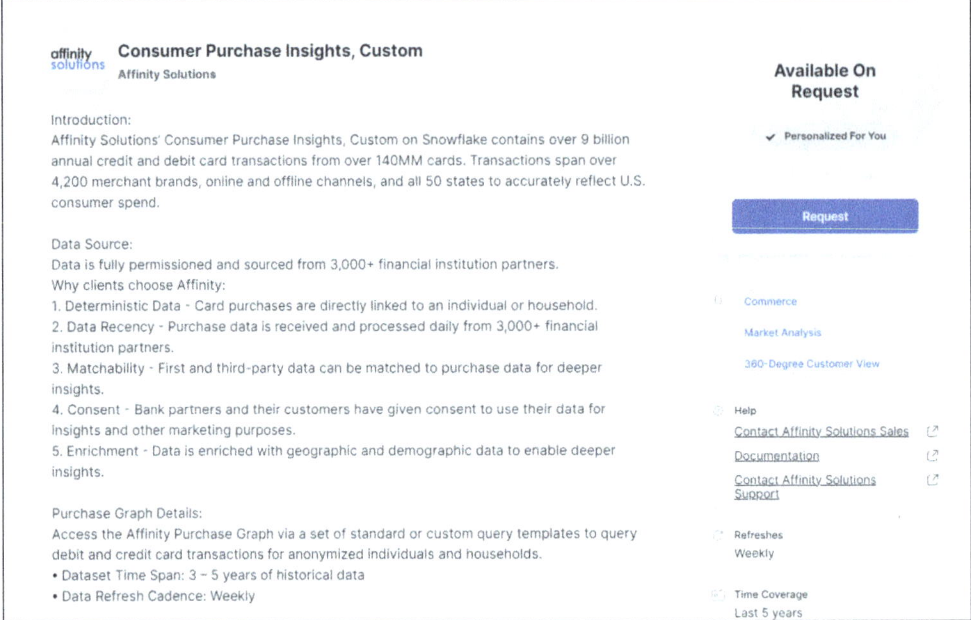

Figure 7-6. *Example of available data in the Snowflake Marketplace*

Evaluating the Data Treasure

- **Free trials**: Many datasets offer free trials, allowing you to test the data before committing.
- **Data samples**: Download and analyze data samples to assess their quality and relevance to your needs.
- **Documentation review**: Carefully review the data provider's documentation, including data format, refresh frequency, and any limitations.

Based on our research, the data set fulfills our requirements. Since the team is opting for a custom solution, we will initiate the request for data and begin to work directly with Affinity Solutions. While that process is in motion let's also look at some weather data to enrich our solution and identify weather's impacts on sales within our environments.

Securing Your Data Haul

The team wants to look at some data before deciding what weather source we will use moving forward—standard or custom. In that agile spirit, we can get a free standard source of historical weather data from AccuWeather in Figure 7-7.

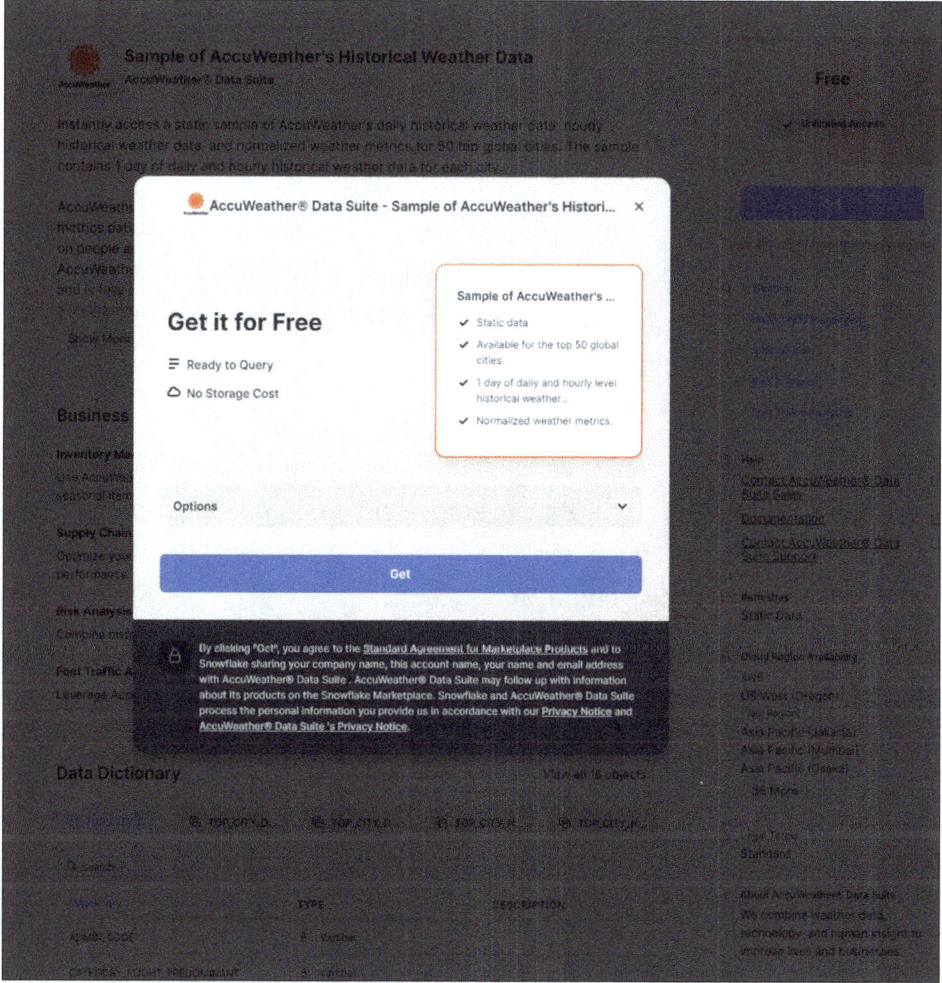

Figure 7-7. *Example of weather data available in the Snowflake Marketplace*

Now the weather data is instantly available for usage in Snowsight worksheets and queryable as seen in Figure 7-8.

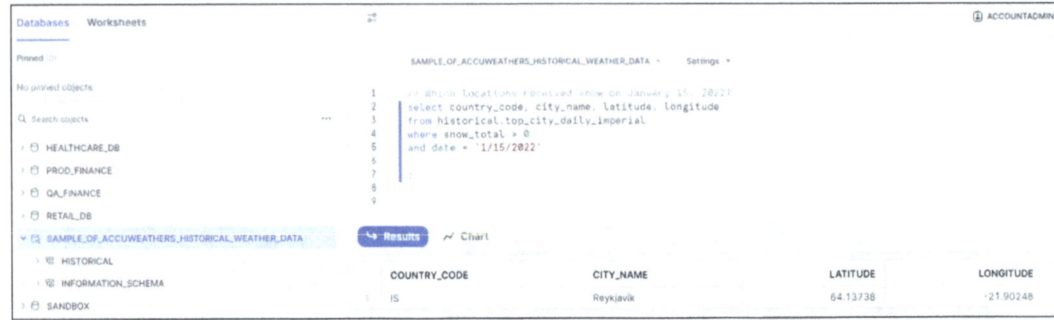

Figure 7-8. Querying weather data share

Tip Stay informed as the Snowflake Marketplace is constantly evolving, so bookmark the page and regularly check for new datasets and updates that may suit your evolving marketing needs.

By following these steps, you can confidently navigate the Snowflake Marketplace and unearth valuable demographic data to empower your marketing campaigns and propel your brand forward. Remember, data is your treasure map—use it wisely to chart a course toward customer engagement and success.

Additional Snowflake Documentation

`https://other-docs.snowflake.com/en/collaboration/consumer-becoming`

`https://app.snowflake.com/marketplace`

Recipe 1-4. Data Monetization

In the digital era, data has transcended its role as a mere byproduct of operations to become a strategic asset with immense economic potential. Data monetization, the practice of converting raw information into revenue, represents a paradigm shift where organizations unlock the hidden value of their data beyond immediate operational needs.

However, successful data monetization requires a meticulous approach. The foundation lies in ensuring high-quality data. Robust data quality management practices are crucial, guaranteeing accuracy, completeness, and reliability. Clean, trustworthy data forms the bedrock for valuable data products that benefit internal analytics and external monetization endeavors.

However, not all data is created equal. Recognizing datasets with intrinsic value is paramount. This requires looking beyond your domain and understanding market demands, consumer needs, and potential cross-industry collaborations. For instance, a construction company's data on worker movement patterns might be valuable for a clothing retailer looking to optimize delivery routes. The key is to identify datasets with unique insights or relevance that can be packaged into compelling data products.

These data products need to be consumable and accessible to target audiences. Packaging data into formats like APIs, curated datasets, or insightful visualizations aligns with user preferences and facilitates value extraction. Remember, value comes from adoption, so make data easily digestible and relevant to your target customers.

Of course, security and compliance are non-negotiable. Data products must adhere to stringent data protection regulations, privacy standards, and industry-specific compliance requirements. Building trust through robust security measures is essential for successful data monetization.

Once data is packaged and secured, monetization strategies come into play. Organizations must determine effective pricing models, whether per-use, subscription-based, or other innovative approaches. The pricing strategy should reflect the perceived value of the data and align with market expectations.

But pricing isn't everything. Distribution channels are crucial for reaching the right audience. Data marketplaces, partnerships with data brokers, or direct customer engagement are all viable options, depending on your target market and data product. Remember, the Snowflake Marketplace is just one example of a channel; explore and identify the most effective route to reach your data consumers.

Ethical considerations are paramount throughout the process. Data governance ensures responsible data practices that address privacy, consent, and ethical use. Transparency and clear guidelines build trust and mitigate potential legal and reputational risks.

To govern data effectively, you need to know it. Data literacy is essential for both internal and external stakeholders. Organizations should invest in training programs to empower their workforce and provide educational resources for external users. This ensures everyone understands and utilizes the value of data responsibly.

Finally, data monetization is an evolving landscape. Continuous innovation is key to success. Organizations must stay abreast of technological advancements, market trends, and emerging use cases to constantly refine their data monetization strategies. This requires a flexible and scalable data infrastructure that can seamlessly adapt to changing demands and integrate new data sources. Cloud-based solutions, modern architectures, and agile DataOps practices provide the agility to navigate this dynamic field.

By carefully considering data quality, identification of valuable data, packaging, security, compliance, monetization strategies, distribution channels, governance, ethics, data literacy, infrastructure, and continuous innovation, organizations can unlock new revenue streams and capitalize on their data assets' full potential. As you explore alternative revenue streams, remember that the key lies in understanding how your organization's posture on each element contributes to successful data monetization.

Problem

XYZ Equipment Rentals stands apart in the rugged, data-sparse landscape of construction. As a leading provider of heavy machinery, they don't just build the future; they generate insights. The vast fleet possesses hidden truths, captured in data on equipment usage, maintenance history, and other administrative sources. Yet, this valuable intelligence remains largely underutilized, a missed opportunity for new revenue streams.

Solution

Transforming these data-driven insights into tangible profit is the crux of XYZ's challenge, shared by many B2B firms grappling with the unlockable potential of their information assets. The key lies in understanding the unique dimensions of this data. XYZ's data delves deep into the operational lifecycles of their machines, painting a granular picture of fuel consumption patterns, predictive maintenance needs, and performance metrics. These detailed records hold immense potential for forging lucrative B2B partnerships. Consider how this data can transcend the confines of individual machines, offering a window into broader construction workflows, equipment utilization trends, and even safety hazards. This contextual richness unlocks additional avenues for monetization beyond pure equipment rental.

Unlocking the hidden power of this data demands a strategic approach to data monetization. Envision the possibilities.

- **Predictive maintenance as a service**: Partnering with construction firms to leverage XYZ's data-driven insights for proactive maintenance, preventing downtime, and maximizing equipment utilization.
- **Construction workflow optimization**: Anonymized data sets could be sold to software developers, empowering them to build tools that streamline construction processes and boost efficiency across the industry.
- **Safety insights marketplace**: XYZ could create a B2B marketplace where construction companies access curated data sets on equipment-related safety risks and preventive measures, fostering a safer and more efficient work environment.

By capitalizing on both the granularity and contextual richness of its data, XYZ can transform its information assets into a potent engine for growth. They evolve from mere equipment providers to data partners, offering invaluable insights and optimizing workflows for the entire construction ecosystem. This data-driven transformation promises financial rewards and deeper integration into the industry, forging lasting partnerships and solidifying their position as data-driven innovators.

XYZ's journey serves as a guiding light for many B2B organizations navigating the data-driven age. It compels us to reimagine the value hidden within our data, unlock its potential, and forge new paths to sustainable growth in the ever-evolving landscape of information-driven business. Now, let's look at a practical solution on how XYZ can leverage Snowflake in their monetization strategy.

How It Works

XYZ has defined its data product and chosen Snowflake Data Cloud as its ideal platform for data monetization.

Data Preparation

- **Curation and cleansing**: XYZ implements data governance practices to ensure the accuracy, completeness, and consistency of its data. This involves addressing missing values, outliers, and potential duplicates while complying with relevant privacy regulations.
- **Data enrichment**: Additional data sources can be integrated (e.g., weather data, construction project timelines) to further enrich the dataset and unlock new analytical possibilities.

Data Asset Identification

- **Predictive maintenance models**: XYZ leverages machine learning algorithms on historical equipment data to develop predictive maintenance models. These models estimate equipment failure probabilities and recommend preventive maintenance schedules, benefiting equipment manufacturers and other rental companies.
- **Construction project efficiency benchmarks**: By analyzing factors like equipment utilization, downtime, and project completion times, XYZ can establish industry benchmarks for construction project efficiency. This valuable data is attractive to project managers and construction firms seeking to optimize their workflows.
- **Risk assessment models**: Combining equipment performance data with accident reports and environmental factors, XYZ can create risk assessment models for insurance companies and safety contractors. These models identify high-risk scenarios and inform preventive measures, enhancing workplace safety and reducing insurance costs.

Impact

- **New revenue stream**: XYZ diversifies its income by unlocking a significant new revenue stream generated through data monetization, reducing reliance solely on equipment rentals.
- **Competitive advantage**: XYZ establishes itself as a data-driven leader in the construction industry, attracting new customers and partnerships through its unique data assets and analytical capabilities.
- **Industry insights**: XYZ's data-driven initiatives contribute to advancing the entire construction ecosystem by providing valuable insights that improve efficiency, safety, and risk management practices across the industry.

Snowflake Data Cloud: The Ideal Platform for Data Monetization

Snowflake's secure and scalable cloud platform presents the ideal solution for XYZ's data monetization journey. Here's why.

- **Secure data management**: Snowflake ensures the highest data privacy and security standards, adhering to industry regulations and protecting sensitive information throughout the data lifecycle.
- **Seamless data sharing and collaboration**: Snowflake Marketplace facilitates creating and distributing data products like those envisioned earlier, enabling secure data sharing and collaboration with partners and customers.
- **Flexible pricing models**: Snowflake provides diverse pricing models for subscriptions, API access, and data usage, allowing XYZ to customize monetization strategies and cater to different customer needs.
- **Scalability and agility**: Snowflake effortlessly scales to accommodate XYZ's growing data volume and user base, ensuring uninterrupted data operations and future-proof growth.

The data product share is ready to publish, and assuming XYZ's product owner has the correct permissions, it's time to start publishing data to the Snowflake Marketplace.

Similar to the consumer discussed in the last section, a provider has its own terms of service. The terms of service must be accepted through an ORGADMIN as shown in Figure 7-9.

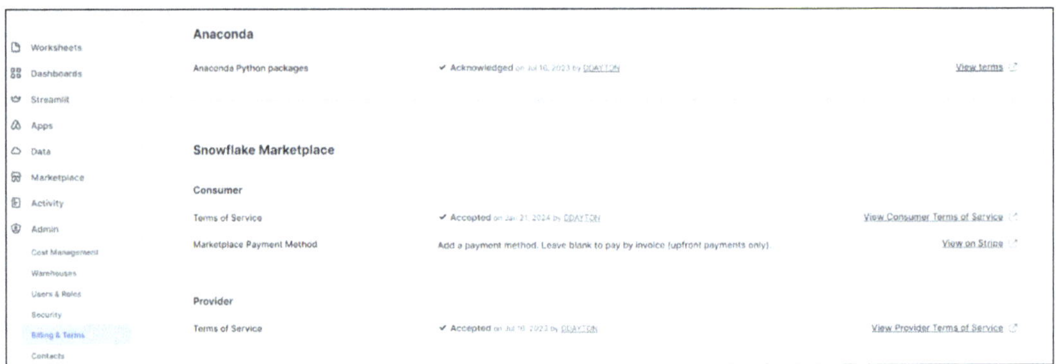

***Figure 7-9.** Snowflake consumer, provider, and Anaconda terms of service*

Next, a provider profile in Figure 7-10 is needed to help consumers learn more about the data product. Once the profile is complete and approved by Snowflake, the product listings are available to anyone on the marketplace.

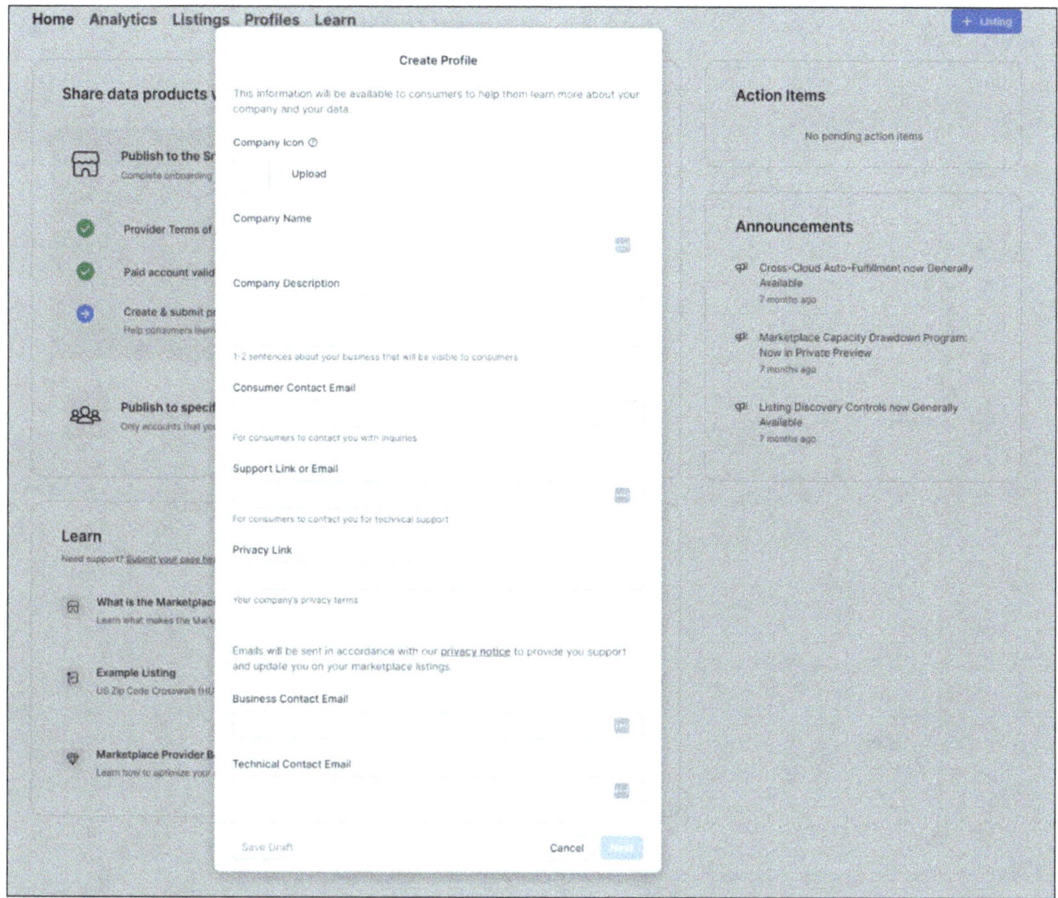

Figure 7-10. *Provider profile*

Until that approval, there is still an option to publish directly to consumers, which has been specified. Here in Figure 7-11, you can see an example of sharing a specific object from one account to another.

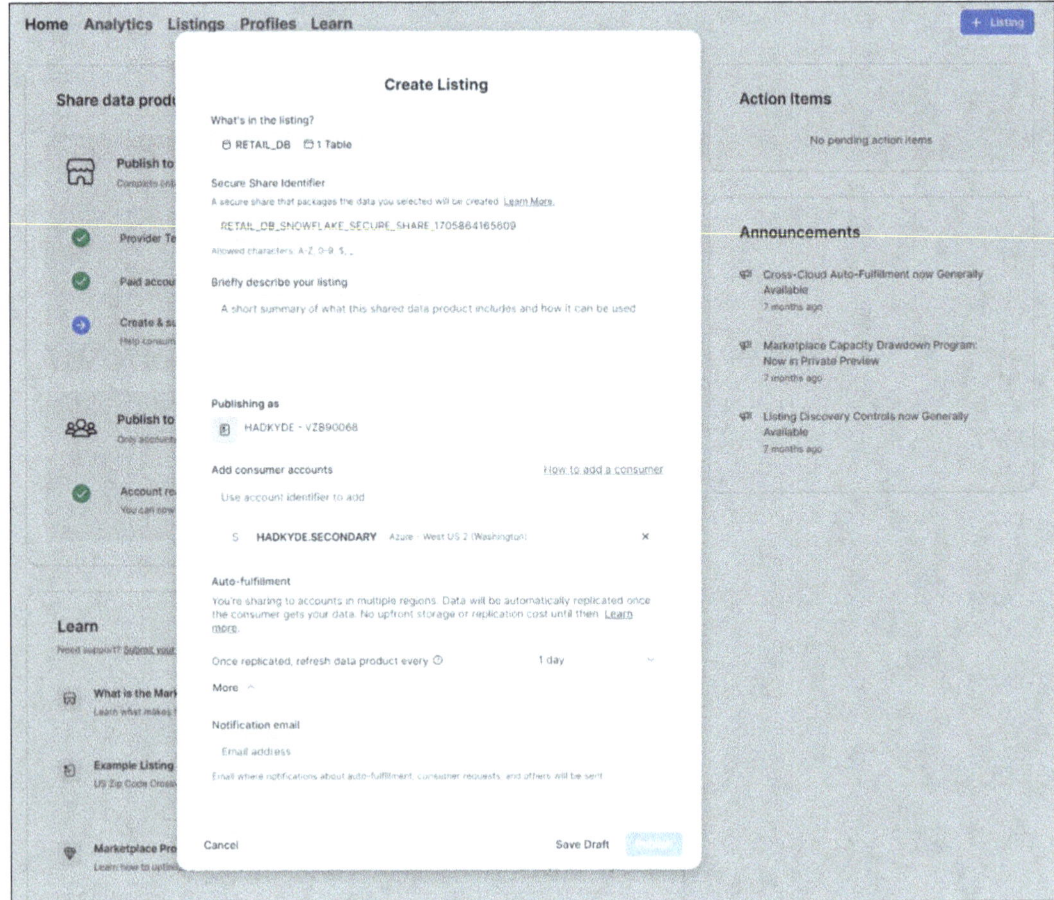

Figure 7-11. Creating a listing on the Snowflake Marketplace

XYZ now has its first provider listing and is officially on its monetization journey.

Tip Did you know Snowflake also supports Snowflake Native Apps on the Snowflake Marketplace? Products and monetization can span far beyond just data. How can you contribute to the marketplace?

XYZ Equipment Rentals demonstrates how B2B data monetization unlocks new possibilities for companies with rich data resources. By leveraging Snowflake's capabilities, construction enterprises can transform their data from a hidden asset into a powerful engine for growth, enhancing their operations and contributing to the advancement of the entire industry. As data takes center stage in the construction landscape, XYZ's pioneering journey serves as a roadmap for embracing the transformative potential of data monetization.

This revised version emphasizes technical details, avoids metaphors and colorful language, and incorporates a dedicated section highlighting the specific benefits of the Snowflake Data Cloud for XYZ's data monetization strategy.

Additional Snowflake Documentation

`https://www.snowflake.com/en/data-cloud/marketplace/`

`https://www.snowflake.com/en/data-cloud/overview/marketplace/snowflake-marketplace-for-data-and-application-partners/`

`https://docs.snowflake.com/en/user-guide/data-exchange-becoming-a-provider`

CHAPTER 8

Data Recovery and Protection

Overview

Business continuity and disaster recovery (BC/DR) planning is not just a formality but a critical process that holds immense value for organizations seeking long-term stability and success. In today's unpredictable business landscape, where unforeseen events like natural disasters, cyberattacks, or emergencies can strike at any moment, a robust BC/DR plan becomes paramount.

The true value of a well-designed BC/DR plan cannot be emphasized enough. Its absence leaves companies vulnerable to the loss of critical data, substantial revenue setbacks, and potential damage to their hard-earned reputation. By implementing an effective BC/DR strategy, organizations can significantly minimize downtime, mitigate the risk of data loss, and ensure a seamless experience for their customers. Ultimately, these factors directly impact the company's bottom line and sustainability in the long run.

The upcoming chapters delve into the intricacies of BC/DR planning within the Snowflake platform. We will uncover key concepts and shed light on how organizations can tailor their BC/DR plans to address specific needs. By the time you reach the end of this comprehensive exploration, you should have a profound understanding of the significance of BC/DR planning and be equipped with the necessary knowledge and steps to develop a robust plan within the Snowflake data cloud. With such a plan in place, organizations can confidently navigate through times of crisis while ensuring uninterrupted operations and continued success.

D. Dayton and J. Eipe, *Snowflake Recipes*, https://doi.org/10.1007/979-8-8688-0938-5_8

Recipe 8-1. Fail-safe and Time Travel

Snowflake is a powerful and versatile cloud data platform that enables businesses to process and analyze massive amounts of data quickly and efficiently. However, even the most reliable platforms can experience disruptions, and data can become compromised or lost. This is why Snowflake has designed a suite of core features to ensure businesses have robust data protection and recovery mechanisms. Among these are Snowflake's Fail-safe and Time Travel features.

It is important to note that the seven-day fail-safe period requires a Snowflake Support Ticket to recover and is non-configurable. Once the Time Travel retention period concludes, the fail-safe period begins, enabling potential data recovery. However, it is worth considering that if there is an ongoing Time Travel statement, the movement of data and objects (such as tables, schemas, and databases) into the fail-safe period may be delayed until the query is completed.

Snowflake Fail-safe is a crucial component of data resilience and disaster recovery strategies, ensuring that organizations have a safety net to rely on in case of unexpected disruptions. It provides businesses with peace of mind, knowing that their data is protected, recoverable, and available, reinforcing the reliability and dependability of the Snowflake platform. Consider the following notes from Snowflake regarding Fail-safe.

> *Fail-safe is a data recovery service that is provided on a best effort basis and is intended only for use when all other recovery options have been attempted.*
>
> *Fail-safe is not provided as a means for accessing historical data after the Time Travel retention period has ended. It is for use only by Snowflake to recover data that may have been lost or damaged due to extreme operational failures.*
>
> *Data recovery through Fail-safe may take several hours to days to complete.*[2]

Meanwhile, Snowflake's Time Travel enables businesses to restore their data to any point in time, up to 90 days in the past (Enterprise Edition or higher), without relying on traditional backup and restore processes with their long execution times and costly processes. This self-service feature provides businesses with peace of mind. It offers the flexibility to explore and analyze historical data, restore data objects quickly, and create data object replicas from historical data in the past. Consider the table activity and volume, which could increase overall storage consumption. It is important to have a cohesive monitoring strategy to help manage and reduce costs if needed.

This chapter delves into these features and explores how they can be used to help businesses protect their data and maintain business continuity in the face of unexpected disruptions. Together, these two features help ensure high availability, data durability, and business continuity in case of disasters or system failures.

Problem

A retail company has relied on Snowflake as its primary data warehousing solution for several years. Unfortunately, an accidental command was executed during a recent code release, leading to incorrect or "bad" data in a crucial table from their Snowflake account. This table holds vital sales data that plays a critical role in inventory management and forecasting. The absence of this data has the potential to severely disrupt the company's operations, underscoring the urgent need to recover it promptly.

In this challenging situation, the company is desperate to correct the incorrect data and mitigate the impact on its business activities. Timely recovery of the data is imperative to ensure uninterrupted inventory management and accurate sales forecasting. Without the vital sales data, the company may encounter significant challenges in replenishing stock, meeting customer demands, and making informed business decisions.

This incident is a stark reminder of the importance of data protection and the need for robust safeguards to prevent accidental data loss. The company intends to learn from this experience, reinforcing its data governance practices and strengthening data continuity capabilities within Snowflake to ensure the future integrity and availability of its critical data assets.

Solution

Assume we have a transactional fact table called `fact_sales_transactions`, and the pipeline team found a job error that resulted in incorrect records for the last 48 hours. While the job needs to be triaged to determine the root cause and be resolved, the team needs to recover the incorrect data before the business team starts their day.

Evaluating the query history, the team found two pieces of information.

The root cause was due to a lookup table being dropped during the last deployment cycle and it was missed by the DataOps unit tests. Since this lookup table was dropped, all tax amounts were calculated using a generic rate instead of the corresponding regional rate causing the fact table to be incorrect.

Let's tackle the missing lookup table first.

By leveraging the Snowflake query history, Figure 8-1 shows how the team could identify the QUERY_ID of the transaction that dropped the table.

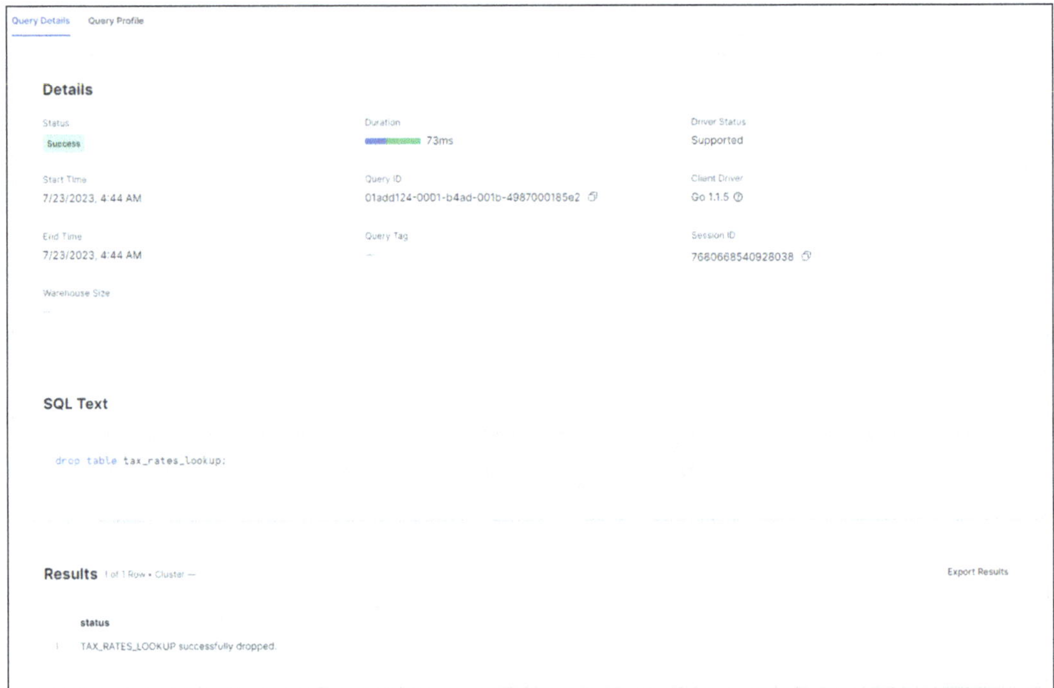

Figure 8-1. *Query Profile in Snowflake Snowsight*

By switching the context to the analyst user, Figure 8-2 shows that this table no longer exists in the database.

Figure 8-2. *Snowflake Snowsight Worksheet*

Using a SHOW TABLES command in Figure 8-3 also helps confirm when the table was dropped. The DROPPED_ON timestamp coincides with the ELT cycle.

```
SHOW TABLES HISTORY LIKE 'tax%' IN retail_db.sales;
```

Results | Chart

	created_on	name	database_name	schema_name	rows	...	dropped_on
1	2023-07-23 03:42:47.626 -0700	TAX_RATES_LOOKUP	RETAIL_DB	SALES	5		2023-07-23 03:44:00.177 -0700

Figure 8-3. *Snowflake Snowsight Worksheet*

The team quickly recovered the object by running the UNDROP command shown in Figure 8-4.

```
use role sysadmin;
use database retail_db;
use schema sales;

UNDROP TABLE tax_rates_lookup;
```

Results | Chart

	status
1	Table TAX_RATES_LOOKUP successfully restored.

Figure 8-4. *Snowflake Snowsight Worksheet*

The analyst role can now view and query `tax_rates_lookup`. Next, the team recovers from the bad data in the FACT table using Time Travel shown in Figure 8-5.

```
use role supply_chain_analyst;

select
    state_code,
    county,
    city,
    tax_rate,
    effective_date,
    expiration_date
from tax_rates_lookup;
```

Results Chart

STATE_CODE	COUNTY	CITY	TAX_RATE	EFFECTIVE_DATE	EXPIRATION_DATE
CA	Los Angeles	Los Angeles	9.50	2023-01-01	2023-12-31
CA	Orange	Irvine	8.75	2023-01-01	2023-06-30
NY	null	null	8.88	2023-01-01	null
TX	Harris	Houston	8.25	2023-01-01	2023-12-31
TX	Dallas	Dallas	8.25	2023-01-01	2023-12-31

Figure 8-5. *Snowflake Snowsight Worksheet*

Figure 8-6 shows that the tax has been generally set to a value of $10 for all transactions in the table. This provides evidence that the `fact_sales_transaction` table is not in a good state.

```
select distinct transaction_id, tax from fact_sales_transaction;
```

Results Chart

TRANSACTION_ID	TAX
1	10.00
2	10.00
3	10.00
4	10.00
5	10.00

Figure 8-6. *Snowflake Snowsight Worksheet*

Time Travel can be used via several methods, such as distinct timestamps, time offsets, and individual statements. Let's identify the `query_id` of the update statement in Figure 8-7 and use it in conjunction with Snowflake clones (discussed in the next section) to create a new cloned table from the original as it was prior to the bad update.

Query Details Query Profile

Details

Status: Success
Duration: 982ms
Driver Status: Supported
Start Time: 7/23/2023, 5:09 AM
Query ID: 01add13d-0001-b4ad-001b-49870001879a
Client Driver: Go 1.1.5
End Time: 7/23/2023, 5:09 AM
Query Tag: —
Session ID: 7680668540928038
Warehouse Size: X-Small

SQL Text

```
update fact_sales_transaction set tax = 10;
```

Results 1 of 1 Row • Cluster 1 Export Results

	number of rows updated	number of multi-joined rows updated
1	5	0

Figure 8-7. *Query Profile in Snowflake Snowsight*

Now that we have query_id from Figure 8-7 let's restore the table to its prior state in Figure 8-8.

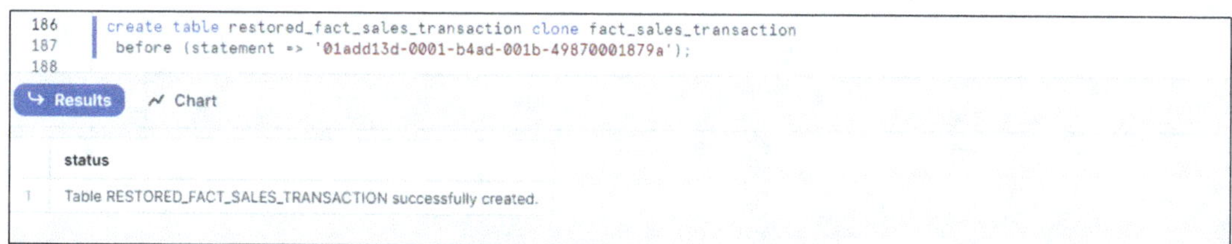
create table restored_fact_sales_transaction clone fact_sales_transaction
before (statement => '01add13d-0001-b4ad-001b-49870001879a');
Results
Chart
status
Table RESTORED_FACT_SALES_TRANSACTION successfully created.

Figure 8-8. *Snowflake Snowsight Worksheet*

Figure 8-9 shows the cloned table has the correct tax information.

```
select distinct transaction_id, tax from restored_fact_sales_transaction;
```

Results Chart

	TRANSACTION_ID	TAX
1	1	3.50
2	2	1.50
3	3	0.50
4	4	2.00
5	5	1.00

***Figure 8-9.** Snowflake Snowsight Worksheet*

Finally, the team is ready to perform a table name swap and restore the system to the correct state prior to the bugged ELT cycle.

```
alter table fact_sales_transaction rename to bad_fact_sales_transaction;
alter table restored_fact_sales_transaction rename to fact_sales_transaction

use role supply_chain_analyst;

select
    TRANSACTION_ID,
    SALE_DATE,
    QUANTITY,
    UNIT_PRICE,
    TOTAL_AMOUNT,
    TAX,
    CREATED_AT,
    UPDATED_AT
from fact_sales_transaction;
```

Results Chart

	TRANSACTION_ID	SALE_DATE	QUANTITY	...	UNIT_PRICE	TOTAL_AMOUNT	TAX	CREATED_AT	
1	1	2023-05-14	3		25.99	77.97	3.50	2023-05-14 10:00:00.000	2023-05-
2	2	2023-05-14	2		19.99	39.98	1.50	2023-05-14 11:30:00.000	2023-05-
3	3	2023-05-15	1		9.99	9.99	0.50	2023-05-15 09:45:00.000	2023-05-
4	4	2023-05-15	5		12.50	62.50	2.00	2023-05-15 14:20:00.000	2023-05-
5	5	2023-05-16	3		8.75	26.25	1.00	2023-05-16 16:05:00.000	2023-05-

***Figure 8-10.** Snowflake Snowsight Worksheet*

Now that the business has accurate data to start its day from Figure 8-10, the DataOps team can focus on resolving the root cause in the deployment process.

Additional Snowflake Documentation

`https://docs.snowflake.com/en/user-guide/data-availability`

`https://docs.snowflake.com/en/user-guide/data-failsafe`

Recipe 8-2. Snowflake Clones

Snowflake clones are a powerful and efficient feature that enables users to create an exact copy of an existing database, table, or schema within seconds. Cloning allows users to work on a replica of data while leaving the original data untouched, as you saw in the last example.

Clones in Snowflake are created almost instantly and require minimal storage resources as they use a zero-copy cloning methodology. They share the underlying data with the original object through metadata pointers in the Snowflake cloud services layer, and any changes made to the clone are only reflected in the clone and, at that point begin using additional storage for the modified data.

In today's world, data is king. Enterprises depend on data for various operations, including decision-making, analytics, and forecasting. As the demand for data-driven insights increases, businesses need a robust and reliable platform to manage their data. This is where Snowflake clones come in.

Cloning is useful in various scenarios, such as creating a test environment, implementing data governance policies, performing data analysis and reporting, and disaster recovery testing. Additionally, cloning provides a fast and efficient way to make a copy of a large dataset without incurring additional storage costs.

Overall, Snowflake clones provide a flexible, cost-effective, and efficient solution to data management challenges, allowing users to work with their data in a safe and scalable manner.

Problem

In a typical data warehouse environment, data is processed and stored for various business purposes, such as analytics, reporting, and data-driven decision-making. To maintain the quality of data and ensure proper testing of processes, it is essential to have a separate testing environment or QA environment that mirrors the production environment as closely as possible. This enables the testing team to perform comprehensive tests without affecting the live data or disrupting business operations.

In the case of a Snowflake data warehouse, creating a QA environment from PROD is a common use case. However, this process can be complex and time-consuming, requiring significant expertise and resources. The challenge is creating a replica of the production environment while ensuring that sensitive data is masked and security is in place. Additionally, the QA environment should be maintained and updated regularly to keep it in sync with production data.

A typical use case for creating a QA environment from PROD in Snowflake could be a financial institution that wants to ensure the accuracy and reliability of its financial reports. To achieve this, they need a testing environment replicating their production environment while maintaining governance and privacy (i.e., QA teams do not need access to PII data for testing). This would involve creating a clone of their existing production database and masking sensitive data. The testing team can perform various unit tests, modeling, and end-to-end tests without impacting the live data. Once testing is complete, the QA environment can be refreshed with updated production data to ensure that it remains current for the next release cycle.

Solution

The data engineering team has just pushed a release branch from DEV to the QA environment for a change to the finance schema. The QA team is now responsible for testing the latest code changes. The QA engineer assigned the story needs to begin with refreshing the QA environment from the PROD environment, specifically the `billing.invoices` table.

Figure 8-11 presents the table to test.

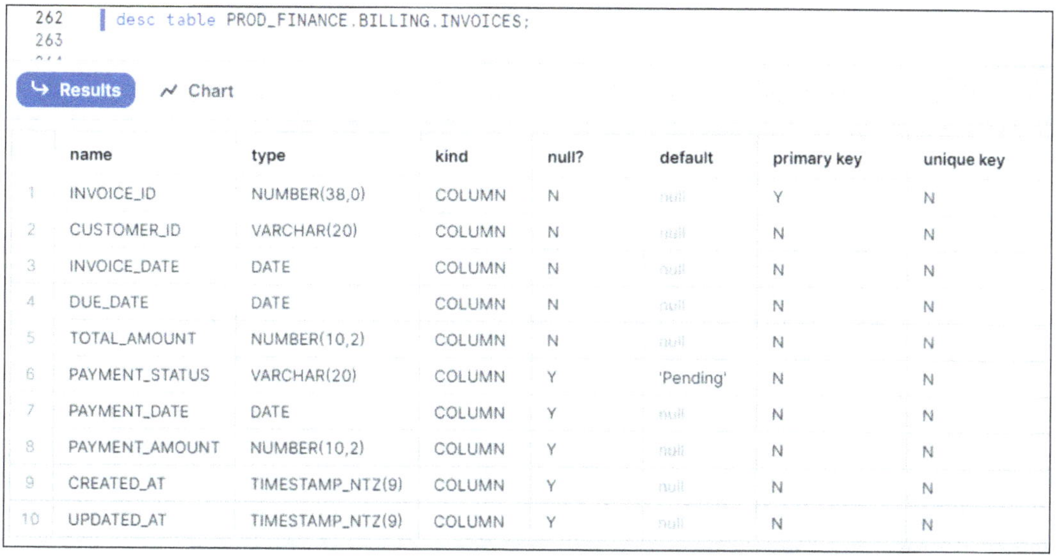

```
desc table PROD_FINANCE.BILLING.INVOICES;
```

Results Chart

	name	type	kind	null?	default	primary key	unique key
1	INVOICE_ID	NUMBER(38,0)	COLUMN	N	null	Y	N
2	CUSTOMER_ID	VARCHAR(20)	COLUMN	N	null	N	N
3	INVOICE_DATE	DATE	COLUMN	N	null	N	N
4	DUE_DATE	DATE	COLUMN	N	null	N	N
5	TOTAL_AMOUNT	NUMBER(10,2)	COLUMN	N	null	N	N
6	PAYMENT_STATUS	VARCHAR(20)	COLUMN	Y	'Pending'	N	N
7	PAYMENT_DATE	DATE	COLUMN	Y	null	N	N
8	PAYMENT_AMOUNT	NUMBER(10,2)	COLUMN	Y	null	N	N
9	CREATED_AT	TIMESTAMP_NTZ(9)	COLUMN	Y	null	N	N
10	UPDATED_AT	TIMESTAMP_NTZ(9)	COLUMN	Y	null	N	N

Figure 8-11. *Snowflake Snowsight Worksheet*

When we query the table in PROD, Figure 8-12 shows a masking policy applied to the PII data.

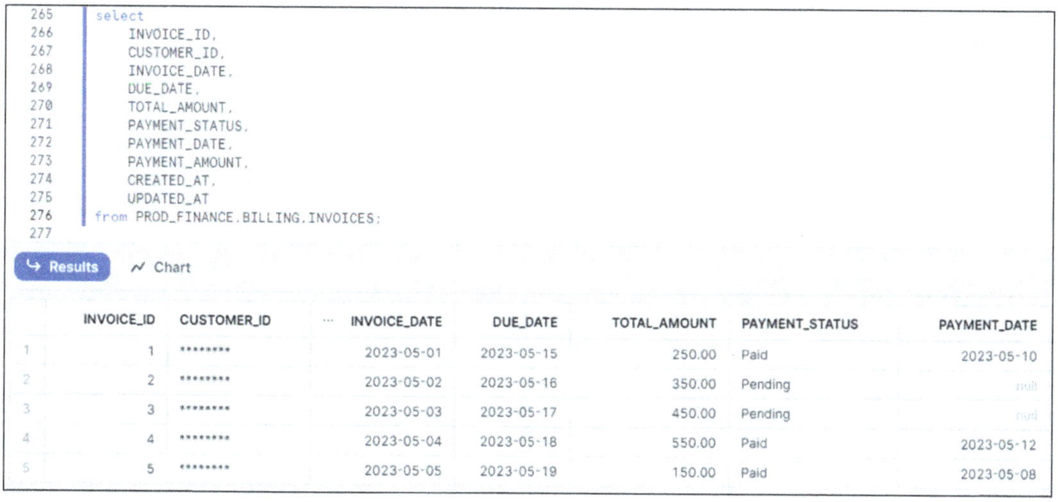

```
select
    INVOICE_ID,
    CUSTOMER_ID,
    INVOICE_DATE,
    DUE_DATE,
    TOTAL_AMOUNT,
    PAYMENT_STATUS,
    PAYMENT_DATE,
    PAYMENT_AMOUNT,
    CREATED_AT,
    UPDATED_AT
from PROD_FINANCE.BILLING.INVOICES;
```

Results Chart

	INVOICE_ID	CUSTOMER_ID	INVOICE_DATE	DUE_DATE	TOTAL_AMOUNT	PAYMENT_STATUS	PAYMENT_DATE
1	1	********	2023-05-01	2023-05-15	250.00	Paid	2023-05-10
2	2	********	2023-05-02	2023-05-16	350.00	Pending	null
3	3	********	2023-05-03	2023-05-17	450.00	Pending	null
4	4	********	2023-05-04	2023-05-18	550.00	Paid	2023-05-12
5	5	********	2023-05-05	2023-05-19	150.00	Paid	2023-05-08

Figure 8-12. *Snowflake Snowsight Worksheet*

Looking at the masking policy, Figure 8-13 shows that only the `finance_admin` has access to this column—a role our QA engineer has not granted them.

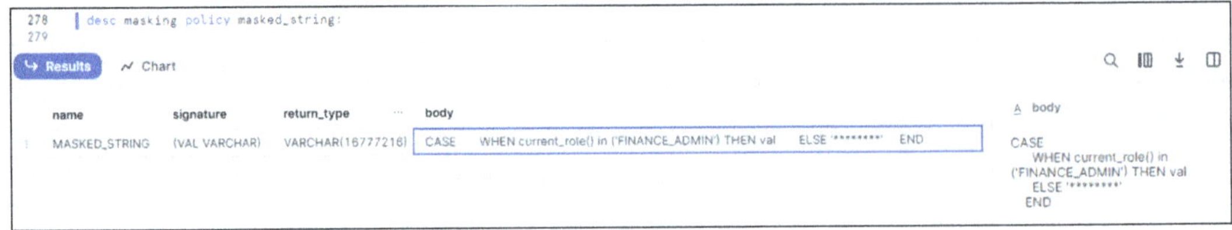

***Figure 8-13.** Snowflake Snowsight Worksheet*

Now, let's refresh the QA environment with a clone in Figure 8-14 from `prod_finance.billing` to `qa_finance.billing`.

***Figure 8-14.** Snowflake Snowsight Worksheet*

The engineer confirms the invoice table is now accessible in the QA environment. Note the source database as `qa_finance` and the masked `customer_id` maintaining privacy.

```
use role qa_engineer;
select
    INVOICE_ID,
    CUSTOMER_ID,
    INVOICE_DATE,
    DUE_DATE,
    TOTAL_AMOUNT,
    PAYMENT_STATUS,
    PAYMENT_DATE,
    PAYMENT_AMOUNT,
    CREATED_AT,
    UPDATED_AT
from QA_FINANCE.BILLING.INVOICES;
```

Results Chart

	INVOICE_ID	CUSTOMER_ID	INVOICE_DATE	DUE_DATE	TOTAL_AMOUNT	PAYMENT_STATUS	PAYMENT_DATE
1	1	********	2023-05-01	2023-05-15	250.00	Paid	2023-05-10
2	2	********	2023-05-02	2023-05-16	350.00	Pending	null
3	3	********	2023-05-03	2023-05-17	450.00	Pending	null
4	4	********	2023-05-04	2023-05-18	550.00	Paid	2023-05-12
5	5	********	2023-05-05	2023-05-19	150.00	Paid	2023-05-08

***Figure 8-15.** Snowflake Snowsight Worksheet*

As previously discussed, the advantage of using Snowflake clones is no additional storage costs and allowing isolated updates to the clone that don't modify or affect the origin table, which is in production in this case. This can be seen in Figure 8-16 by querying the `information_schema.table_storage` metrics table using the `accountadmin` role. Depending on the RBAC strategy, one could also use the table owner role to achieve the same outcome.

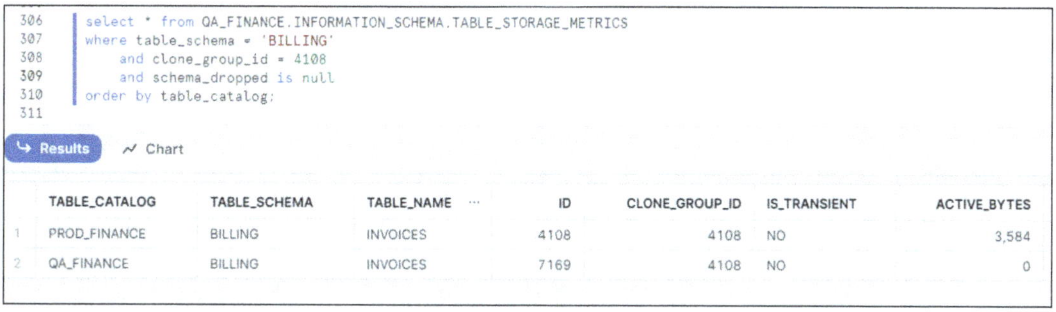

```
select * from QA_FINANCE.INFORMATION_SCHEMA.TABLE_STORAGE_METRICS
where table_schema = 'BILLING'
    and clone_group_id = 4108
    and schema_dropped is null
order by table_catalog;
```

Results Chart

	TABLE_CATALOG	TABLE_SCHEMA	TABLE_NAME ···	ID	CLONE_GROUP_ID	IS_TRANSIENT	ACTIVE_BYTES
1	PROD_FINANCE	BILLING	INVOICES	4108	4108	NO	3,584
2	QA_FINANCE	BILLING	INVOICES	7169	4108	NO	0

Figure 8-16. *Snowflake Snowsight Worksheet*

Note how the QA table is a child of PROD via the `clone_group_id` yet still has its distinct table with zero active bytes identified by the `id` column in Figure 8-17.

The engineer can begin testing after the QA environment has been refreshed. During those tests, it is common to insert, modify, and delete records to simulate an ELT process. The engineer generates test data to modify the table and then shows how the clone is now utilizing storage, and the changes do not apply to the origin table.

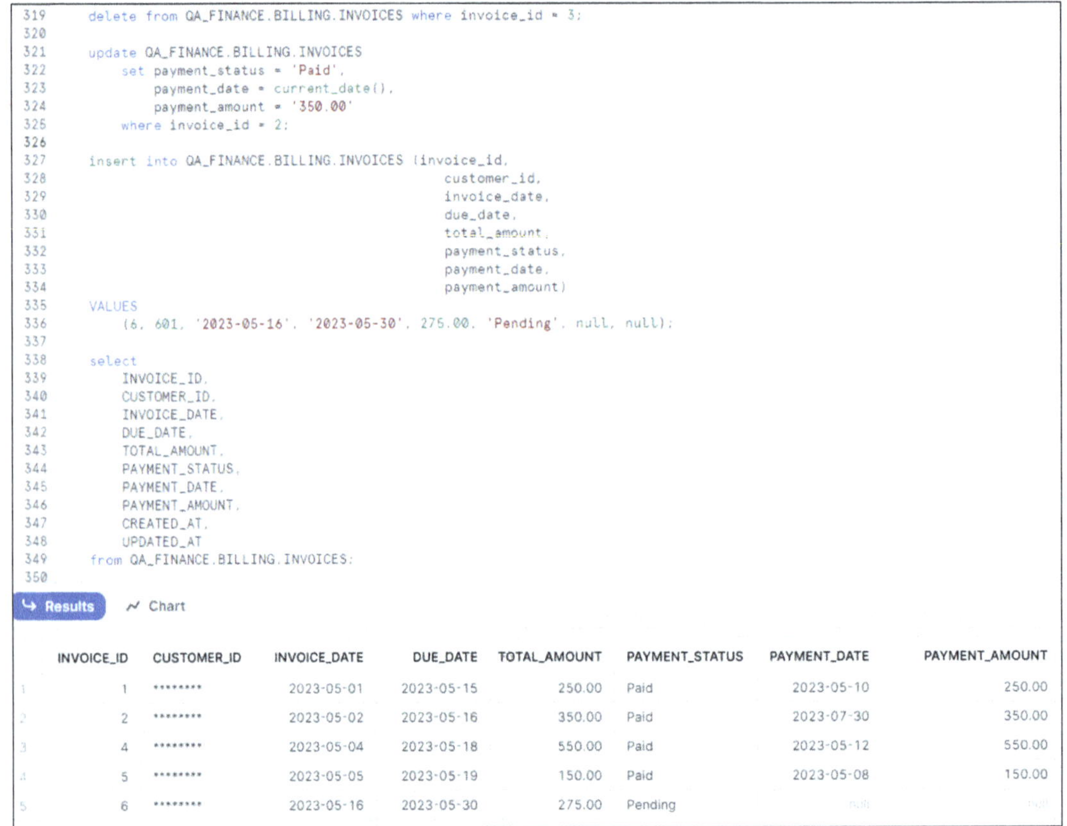

Figure 8-17. *Snowflake Snowsight Worksheet*

If we rerun our query against the `table_storage_metrics` table, Figure 8-18 shows that the clone table in the QA environment is using storage.

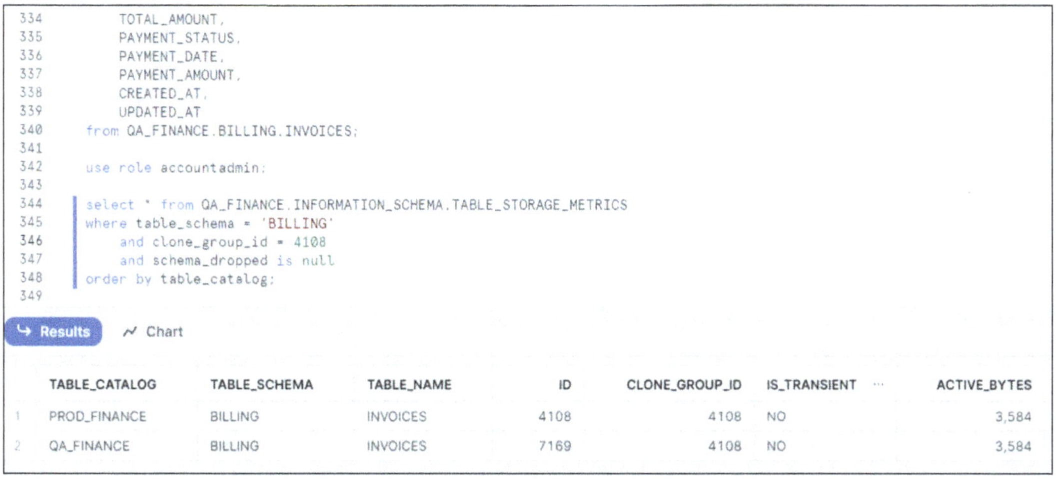

```
        TOTAL_AMOUNT,
        PAYMENT_STATUS,
        PAYMENT_DATE,
        PAYMENT_AMOUNT,
        CREATED_AT,
        UPDATED_AT
    from QA_FINANCE.BILLING.INVOICES;

    use role accountadmin;

    select * from QA_FINANCE.INFORMATION_SCHEMA.TABLE_STORAGE_METRICS
    where table_schema = 'BILLING'
        and clone_group_id = 4108
        and schema_dropped is null
    order by table_catalog;
```

Results Chart

	TABLE_CATALOG	TABLE_SCHEMA	TABLE_NAME	ID	CLONE_GROUP_ID	IS_TRANSIENT	···	ACTIVE_BYTES
1	PROD_FINANCE	BILLING	INVOICES	4108	4108	NO		3,584
2	QA_FINANCE	BILLING	INVOICES	7169	4108	NO		3,584

Figure 8-18. *Snowflake Snowsight Worksheet*

Using Snowflake clones for agile development is a powerful method that can help drive innovation and value for any organization leveraging this feature.

Additional Snowflake Documentation

`https://docs.snowflake.com/en/user-guide/object-clone`

Recipe 8-3. Account Replication and Failover (Disaster Recovery)

In today's data-driven world, it is essential to have a reliable and resilient data platform that can handle unforeseen circumstances and minimize downtime. The Snowflake data cloud offers several features that ensure high availability and disaster recovery for its customers. Two of these critical features are account replication and failover.

Account replication allows Snowflake customers to replicate databases and shares in all accounts and other account objects with Business Critical Edition. This feature enables companies to maintain a replica in another account within their Snowflake organization, in a different region, or even across clouds, providing disaster recovery capabilities in the event of regional outages or disasters. With account replication, companies can have a secondary instance of their data ready to be activated in the event of a primary region outage. It is important to evaluate the overall cost of replication against the cost of not having replication in the BC/DR strategy.

Failover (Business Critical Edition) is another feature that enhances the reliability of Snowflake's data platform. Failover is the process of switching to the secondary account replica in the event of a primary region outage. Snowflake's failover feature ensures that customers experience minimal downtime and quickly switch to the secondary replica. This ensures business continuity and avoids data loss, which is critical in today's data-dependent world.

Overall, Snowflake's account replication and failover features provide customers with a robust disaster recovery solution that minimizes the impact of unexpected outages or disasters. This chapter explores these features in practice and understands how they can help organizations maintain high availability and business continuity.

Problem

A company has implemented Snowflake as its modern data platform and recognizes the crucial need for ensuring uninterrupted business operations and efficient disaster recovery (BC/DR) for its critical data. In the face of potential disasters or outages, the company aims to mitigate the impact on its business and swiftly restore its data to its normal state. To achieve this, the company is determined to establish a robust and dependable BC/DR strategy within Snowflake that guarantees the safety and seamless accessibility of its data, irrespective of any system disruptions or failures.

Recognizing the significance of a comprehensive BC/DR plan, the company understands that it protects against unforeseen events that can jeopardize its data, revenue, and reputation. By implementing an effective BC/DR strategy, the company seeks to minimize downtime, reduce the risk of data loss, and uphold customer satisfaction, which all contribute to maintaining a strong bottom line.

To accomplish their BC/DR objectives within Snowflake, the company recognizes the importance of adhering to industry best practices and leveraging Snowflake's native capabilities. They aim to configure the account replication and failover setup in alignment with their specific business needs and data governance policies. By doing so, the company can confidently navigate any potential disruptions, ensuring the continuity of its operations and safeguarding the integrity of its critical data.

Solution

To address the company's need for uninterrupted business operations and efficient disaster recovery (BC/DR) for their critical data within Snowflake, consider the native functionality of account replication and failover.

The first step is ensuring there is a secondary account under the Snowflake organization. At a minimum, this should be in a different region than the primary account. Ideally, this secondary account would be in a different cloud entirely. In this instance, the organization has decided to have the primary account in AWS US West and the secondary account in Azure US West 2. Additionally, there were some late arriving requirements from the marketing team based on the East Coast. To help support performance it was decided to replicate the primary account to a third account located in AWS US East. This addition hardens the BC/DR strategy while providing the data closer to the marketing end users.

Consider the Snowflake accounts shown in Figure 8-19, which are included in the failover group.

ACCOUNT ↓	EDITION	CLOUD	REGION
VZB90068	Business Critical	AWS	US West (Oregon)
TERTIARY	Business Critical	AWS	US East (N. Virginia)
SECONDARY	Business Critical	Azure	West US 2 (Washington)

Figure 8-19. *Snowflake Admin Accounts page*

If this is the first time setting up replication within the Snowflake organization, an `orgadmin` is required to enable this feature. This is achieved through a global parameter change of `enable_account_database_replication` shown in Figure 8-20.

```
USE ROLE ORGADMIN;

SHOW ORGANIZATION ACCOUNTS;

SELECT SYSTEM$GLOBAL_ACCOUNT_SET_PARAMETER('HADKYDE.VZB90068', 'ENABLE_ACCOUNT_DATABASE_REPLICATION', 'true');
SELECT SYSTEM$GLOBAL_ACCOUNT_SET_PARAMETER('HADKYDE.SECONDARY', 'ENABLE_ACCOUNT_DATABASE_REPLICATION', 'true');
SELECT SYSTEM$GLOBAL_ACCOUNT_SET_PARAMETER('HADKYDE.TERTIARY', 'ENABLE_ACCOUNT_DATABASE_REPLICATION', 'true');
```

Results Chart

	SYSTEM$GLOBAL_ACCOUNT_SET_PARAMETER('HADKYDE.TERTIARY', 'ENABLE_ACCOUNT_DATABASE_REPLICATION', 'TRUE')
1	["SUCCESS"]

Figure 8-20. *Snowflake Snowsight Worksheet*

A failover group must be created now that the account configuration has been updated and the feature is enabled. This create statement identifies which object types will be replicated, the allowed database(s), the allowed target account(s), and the replication schedule.

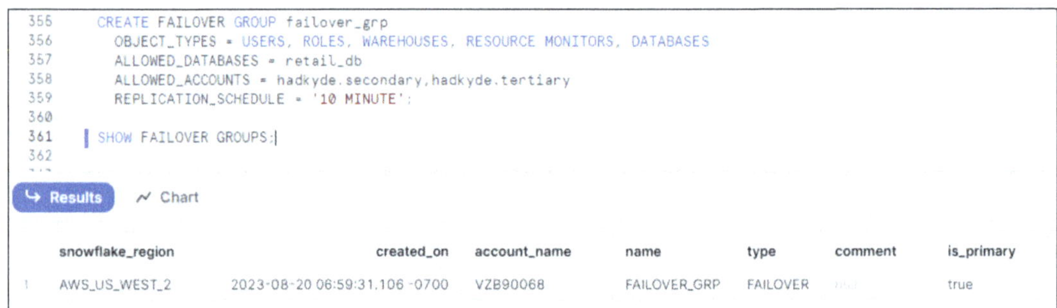

Figure 8-21. *Snowflake Snowsight Worksheet*

Once the failover group has been created in the primary account from Figure 8-21, it must be created as a replica failover group on the target account(s). Remember to adjust the replication schedule to your company's needs to strike an appropriate balance of cost. More frequent, large-volume replications increase storage and egress costs when moving data across regions.

Figure 8-22 shows the failover group created in the Azure secondary account.

```
CREATE FAILOVER GROUP failover_grp
  AS REPLICA OF hadkyde.vzb90068.failover_grp;

SHOW FAILOVER GROUPS;
```

Results Chart

	snowflake_region	created_on	account_name	name	type	comment	is_primary
1	AZURE_WESTUS2	2023-08-20 07:02:54.659 -0700	SECONDARY	FAILOVER_GRP	FAILOVER	null	false
2	AWS_US_WEST_2	2023-08-20 06:59:31.106 -0700	VZB90068	FAILOVER_GRP	FAILOVER	null	true

Figure 8-22. *Snowflake Snowsight Worksheet*

Figure 8-23 shows the failover group created in the AWS tertiary account.

```
CREATE FAILOVER GROUP failover_grp
  AS REPLICA OF hadkyde.vzb90068.failover_grp;

SHOW FAILOVER GROUPS;
```

Results Chart

	snowflake_region	created_on	account_name	name	type	comment	is_primary
1	AZURE_WESTUS2	2023-08-20 07:02:54.659 -0700	SECONDARY	FAILOVER_GRP	FAILOVER	null	false
2	AWS_US_EAST_1	2023-08-20 07:14:50.121 -0700	TERTIARY	FAILOVER_GRP	FAILOVER	null	false
3	AWS_US_WEST_2	2023-08-20 06:59:31.106 -0700	VZB90068	FAILOVER_GRP	FAILOVER	null	true

Figure 8-23. *Snowflake Snowsight Worksheet*

When using a refresh schedule with a failover group, the initial create statement automatically triggers the replication workflow. However, a refresh command can always be manually executed in the target accounts shown in Figure 8-24. Remember that this command needs to be applied to *all* target accounts.

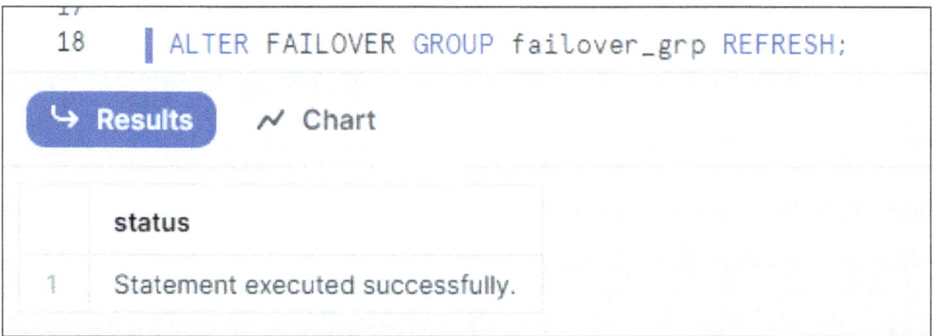

Figure 8-24. *Snowflake Snowsight Worksheet*

Once the initial replication has completed the object types within the scope of the allowed databases are available in the target accounts.

Figure 8-25 shows the `retail_db` schema and tables now exist in the `secondary` target account in Azure US West 2.

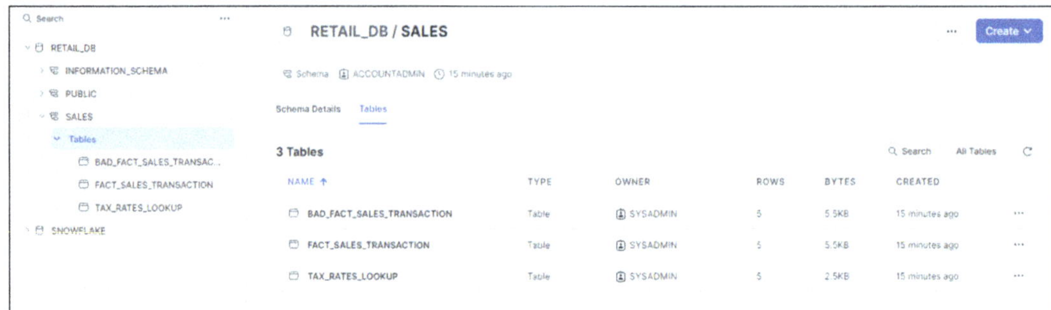

Figure 8-25. Snowflake Snowsight Data View

Figure 8-26 shows the `retail_db` schema and tables now exist in the tertiary target account in AWS US EAST to support the marketing team.

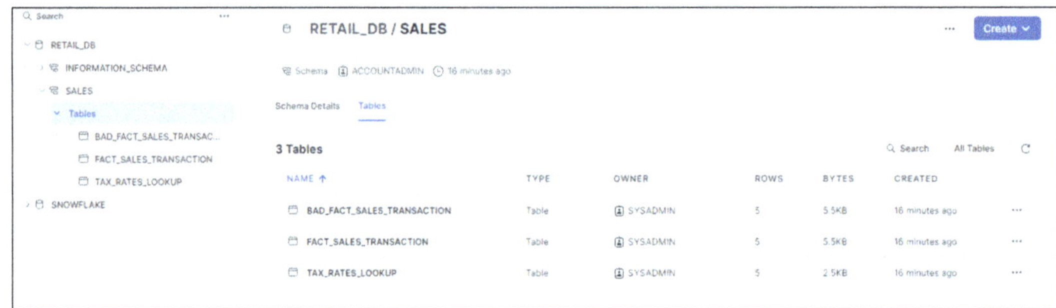

Figure 8-26. Snowflake Snowsight Data View

To demonstrate to the business and security teams that replication is working we can add a test table to the `retail_db` in the source account and then validate it exists on the target accounts after the next scheduled replication.

RETAIL_DB.PUBLIC ▾ Settings ▾

```
-- modify source database to demostrate replication

create or replace table retail_db.sales.repl_test (x varchar);
insert into retail_db.sales.repl_test (x) values ('test');
select x from retail_db.sales.repl_test;
```

Results Chart

	X
1	test

Figure 8-27. *Snowflake Snowsight Worksheet*

The `repl_test` table is now available on the `secondary` account, as shown in Figure 8-28.

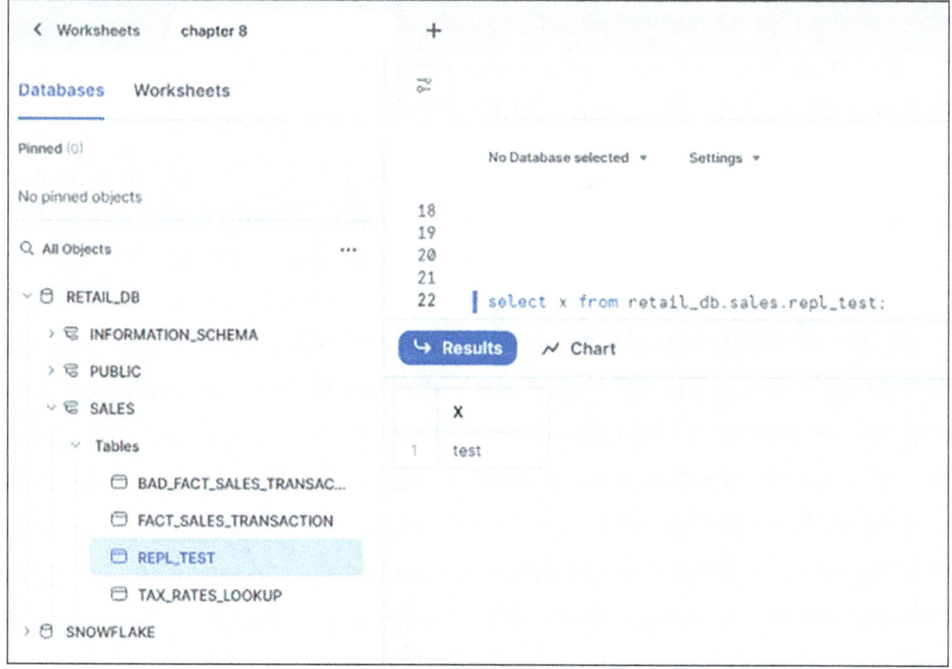

Figure 8-28. *Snowflake Snowsight Worksheet*

The repl_test table is now available on the tertiary account, as shown in Figure 8-29.

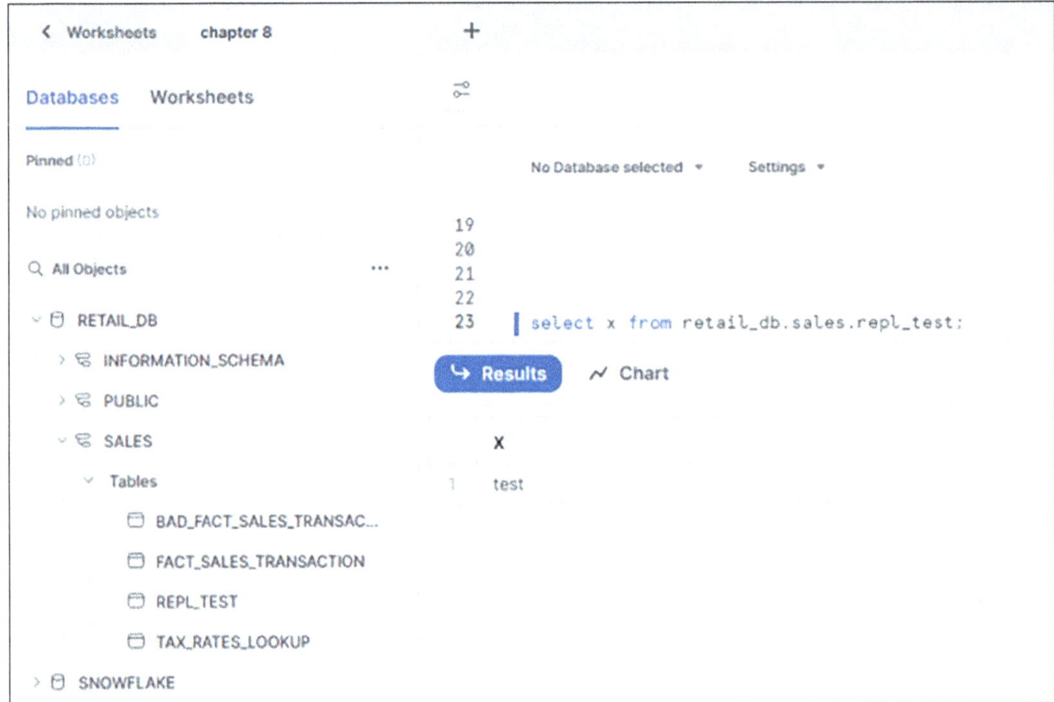

Figure 8-29. *Snowflake Snowsight Worksheet*

Finally, in Figure 8-30 it can be demonstrated how a database can failover to one of the target accounts. The team wants to failover retail_db to the Azure-backed secondary account. An easy test to determine whether the database is a replica is by it being in read-only mode.

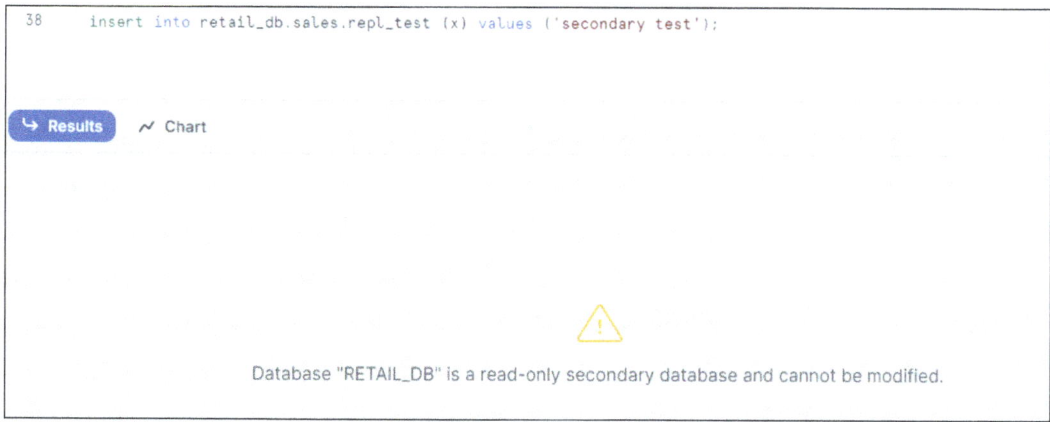

Figure 8-30. *Snowflake Snowsight Worksheet*

Additionally, one could use the `SHOW REPLICATION DATABASES` to validate the primary, secondary, and tertiary accounts.

Now let's set the `secondary` account as the primary in `failover_grp`.

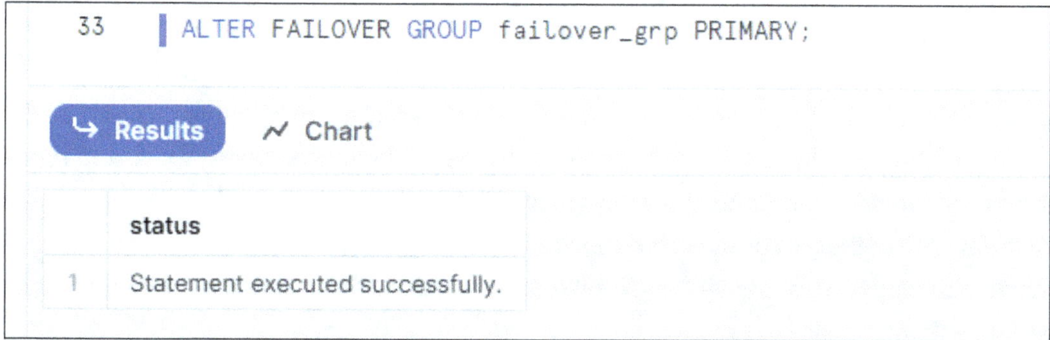

Figure 8-31. *Snowflake Snowsight Worksheet*

Figure 8-32 shows that `retail_db` is no longer in read-only mode and acting as primary. Similarly to the previous step, the `show replication databases` command could be used.

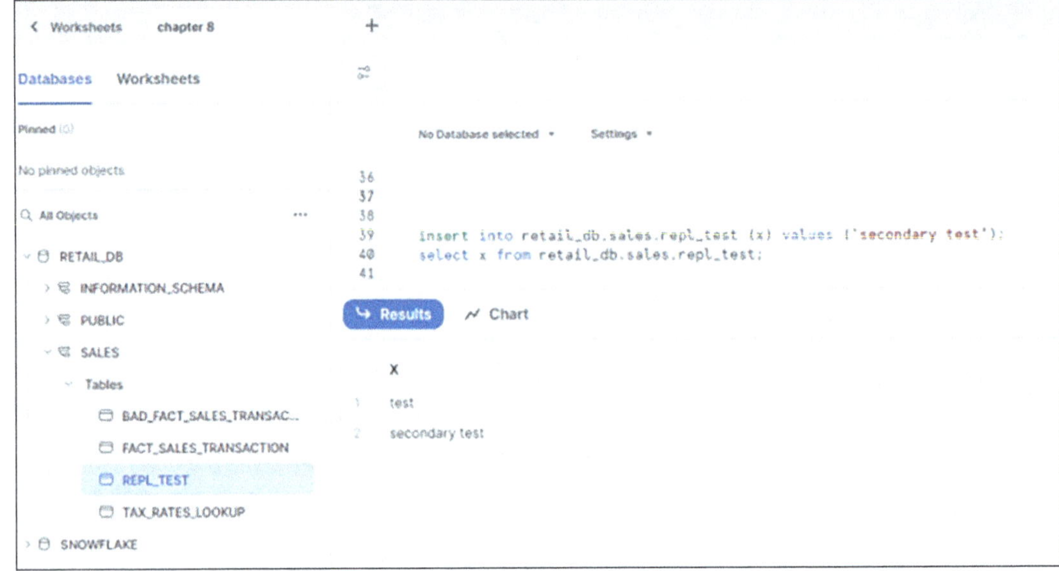

Figure 8-32. Snowflake Snowsight Worksheet

In today's data-driven landscape, having a strong data platform is crucial. The Snowflake data cloud is exceptional in data management and ensuring availability and disaster recovery via key functionality like account replication and failover.

Account replication lets Snowflake users duplicate databases, users, roles, and more across accounts, regions, and cloud providers. This lets organizations keep a copy of their data in another account within the Snowflake organization. This is particularly valuable during disruptions like regional outages; a secondary replica can be activated swiftly, enhancing disaster recovery.

Adding to this, the failover mechanism solidifies Snowflake's strength. Failover enables a smooth shift from a disrupted primary region to a secondary replica, reducing downtime and maintaining data integrity. This aids business continuity by minimizing data loss during unexpected crises.

Additional Snowflake Documentation

https://docs.snowflake.com/en/user-guide/account-replication-intro

https://docs.snowflake.com/en/sql-reference/commands-replication

---

Recipe 8-4. Client Redirect (Business Continuity)

Snowflake's client redirect is a powerful feature designed to seamlessly redirect client traffic, such as Microsoft Power BI, from one Snowflake account to another in the event of a failover scenario, ensuring BC/DR is end-to-end. This capability becomes particularly valuable when the original Snowflake account becomes inaccessible or encounters performance issues. In the last section, we configured a failover group, which ensures data is another account, but what about end users? By leveraging the client redirect feature, users can ensure uninterrupted business operations and effectively minimize potential downtime by effortlessly rerouting traffic to a healthy Snowflake account replica.

When the primary Snowflake account becomes unavailable in critical situations, the client redirect feature automatically redirects users to a replica account. This ensures that users can continue accessing their data and performing tasks without significant disruption. Whether running important queries, generating reports, or analyzing real-time data, the client redirects feature guarantees a seamless transition, allowing users to maintain their productivity and meet their business needs.

This chapter delves into an exploration of Snowflake's client redirect feature, uncovering its numerous benefits, robust functionality, and best practices for implementation. By understanding the full potential of this feature, users can optimize their failover strategies and maximize the resiliency of their Snowflake deployments. Furthermore, we examine how Snowflake's failover and client redirect capabilities work in harmony, delivering a cohesive and uninterrupted user experience even during challenging circumstances.

Problem

In a highly interconnected and data-driven world, businesses heavily rely on cloud-based services to ensure the continuity and efficiency of their operations. However, even with robust cloud infrastructure, unforeseen events such as rare cloud outages can disrupt critical services and significantly impact an organization's ability to function.

Consider a situation where an organization depends on its primary Snowflake account deployed in AWS US West 2 to power various pipelines, analytics processes, and downstream jobs. These critical components are pivotal in the organization's day-to-day operations, enabling data-driven decision-making, customer insights, and business growth. However, an unforeseen and rare AWS cloud outage occurs, rendering the primary Snowflake account inaccessible and disrupting the entire data ecosystem.

In this scenario, the organization's preparedness and proactive measures become crucial in mitigating the impact of the outage. Fortunately, the organization had previously implemented Snowflake account replication and failover, a strategic approach to its disaster recovery plan. This entails maintaining a secondary replica account located in Azure US West 2, another cloud provider.

In this scenario, the organization's primary objective is to maintain uninterrupted business operations and minimize the impact of the AWS cloud outage on their critical data workflows. The organization can redirect its pipelines, analytics processes, and downstream by successfully leveraging the secondary replica account. This ensures data availability, reduces downtime, and allows continued productivity and business continuity.

Solution

At the core, Snowflake's client redirect is a logical connection string that routes traffic to the current primary connection. The first step in Figure 8-33 to enable this feature is to create a `connection` on the `primary` account and enable the `secondary` account as a replica of that specific `connection`.

Figure 8-33. Snowflake Snowsight Worksheet

Next, the admin needs to navigate to the secondary account and create a `replica of the connection` created on the primary account shown in Figure 8-34.

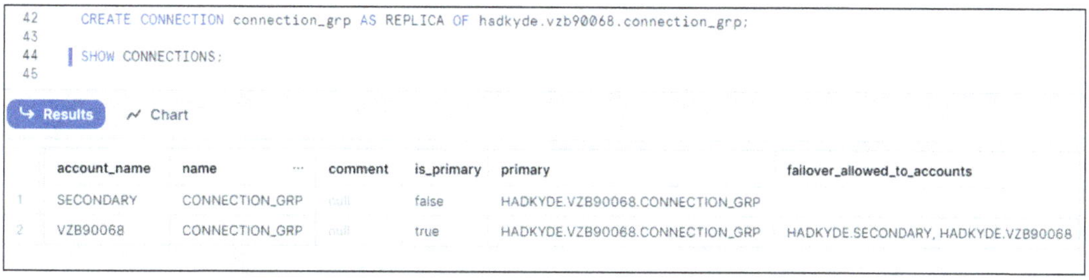

```
CREATE CONNECTION connection_grp AS REPLICA OF hadkyde.vzb90068.connection_grp;

SHOW CONNECTIONS;
```

Results Chart

	account_name	name	comment	is_primary	primary	failover_allowed_to_accounts
1	SECONDARY	CONNECTION_GRP	null	false	HADKYDE.VZB90068.CONNECTION_GRP	
2	VZB90068	CONNECTION_GRP	null	true	HADKYDE.VZB90068.CONNECTION_GRP	HADKYDE.SECONDARY, HADKYDE.VZB90068

Figure 8-34. *Snowflake Snowsight Worksheet*

To demonstrate how client redirect functions, lets mimic an end user using Microsoft Power BI in Figure 8-35 and 8-36 and connect to the `fact_sales_transactions` fact table in `retail_db` to create a simple chart. Remember that instead of using the account-specific `connection_url`, use `connection_url` from `connection_grp`. In this case, `hadkyde-connection_grp.snowflakecomputing.com`.

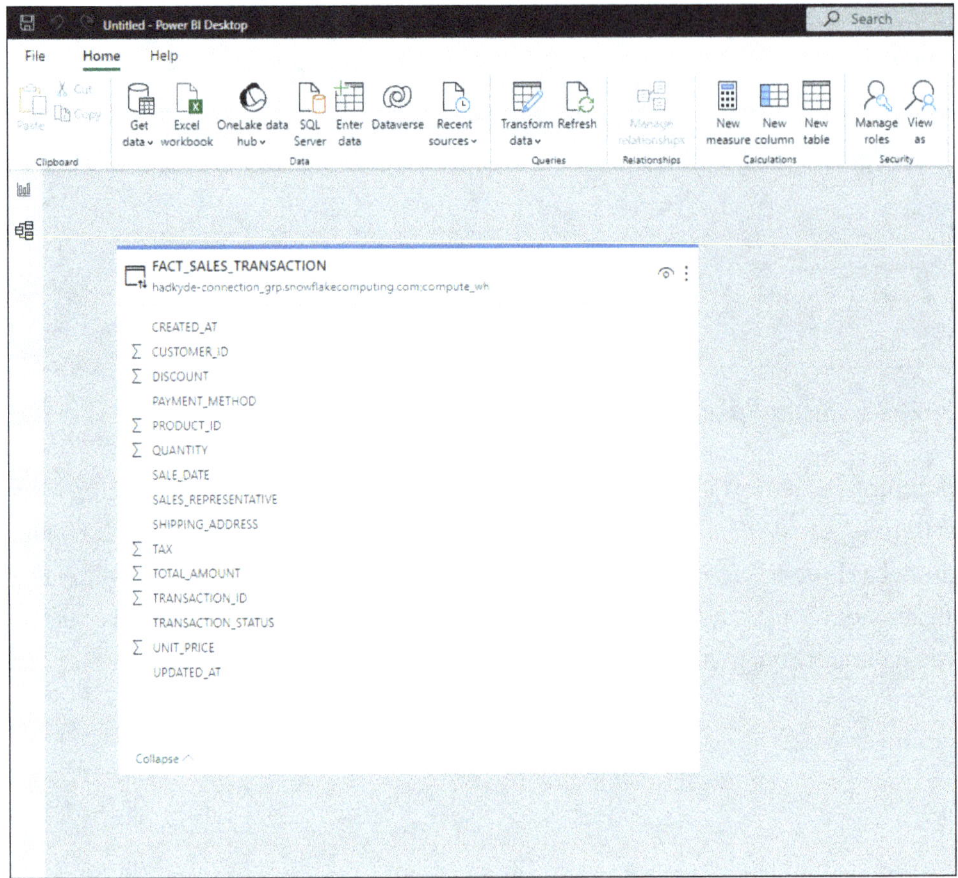

Figure 8-35. *Microsoft PowerBI Desktop Relationships*

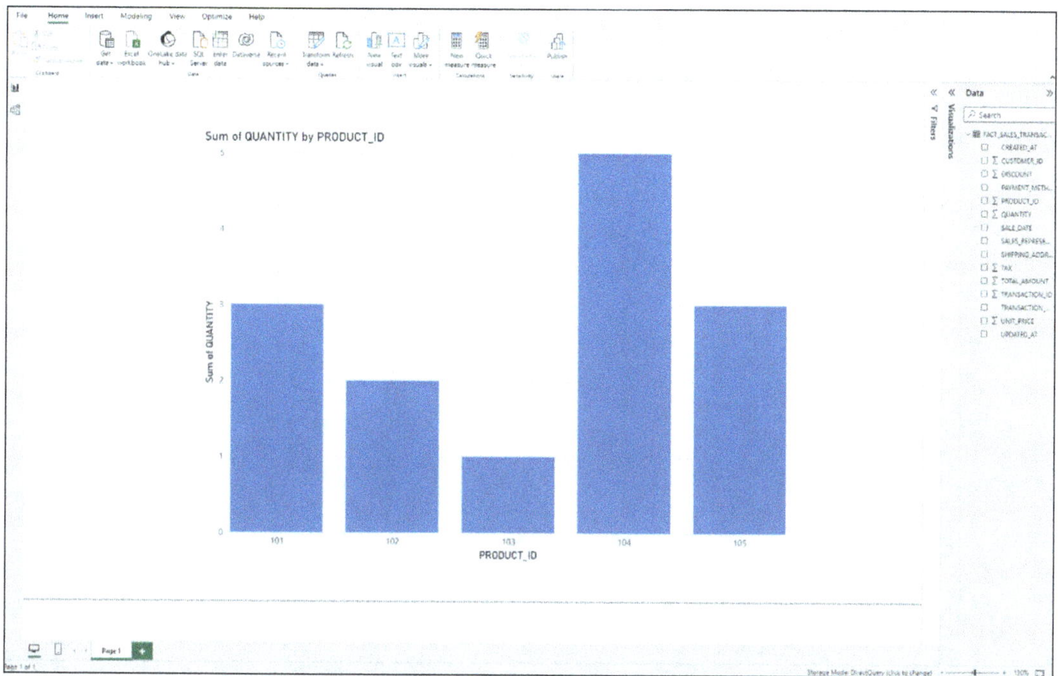

Figure 8-36. *Microsoft PowerBI Report*

Navigating to the primary account in `connection_grp`, the query history shows our connections and queries from Power BI.

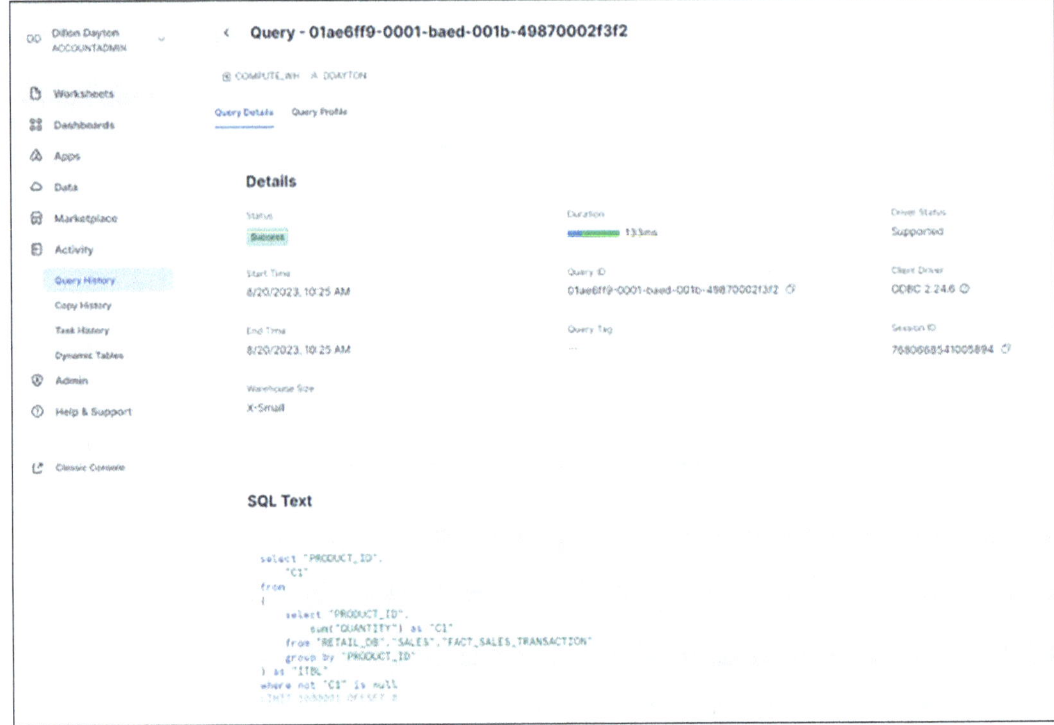

Figure 8-37. *Snowflake Snowsight Query History*

Now, to trigger a client redirect, one would navigate to the secondary or replica and alter the connection to be the primary. Figure 8-38 shows that in the `is_primary` column, the secondary account is now the primary in `connection_grp`.

```
ALTER CONNECTION connection_grp PRIMARY;

SHOW CONNECTIONS;
```

Results Chart

	snowflake_region	...	created_on	account_name	name	comment	is_primary
1	AZURE_WESTUS2		2023-08-20 09:18:43.767 -0700	SECONDARY	CONNECTION_GRP	null	true
2	AWS_US_WEST_2		2023-08-20 09:16:49.622 -0700	VZB90068	CONNECTION_GRP	null	false

Figure 8-38. *Snowflake Snowsight Worksheet*

With the redirect completed, let's make a small modification to our chart in Figure 8-39 and check out the query history on the secondary account while using the exact same `connection_url`, eliminating the need to modify end-user tools or any connection strings being used downstream.

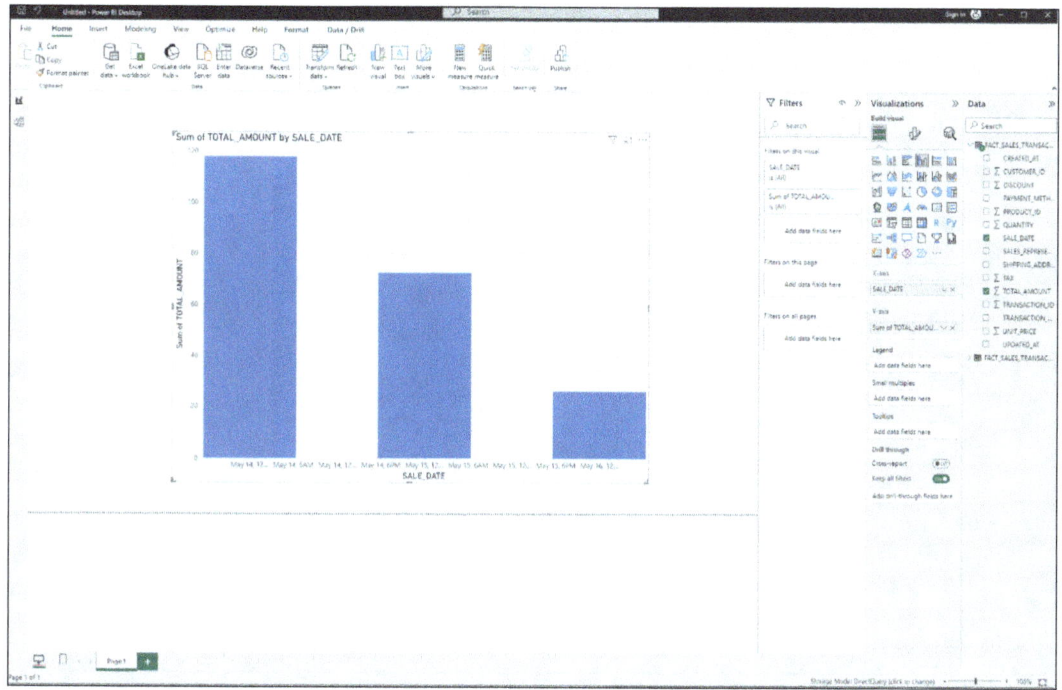

***Figure 8-39.** Microsoft PowerBI Report*

Note the updated SQL showing the change made in the original PBI but via the secondary account.

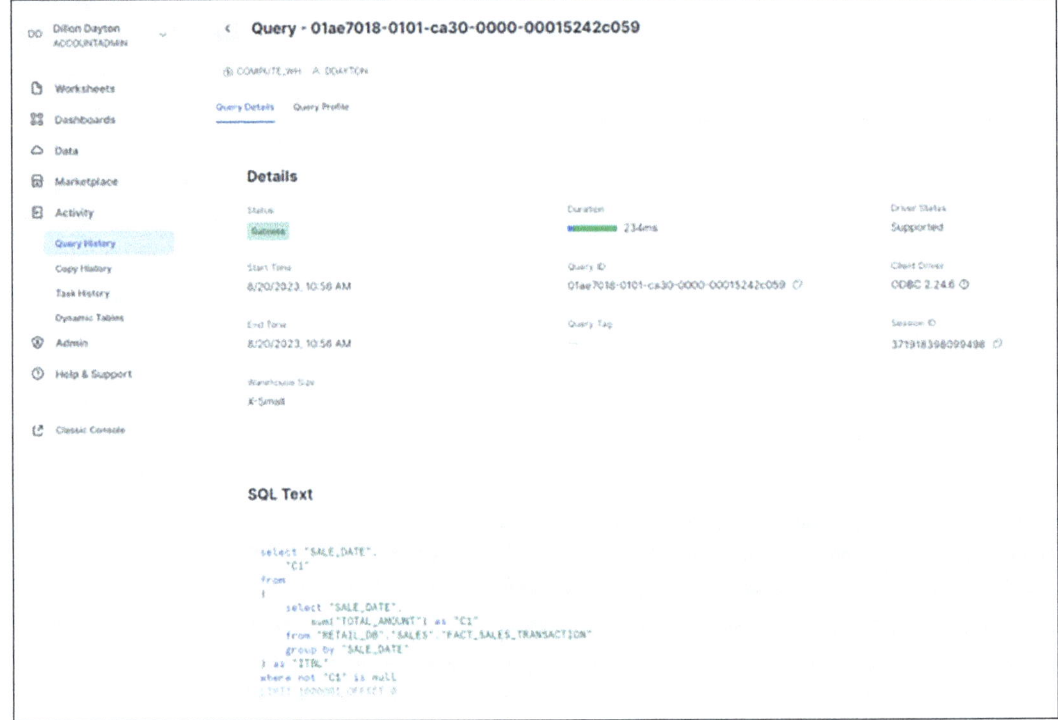

Figure 8-40. *Snowflake Snowsight Query History*

Additional Snowflake Documentation

`https://docs.snowflake.com/en/user-guide/client-redirect`

CHAPTER 9

Application Integration

Overview

Today's businesses demand agility, efficiency, and data-driven decision-making. Application integration within the Snowflake Data Cloud unlocks a wealth of possibilities. By seamlessly connecting applications with centralized data in Snowflake, companies can drive real-time insights, streamline business processes, and ultimately foster a more robust, data-driven ecosystem.

The benefits of integrating applications with Snowflake are far-reaching. From eliminating data silos and reducing costs to enabling faster insights and innovation, it represents a significant competitive advantage. Building applications directly within the Snowflake environment eliminates the burdens of complex data movement and management, leading to operational efficiencies many organizations would struggle to achieve with traditional architectures.

This chapter examines some compelling use cases for application integration with the Snowflake Data Cloud.

- **Connecting applications**: Explore how integrating Snowflake with tools like Microsoft Power BI empowers the creation of dynamic and interactive dashboards that facilitate strategic decision-making.
- **Snowflake Unistore**: How Snowflake enables organizations to efficiently work with transactional and analytical data within a single platform.
- **Streamlit and Snowflake**: Explore how Streamlit provides a flexible framework to build custom front-end applications tailored to your specific business needs, all backed by the power of Snowflake data.

D. Dayton and J. Eipe, *Snowflake Recipes*, https://doi.org/10.1007/979-8-8688-0938-5_9

- **Snowflake for applications**: Understand Snowflake's revolutionary Native App Framework, Snowpark Container Services, and how they streamline app development and delivery on the data cloud, paving the way for data-centric app marketplaces and secure collaboration.

With its scalability, elasticity, and seamless integration capabilities, Snowflake streamlines the process of building data-centric applications. By the end of this chapter, you'll have developed an understanding of this powerful approach, learn about available tools and frameworks, and acquire insights to guide application integration within the Snowflake Data Cloud and gain an advantage for your organization.

Recipe 9-1. Connecting Applications

In the modern data landscape, the seamless integration of applications with robust data platforms is key to innovation and optimized business operations. The Snowflake Data Cloud has emerged as a central hub for organizations, offering an exceptionally scalable, secure, and performant foundation for housing and managing vast datasets.

However, the true value of Snowflake lies in its ability to empower a wide range of powerful applications—a synergy that amplifies efficiency and unlocks deeper insights.

By effectively connecting applications to Snowflake, organizations pave the way for the following.

- **Data-driven decision-making**: Business intelligence tools like Tableau and Power BI, when fueled by the data within Snowflake, transform raw information into stunning, interactive dashboards. These dashboards become invaluable tools for executives and stakeholders to identify trends, monitor KPIs, and make informed decisions supported by real-time data.
- **Streamlined workflows**: Applications designed for automation, task management, and collaboration integrate seamlessly with Snowflake. This means data workflows can be triggered efficiently, data can be securely shared across teams, and processes operate smoothly based on the insights contained within your data warehouse.
- **Business processes and tools**: Applications powering recommendation engines, customer support, or marketing campaigns draw on the rich customer data stored within Snowflake.

It enables organizations to cater to individual preferences, anticipate needs, and deliver tailored experiences that foster loyalty and drive growth. This is only one example, and the possibilities are endless.

Snowflake supports this expansive world of applications with exceptional connectivity options, developer-friendly interfaces, and powerful features.

- **ODBC/JDBC drivers**: Ensure broad compatibility with a multitude of existing applications and systems.
- **Native connectors**: Optimize performance for popular tools like Python, Spark, and Kafka.
- **REST API**: Enables custom integrations for unique business requirements and specialized applications.

Let's delve deeper into specific integration patterns, use cases, and best practices for connecting apps to Snowflake. Whether you're looking to optimize existing applications or deploy cutting-edge new ones, you'll learn how Snowflake can be the launchpad for your business innovations.

Problem

A large retail company has amassed a wealth of valuable sales data within its Snowflake Data Cloud. Data flows steadily from point-of-sale systems, e-commerce transactions, and integrated inventory management tools, encompassing crucial insights on customer behavior, product performance, and operational metrics. However, the company's primary reliance on spreadsheets and basic charting for sales analysis severely restricts its ability to unlock the true value of this data. This approach hampers strategic decision-making, efficiency, and organization-wide collaboration in several ways.

- **Crippled analytics**: Traditional spreadsheets fall woefully short when handling the multi-dimensional, complex nature of sales data. Analysts struggle to visualize emerging trends, spot correlations between product categories and customer demographics, and pinpoint geographic variations in real time. Key patterns that could reveal growth opportunities or mitigate risks remain hidden within the rows and columns.

- **Cumbersome and error-prone workflows**: Manually exporting data from Snowflake into spreadsheets introduces a bottleneck in the analysis process. Each data refresh is time-consuming, and repetitive data manipulation tasks add significant overhead. Manual processes inevitably increase the risk of errors and version control conflicts, undermining the reliability of the insights derived.
- **Missed optimization potential**: The limitations of spreadsheets prevent analysis from extending beyond basic reporting. Identifying ways to optimize pricing, inventory allocation, and marketing spending demands the ability to easily integrate sales data with external factors (like seasonality, promotional activities, and competitor pricing), which would be readily accessible within the Snowflake ecosystem.
- **Siloed decision-making**: Confining analysis to spreadsheets restricts the impact of valuable insights, trapping them within the sales team. Other departments, such as marketing, supply chain, and finance, cannot benefit from timely sales data to inform their decisions. This lack of cross-functional visibility limits the organization's ability to operate in a fully data-driven and responsive manner.

The retail company can revolutionize its data analytics capabilities by strategically connecting its Snowflake Data Cloud with powerful applications such as business intelligence tools and workflow automation platforms. This integration unlocks actionable insights through visual analytics (exposing trends and customer preferences), streamlines operations with automated workflows, fosters cross-functional collaboration through shared insights and ultimately creates a competitive edge by positioning the company to capitalize on data-driven decision-making opportunities that legacy approaches would miss.

Solution

The retailer's reliance on outdated tools like spreadsheets is holding them back. The company has decided to implement Power BI to help extract true value and actionable insights from the data. Spreadsheets cripple complex analysis, introduce inefficiencies, and isolate data within departmental silos. The shift to Power BI empowers them with robust dashboards, streamlined processes, and organization-wide collaboration, ultimately accelerating data-driven decision-making and fueling a competitive advantage.

Prerequisites

- **Snowflake data cloud account**: Have your account access credentials handy. This includes your account name (often within the URL), username, and password.
- **Power BI desktop or service access**: You'll need a Power BI Desktop installed, or you can access the Power BI service through a web browser.
- **Snowflake ODBC driver**: Download and install the ODBC driver compatible with your system, ensuring connectivity between Power BI and Snowflake. You can obtain the driver from Snowflake's website.

The Steps to Connect

1. **Open Power BI Desktop.** You'll find it in your program list or application directory.

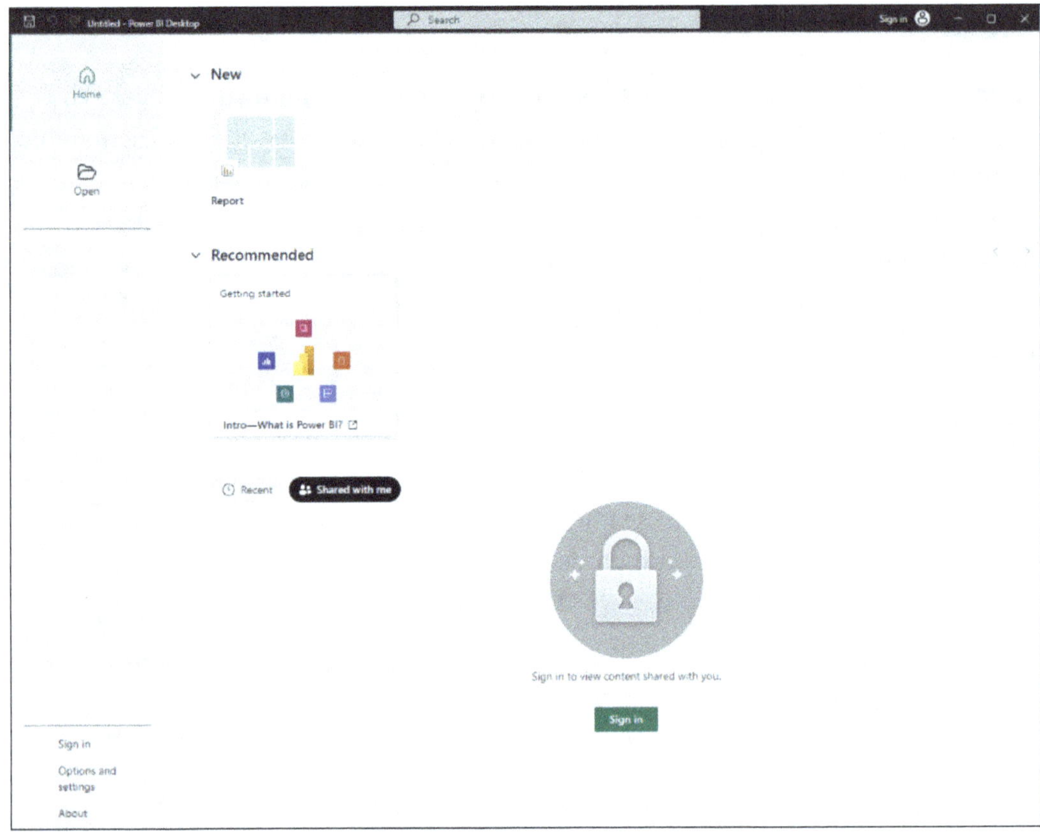

Figure 9-1. *Microsoft PowerBI Desktop*

2. **Initiate a data connection.** Once PowerBI is open as seen in Figure 9-1, click the Get Data button shown in Figure 9-2 within the Home ribbon. This opens a dialog box with a wide selection of data source options.

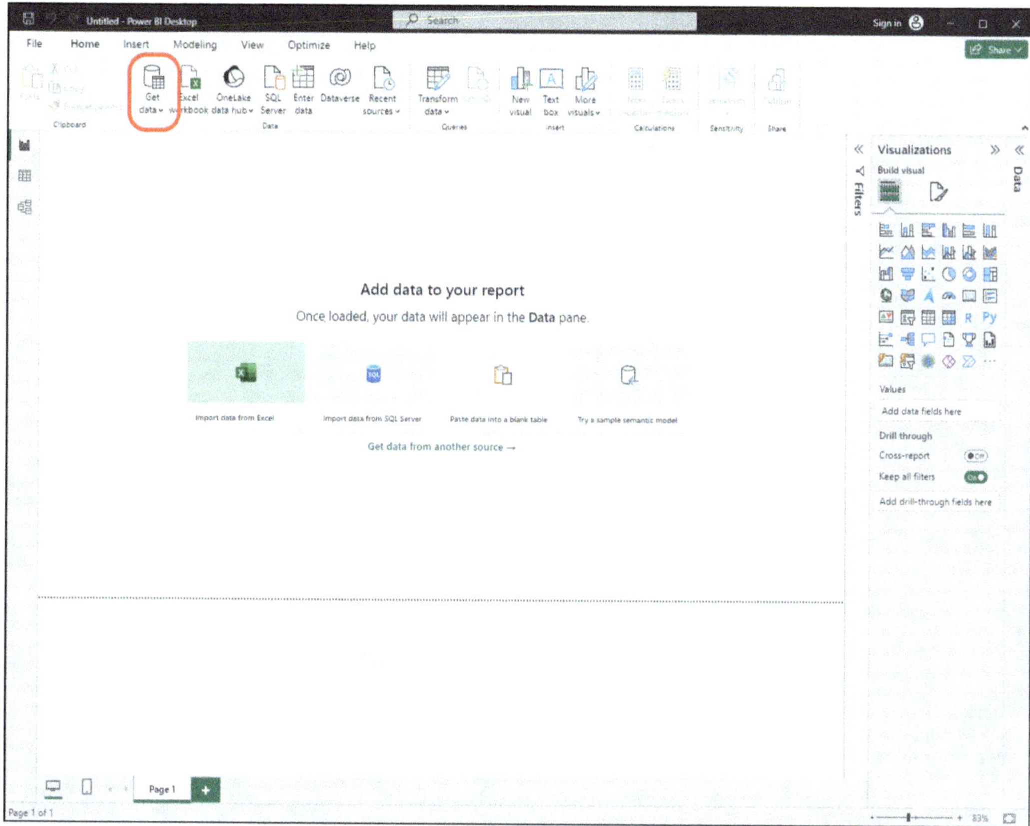

***Figure 9-2.** Microsoft PowerBI Desktop*

3. **Select Snowflake.** Within the Database category seen in Figure 9-3, locate and select the Snowflake connector. A connection dialog box appears.

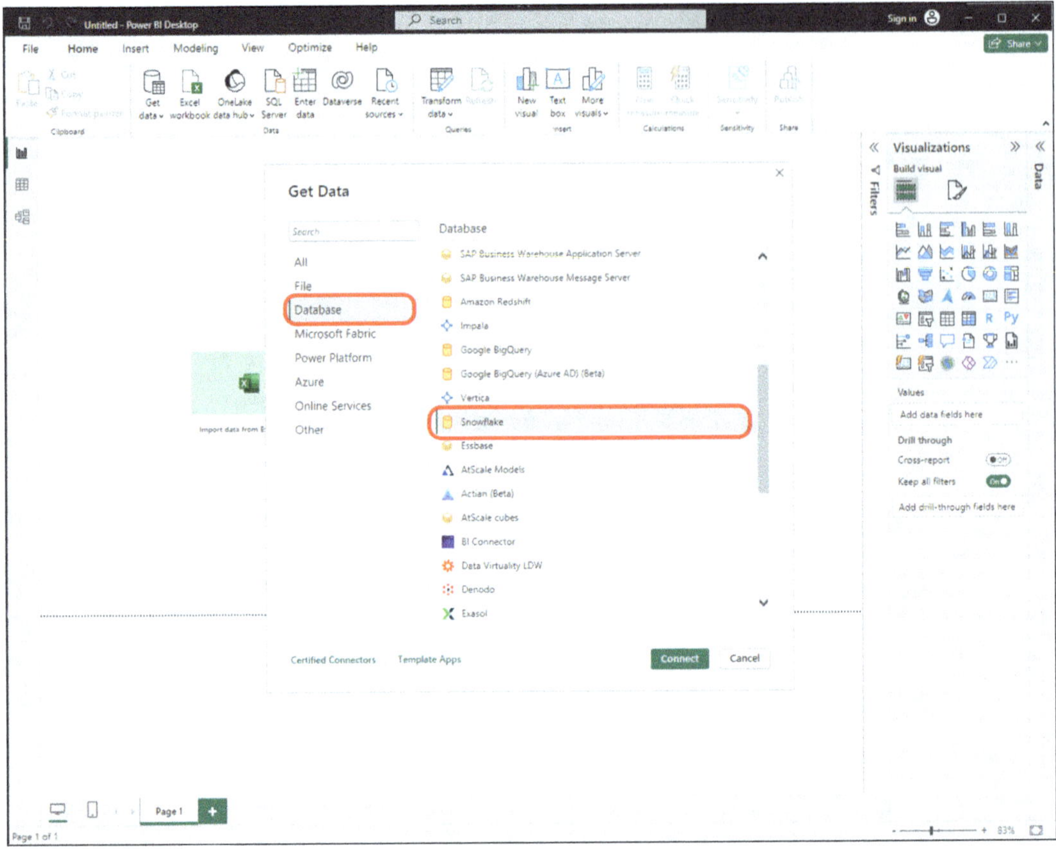

Figure 9-3. *Microsoft PowerBI Desktop Data Source*

4. **Enter your Snowflake credentials.**

   a. **Server**: Carefully enter your Snowflake account URL (e.g., xxx00000.snowflakecomputing.com)

   b. **Warehouse**: Specify the Snowflake warehouse where your relevant data is stored. (e.g., COMPUTE_WH).

   c. **Database (optional)**: Provide the specific database within your Snowflake account if needed.

   d. **Authentication**: Choose your preferred authentication method (e.g., basic with username/password or OAuth2).

   e. Once your credentials are in place, click Connect. The Navigator should prompt you to select which data to start with.

5. **Select Sales Data.** The PowerBI Data Navigator appears in Figure 9-4.

   a. Browse your schemas and locate the tables or views containing your sales, inventory, customer, and other relevant datasets.

   b. Select the necessary tables/views and click the Load button.

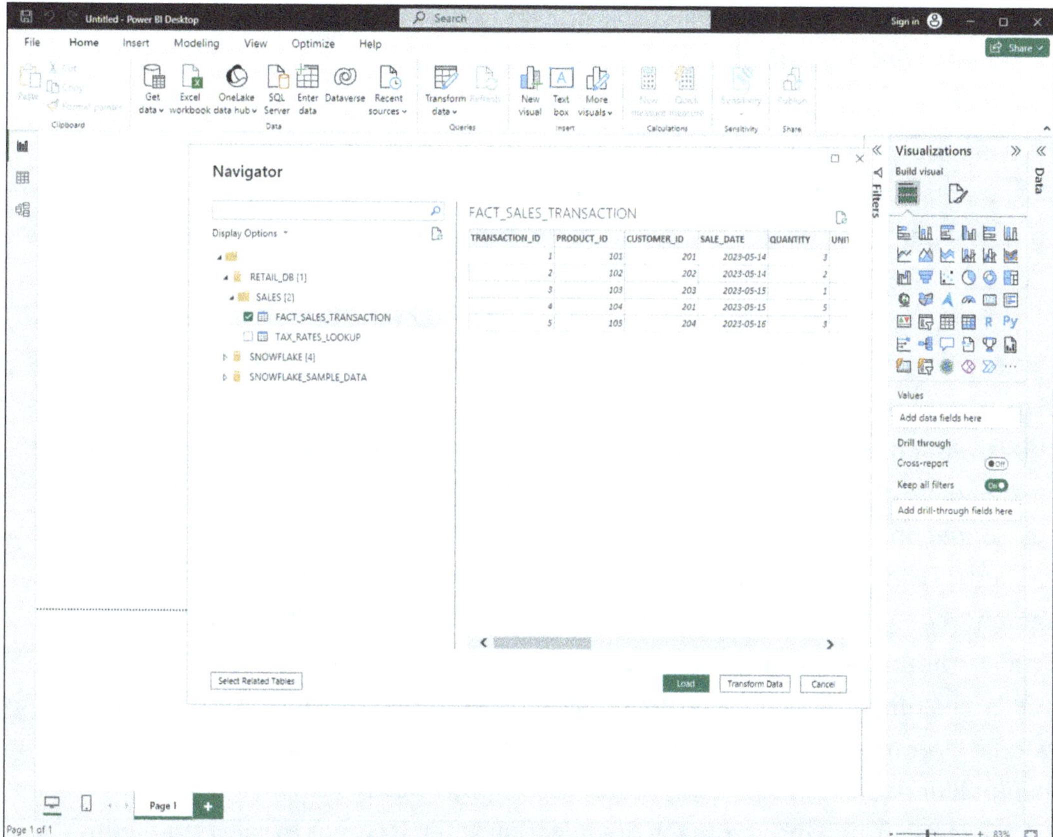

Figure 9-4. Microsoft PowerBI Desktop Data Navigator

6. **Choose the Data Import mode.**

- **Import**: This option copies a snapshot of your selected data into Power BI, allowing offline analysis. Data updates then require manual refreshes.

- **DirectQuery**: This option queries Snowflake directly with each interaction in your Power BI reports and dashboards, ensuring you always work with the most up-to-date data.

> **Tip** If your organization is leveraging MFA, consider enabling ALLOW_CLIENT_MFA_CACHING so Power BI or other applications do not continuously prompt for an MFA validation upon every Snowflake call. This is especially useful when using DirectQuery. For more information, check out the Snowflake documentation at `https://docs.snowflake.com/en/user-guide/security-mfa#using-mfa-token-caching-to-minimize-the-number-of-prompts-during-authentication-optional`.

After successfully connecting Power BI to Snowflake, it's time to unleash the power of your data! Start by experimenting with Power BI's rich library of visualizations. Choose from bar charts, line charts, maps, and more to find the best ways to represent your sales data. Use filters and slicers to empower users to drill down into specific product categories, regions, or dates. Next, consolidate these compelling visuals into dashboards. Create tailored dashboards with key sales metrics that resonate with different teams like sales, marketing, and executives. Finally, to make these insights widely accessible, publish your dashboards to the Power BI service. Implement role-based permissions to ensure secure access and that everyone sees only the data relevant to them.

Connecting Power BI to Snowflake dramatically amplifies the capabilities of the retailer's business intelligence efforts. Snowflake acts as a powerful foundation, easily handling complex sales data and ensuring that Power BI isn't hindered by common database limitations such as scalability. Visualizations reveal crucial insights about trends, customer segments, and campaign effectiveness. Snowflake's streamlined workflows empower analysts to focus on interpreting insights rather than tedious data manipulation. Clear, shared dashboards built upon Snowflake data ignite data-driven decision-making across the organization when it comes to sales, marketing, and inventory strategy. In a competitive landscape, harnessing synergy between any application, not just Power BI and Snowflake, allows organizations to react swiftly to market shifts and capitalize on opportunities fueled by real-time, accurate data.

Additional Snowflake Documentation

`https://docs.snowflake.com/en/user-guide/security-mfa#using-mfa-token-caching-to-minimize-the-number-of-prompts-during-authentication-optional`

Recipe 9-2. Snowflake Unistore

Snowflake Unistore revolutionizes the way applications interact with data. By bridging the divide between transactional workloads (like point-of-sale updates or website interactions) and analytical processing (trend analysis or forecasting), it empowers applications to be more efficient and responsive. Traditionally, these workloads required separate, specialized databases, creating a bottleneck as data moved between them, leading to outdated analysis and delays. Unistore removes this obstacle; applications can directly query and update data within Snowflake, ensuring they work with the most current, accurate snapshot available and optimizing for OLTP workloads.

Consider some of the possibilities.

- **Operational agility**: Organizations can respond in near real-time to changing conditions. For example, inventory systems instantly react to surging demand, or financial risk models constantly incorporate the latest market data.
- **Improved user experiences**: Customers interact with applications backed by near real-time data. Personalized recommendations reflect recent purchases and support inquiries gain instant context from a fully integrated transaction history.
- **Accelerated innovation**: Development teams no longer need to engineer complex data pipelines between systems. The focus shifts to building new features and improving analytics, enabling faster delivery of value to the business.
- **Reduced latency**: Decision-makers throughout the organization have access to insights without the wait. This ability to leverage the freshest possible data is a catalyst for seizing new opportunities before they're lost.

With Snowflake Unistore, the benefits of a data-driven organization, greater customer satisfaction, streamlined operations, and informed strategic choices can be fully realized.

Problem

The agricultural company values its network of dedicated field technicians. These technicians are the backbone of client relationships, diligently collecting data and providing crucial on-the-ground support. However, their current workflow is mired in outdated and cumbersome processes that are hindering both the technicians and the business.

Right now, it's all about paperwork and delayed updates. Field technicians carefully record test results, snap photos of crop conditions and take detailed notes on client supply orders. It's trapped in notebooks and on camera memory cards until the end of a long workday, or worse, the next day, before it can be manually entered into the company's CRM and data lakehouse.

Technicians lose precious time to repetitive data entry, time better spent in the field. Frustration grows when simple client information requests can't be answered promptly due to outdated systems.

Meanwhile, back at headquarters, insights into emerging crop health issues, supply trends, and potential sales opportunities are stale before they arrive. This puts the company at a disadvantage, unable to act quickly or spot patterns that could boost yields and customer satisfaction.

The company's small, agile IT team sympathizes but is already stretched thin. Building and maintaining a full-stack mobile application to solve this feels daunting and likely to divert resources from other critical projects.

The company knows something needs to change. They need a lightweight, easily accessible solution that empowers their field technicians to share data directly from the field. This real-time data would transform decision-making, streamline workflows, and let their technicians focus on what they do best—aiding their clients. There's a sense that the right technology platform exists; they just need to find it. A proof-of-concept seems like the perfect way to find out.

Let's explore how Snowflake Unistore and hybrid tables could support a lightweight application the team is looking to build.

Solution

The first step in this journey is determining whether Snowflake's capabilities align with your data application's needs. For applications demanding low latency and high throughput for transactional workloads (OLTP), Snowflake's Hybrid tables offer a compelling solution.

Note At the time of this writing, Snowflake hybrid tables are a feature in public preview for AWS only.

Hybrid tables seamlessly integrate into Snowflake's architecture, allowing your teams to fully leverage Snowflake's strengths. These tables offer a unique blend of features.

- **Data integrity**: They enforce uniqueness and referential integrity, ensuring your data remains consistent and reliable.
- **Optimized performance**: Indexes enhance query efficiency, making data retrieval faster.
- **Concurrency control**: Row locking allows multiple users to interact with the data safely, preventing conflicts.

Unlike Snowflake's standard tables, Hybrid tables offer these distinct advantages. This combination of data integrity, performance, and concurrency control makes them a powerful choice for applications that demand reliable and responsive data handling.

Let's consider the small data model shown in Figure 9-5. It consists of a PRODUCT, CUSTOMER, SALES, and REPORT hybrid tables with primary keys, foreign keys, and indexes.

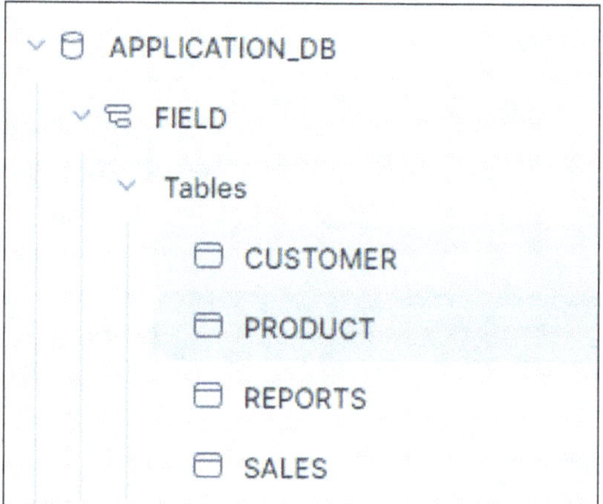

***Figure 9-5.** Snowflake Snowsight Data*

```
CREATE HYBRID TABLE IF NOT EXISTS application_db.field.product(
   product_id number autoincrement primary key,
   product varchar,
   product_desc varchar,
   price decimal(10,2),
   created_at timestamp_ntz,
   updated_at timestamp_ntz,
INDEX product_idx (product)
);

CREATE HYBRID TABLE IF NOT EXISTS application_db.field.customer(
   customer_id number autoincrement primary key,
   fname varchar,
   lname varchar,
   email varchar,
   address varchar,
   city varchar,
   state varchar,
   zip varchar,
   created_at timestamp_ntz,
   updated_at timestamp_ntz,
INDEX customer_idx (lname)
);

CREATE HYBRID TABLE IF NOT EXISTS application_db.field.sales (
   sale_id number autoincrement primary key,
   product_id number,
   customer_id number,
   sale_date timestamp,
   quantity number,
   sub_total decimal(10,2),
   created_at timestamp_ntz,
   updated_at timestamp_ntz,
   FOREIGN KEY (product_id) REFERENCES application_db.field.
   product(product_id),
```

```
    FOREIGN KEY (customer_id) REFERENCES application_db.field.
    customer(customer_id)
);

CREATE HYBRID TABLE IF NOT EXISTS application_db.field.reports (
    report_id number autoincrement primary key,
    technician_id number,
    customer_id number,
    product_id number,
    report_date timestamp_ntz,
    issue_description varchar,
    resolution varchar,
    time_spent number,
    report_status varchar default 'open',
    created_at timestamp_ntz,
    updated_at timestamp_ntz,
    FOREIGN KEY (customer_id) REFERENCES application_db.field.
customer(customer_id),
    FOREIGN KEY (product_id) REFERENCES application_db.field.
product(product_id)
);
```

Tip Don't forget to leverage dynamic data masking to mask PII columns in the CUSTOMER table.

Now that our tables are set, we can insert synthetic data to test functionality. The query shown in Figure 9-6 represents a common query that might be seen in an application. Here, we are joining our tables to get a holistic picture of the order, customer, and report leading to the sale. Notice how the customer's name and email are masked? We executed this query using a role that does not have access to view the PII.

```
select
    concat(c.lname,', ', c.fname) as customer_name,
    c.email as customer_email,
    p.product_desc as product,
    s.quantity,
    s.sub_total,
    s.sale_date,
    r.technician_id
from application_db.field.sales s
left join application_db.field.product p
    on s.product_id = p.product_id
left join application_db.field.customer c
    on s.customer_id = c.customer_id
left join application_db.field.reports r
    on s.customer_id = r.customer_id
where sale_id = 101;
```

Results Chart

	CUSTOMER_NAME	CUSTOMER_EMAIL	PRODUCT	QUANTITY	SUB_TOTAL	SALE_DATE	TECHNICIAN_ID
1	*********, *********	*********	Golden Bantam - 5lb	3	519.87	2024-03-01 05:29:22.138	12

Figure 9-6. *Snowflake Snowsight Worksheet*

If you switch to the ANALYST role and run the same query, you see the PII in Figure 9-7 (in this case, synthetic data).

Figure 9-7. *Snowflake Snowsight Worksheet*

While creating hybrid tables, we used an autoincrement sequence for primary key generation. This ensures that each inserted record receives a unique key. A robust key strategy in production environments is crucial for data integrity, facilitating reliable backups and recovery.

Let's test the enforcement of primary key uniqueness by inserting a new record with a pre-existing PRODUCT_ID of '1.'

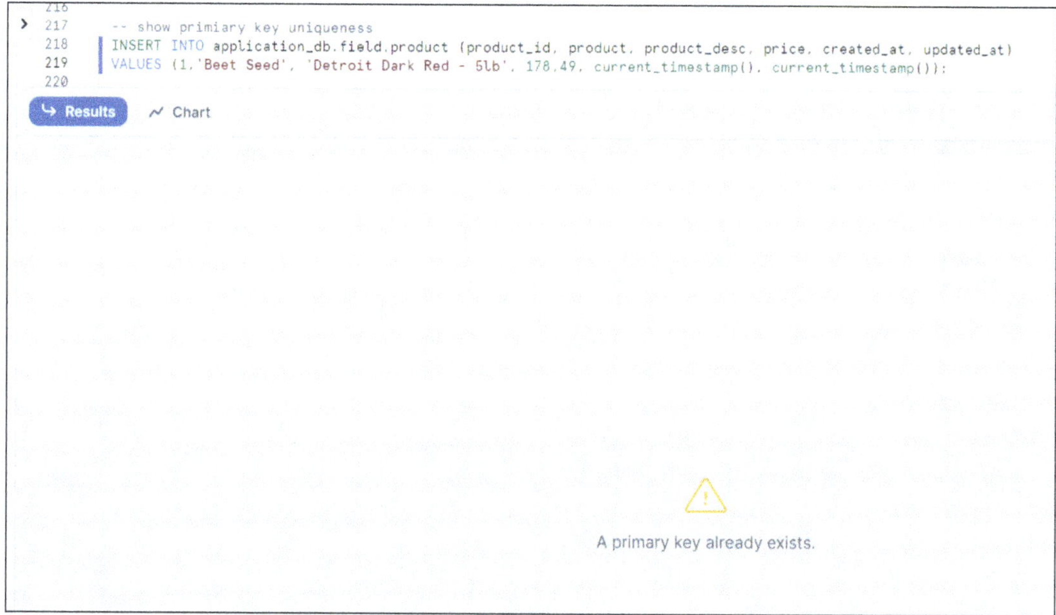

Figure 9-8. *Snowflake Snowsight Worksheet*

As seen in Figure 9-8, Snowflake prevented the insert due to a non-unique primary key.

Now, let's insert the new product by allowing the primary key to autoincrement the ID.

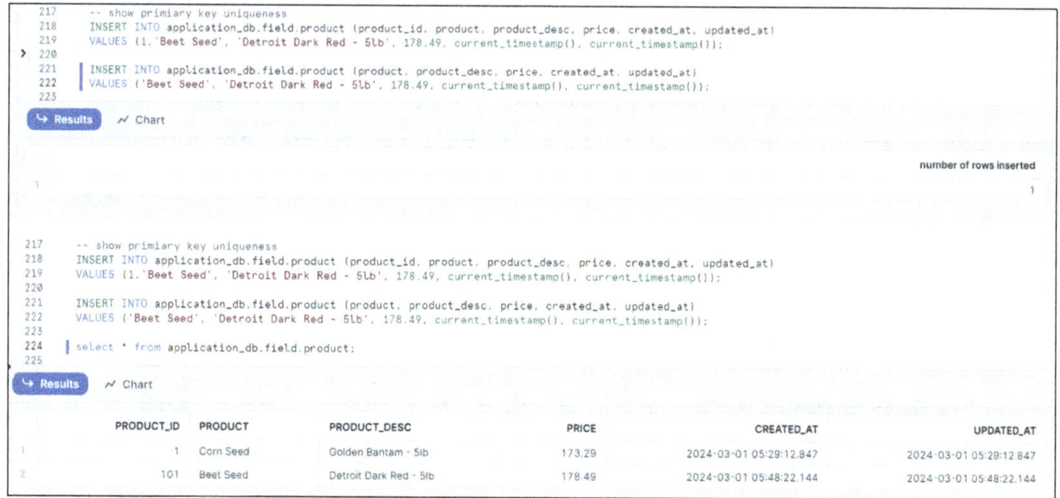

Figure 9-9. *Snowflake Snowsight Worksheet*

Now, as seen in Figure 9-9, we know hybrid tables enforce primary key uniqueness, on to testing referential integrity with foreign keys. This scenario attempts to delete PRODUCT_ID = '201,' which is referenced in the SALES and REPORTS tables.

```
-- show referential integrity
delete from application_db.field.product where product_id = 1;
```

Results Chart

Foreign keys that reference key values still exist.

***Figure 9-10.** Snowflake Snowsight Worksheet*

Again, shown in Figure 9-10 Snowflake enforced referential integrity by not allowing the DELETE on PRODUCT because PRODUCT_ID = 1 is referenced in other tables as a foreign key.

Snowflake's hybrid tables aren't designed to outperform standard tables for complex analytics on massive datasets. Instead, they excel in delivering quick responses to operational queries common in transactional applications. Hybrid tables are ideal for scenarios like the following.

- **Frequent lookups**: Retrieving individual records based on their unique key (e.g., fetching a specific sales record)
- **Rapid updates**: Making frequent modifications to individual records
- **Focused read operations**: Extracting a small set of complete records rather than running aggregations over large chunks of data

The value in Snowflake hybrid tables has transactional data collocated with your analytic data, leading to streamlined workflows and lower operational overhead, allowing your teams to focus on driving innovative value instead of maintaining the status quo. Our agricultural company sees that value and confidence in Snowflake Unistore to solve their needs moving forward.

Additional Snowflake Documentation

https://docs.snowflake.com/en/user-guide/tables-hybrid

https://docs.snowflake.com/en/sql-reference/sql/create-hybrid-table

---

Recipe 9-3. Streamlit

In a modern, rapidly evolving business landscape, building custom, data-centric applications is essential for extracting actionable insights, streamlining decision-making, and creating user-friendly data experiences. Streamlit, an open source Python framework paired with the power of Snowflake Data Cloud, offers a compelling solution for rapid application development.

Streamlit and Snowflake form a powerful partnership for accelerating data-driven application development. Streamlit's streamlined approach to web front-end creation and intuitive syntax lets developers focus on rapid prototyping and extracting insights instead of complex web coding. A seamless connection to Snowflake provides on-demand access to vast datasets for flexible exploration. This enables the crafting of custom applications with tailored visualizations and interactive elements, all powered by the data in Snowflake. Importantly, Streamlit lowers the technical barrier to understanding data, making dashboards and tools accessible to everyone in the organization. This promotes democratized data analysis and collaborative decision-making. Backed by Snowflake's scalability and performance, Streamlit applications can tackle demanding workloads and meet increasing user needs within a cloud-native environment.

We dive into a real-world use case to demonstrate leveraging various Streamlit libraries, components, and charting options to create innovative data applications backed by Snowflake.

Problem

The distribution company has a mutually beneficial arrangement with the trucking union. By outsourcing shipping, they avoid the substantial costs associated with owning and maintaining a fleet of trucks while ensuring reliable delivery of their goods. However, this streamlined operation has a hidden complexity: managing the dynamic quarterly union rate updates.

These updates arrive via an unassuming email attachment—a CSV file packed with new numbers. This triggers a cascade of manual work, typically falling on the shoulders of an already-burdened team member—all leading to several issues noticed by management.

- **Data inaccuracy risk**: Manual data entry from the CSV is time-consuming and prone to human errors, potentially compromising the accuracy of bids, profitability analysis, and overall financial estimations.
- **Operational bottleneck**: The existing process introduces a delay between receiving the union rates and those rates being reflected in the data lakehouse. This creates operational friction and impairs real-time decision-making.
- **Limited visibility**: Unless the data is manually transformed for reporting and analysis, business users and analysts struggle to access, understand, and work with updated trucking union rates seamlessly.
- **Inability to scale**: As the distribution company expands, the manual approach becomes increasingly cumbersome, leading to further delays, errors, and operational overhead.

The downstream effects are significant. The delay in integrating the updated rates hinders the sales team's ability to produce accurate, timely bids. They either rely on potentially outdated information or are forced to wait as updated rates make their way through the bottleneck of manual entry. Meanwhile, analysts aiming to provide insights into project profitability or overall trends know their reports are only as fresh as the last rate update, leading to incomplete and potentially misleading insights. These inefficiencies and the potential for errors directly impact the company.

- **Profit margins**: Inaccurate bids risk undercutting profitability or overpricing projects, leading to lost business opportunities.
- **Resource utilization**: Valuable time spent on manual data manipulation subtracts from strategic analysis and proactive planning.
- **Agility and competitiveness**: Delays in incorporating updated rates slow the company's ability to adapt to changing market conditions, impacting its ability to compete effectively.

This bottleneck highlights a missed opportunity. If the process were automated and the data seamlessly integrated into the company's systems, the distribution company could operate with a level of agility that is currently unachievable. Sales could react swiftly to rate changes, optimizing bids to safeguard profit margins, while ongoing profitability analysis could reveal hidden trends. However, their current reliance on manual processes obscures these potential gains and hinders the company's ability to compete at its true potential. Let's look at Streamlit in Snowflake and see how small teams can quickly build business value applications.

Solution

The team was able to rapidly build a POC of a simple input form that is built leveraging Streamlit in Snowflake. The application provides a user-friendly interface for entering, updating, and viewing union role rates in a Snowflake hybrid table. Let's look at the major sections of the POC code and then discuss potential enhancements that could be considered when designing the project.

Key Components

Import and Setup

- `import streamlit as st` imports the Streamlit library for building web interfaces.
- `from snowflake.snowpark.context import get_active_session` enables interaction with Snowflake from within a Streamlit deployment in Snowflake.

- session = get_active_session() establishes the Snowflake session, leveraging pre-configured credentials within the Snowflake environment.

```
## import required packages
import streamlit as st
from snowflake.snowpark.context import get_active_session

## create a session.
## Using Streamlit in Snowflake so creds are already sourced
session = get_active_session()

st.title("Rates - Data Entry Form")
```

Input Form (Figure 9-11)

- with st.form("rate_update", clear_on_submit=True) creates a form titled "rate_update" that clears its input fields after submission.
- Input fields (st.text_input, st.number_input, etc.) collect various pieces of rate data.
    - Company Code and Name
    - County and City
    - Role
    - Hourly Rate
    - Prevailing Wage Indicator
    - Effective and Expiration Dates

```
## entry form
with st.form("rate_update", clear_on_submit=True):
   company_code = st.text_input('Company Code')
   company_name = st.text_input('Company Name')
   county = st.text_input('County')
   city = st.text_input('City')
   role = st.text_input('Role')
   hourly_rate = st.number_input('Hourly Rate')
```

```
    is_prevailing = st.checkbox('Prevailing Wage?')
    effective_date = st.date_input('Effective Date')
    expiration_date = st.date_input('Expiration Date')
    sub_comment = st.form_submit_button('Submit')
```

Form Submission and Basic Validation

- The Submit button triggers actions using `st.form_submit_button('Submit')`.
- An if loop executes the code if the Submit button has been pressed and all validations have passed.
- Validation checks ensure all required fields are populated, displaying warnings to the user if there are omissions and stopping the insertion of incomplete or malformed data.

```
# form validation and merge to snowflake
if sub_comment:
    if not company_code:
        st.warning('Missing company code value',icon='🚨')
    elif not company_name:
        st.warning('Missing company name value',icon='🚨')
    elif not county:
        st.warning('Missing county value',icon='🚨')
    elif not city:
        st.warning('Missing city value',icon='🚨')
    elif not role:
        st.warning('Missing role value',icon='🚨')
    elif not hourly_rate:
        st.warning('Missing hourly rate value',icon='🚨')
    elif not effective_date:
        st.warning('Missing effective date value',icon='🚨')
    elif not expiration_date:
        st.warning('Missing expriation date value',icon='🚨')
```

```
    else:
        session.sql(<merge statement>
);""").collect()
        st.success('Success!', icon="✅")
```

Snowflake Merge

- The SQL code block can be executed once the Submit button has been pressed and all validation has passed.
- `session.sql(...)` constructs and executes a MERGE SQL statement to either insert a new rate record or update an existing one if the company code matches.

```
session.sql(f"""merge into retail_db.distribution.rates_lookup as
tgt using (select

'{company_code}' as company_code,
'{company_name}' as company_name,
'{county}' as county,
'{city}' as city,
'{role}' as role,
'{hourly_rate}' as hourly_rate,
'{is_prevailing}' as is_prevailing,
'{effective_date}' as effective_date,
'{expiration_date}' as expiration_date) as src
                        on tgt.company_code = src.company_code
                        when matched then update set tgt.company_
                        name = src.company_name,
tgt.county = src.county,
tgt.city = src.city,
tgt.role = src.role,
tgt.hourly_rate = src.hourly_rate,
tgt.is_prevailing = src.is_prevailing,
tgt.effective_date = src.effective_date,
tgt.expiration_date = src.expiration_date
```

```
                                    when not matched then insert
        (company_code,
        company_name,
        county,
        city,
        role,
        hourly_rate,
        is_prevailing,
        effective_date,
        expiration_date)
                                    values
        ('{company_code}',
        '{company_name}',
        '{county}',
        '{city}',
        '{role}',
        '{hourly_rate}',
        '{is_prevailing}',
        '{effective_date}',
        '{expiration_date}');""").collect()
```

- st.success('Success!', icon="✅"): Displays a success message

Data Display

As part of the POC, the team added a display back to the Streamlit application to see records as they are inserted.

```
#return the current data for that company
rates_lookup = f"""select company_code,
                        company_name,
                        county,
                        city,
                        role,
                        hourly_rate,
                        is_prevailing,
                        effective_date,
```

```
                        expiration_date
                from retail_db.distribution.rates_lookup
                where company_code = '{company_code}';"""
df_comments = session.sql(rates_lookup).to_pandas()
st.dataframe(df_comments, use_container_width=True)
```

Now that you have seen the code, let's look at the Steamlit application in Snowflake.

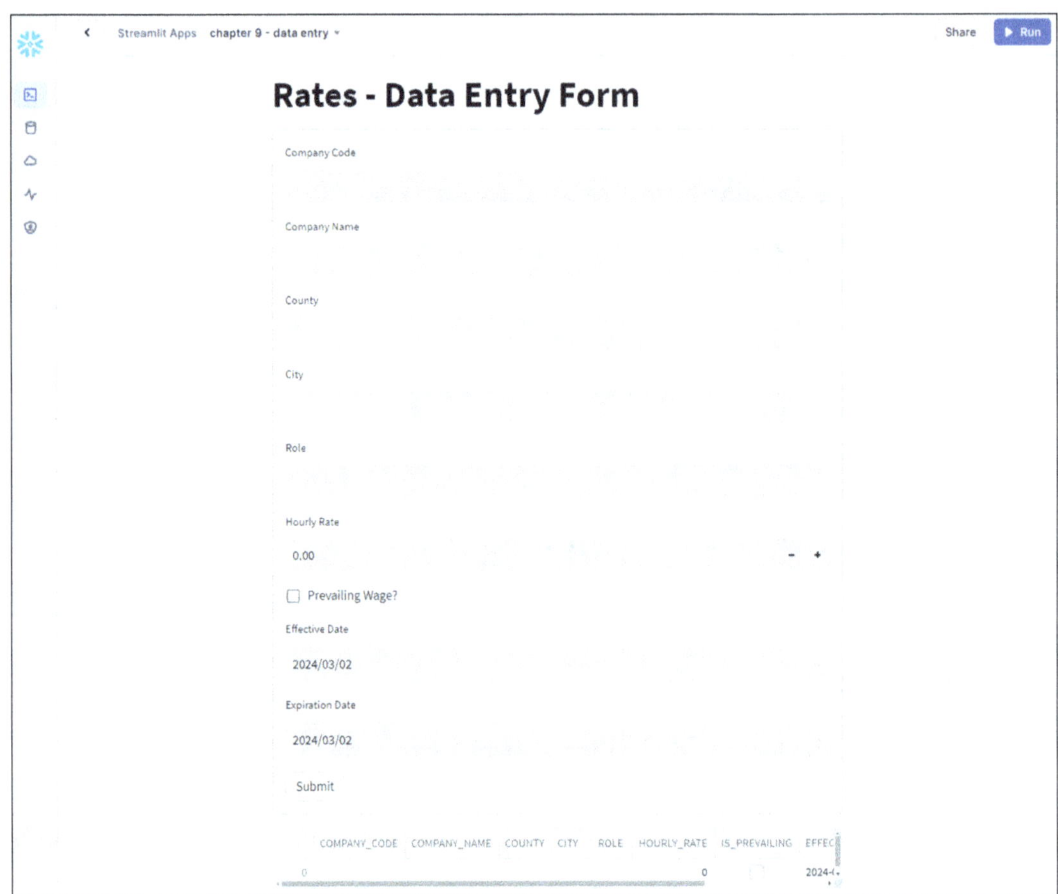

Figure 9-11. *Streamlit on Snowflake (SoS)*

Next, test the app by adding a new record.

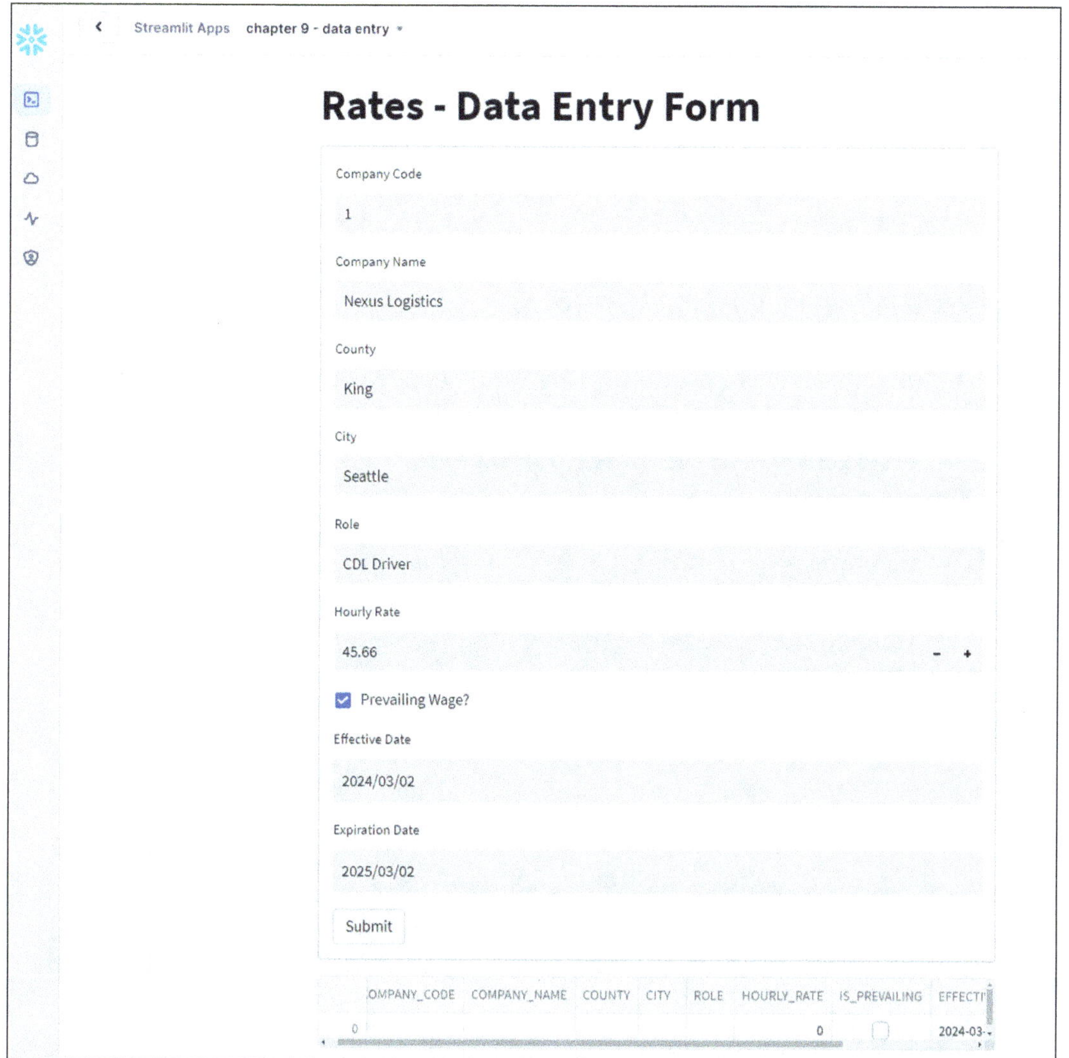

Figure 9-12. *Streamlit on Snowflake (SoS)*

Upon submission, the record inserted in Figure 9-12 was successfully merged to the Snowflake table.

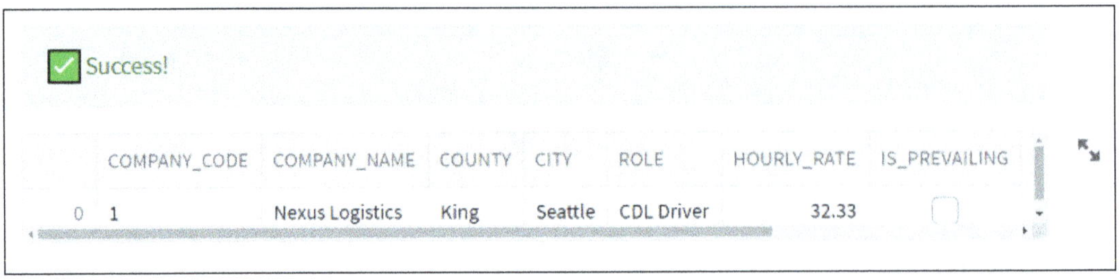

Success!

	COMPANY_CODE	COMPANY_NAME	COUNTY	CITY	ROLE	HOURLY_RATE	IS_PREVAILING
0	1	Nexus Logistics	King	Seattle	CDL Driver	32.33	

***Figure 9-13.** Streamlit on Snowflake (SoS)*

Let's assume this role is a prevailing wage and needs to be updated.

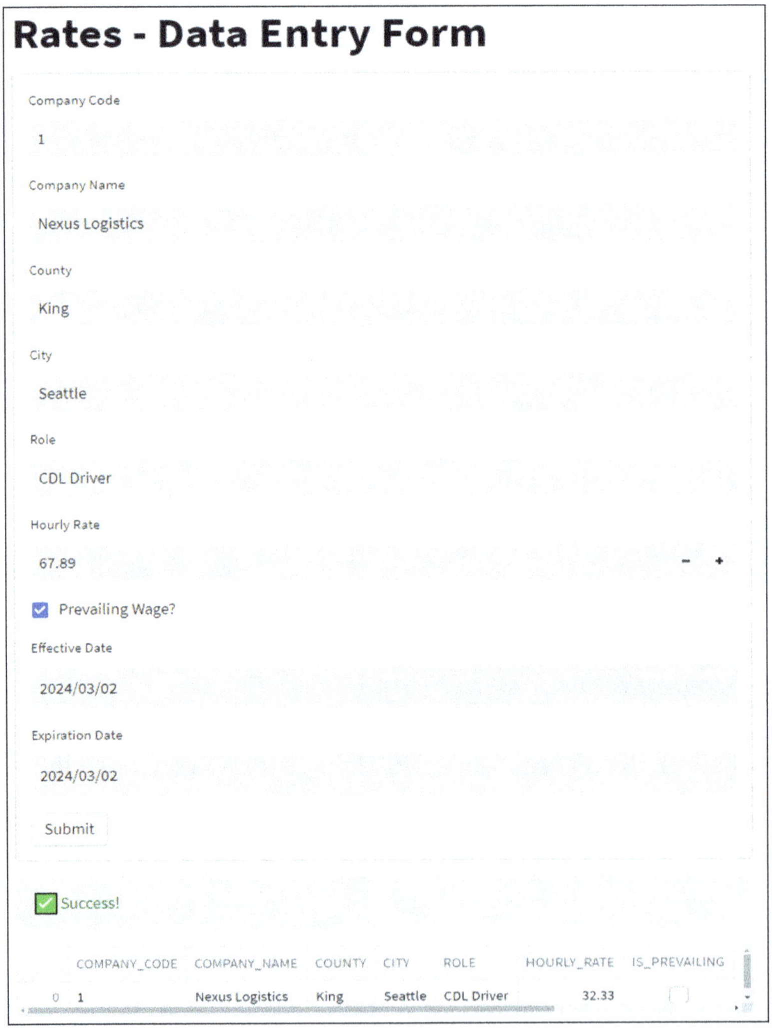

Rates - Data Entry Form

Company Code

1

Company Name

Nexus Logistics

County

King

City

Seattle

Role

CDL Driver

Hourly Rate

67.89

Prevailing Wage?

Effective Date

2024/03/02

Expiration Date

2024/03/02

Submit

Success!

	COMPANY_CODE	COMPANY_NAME	COUNTY	CITY	ROLE	HOURLY_RATE	IS_PREVAILING
0	1	Nexus Logistics	King	Seattle	CDL Driver	32.33	

***Figure 9-14.** Streamlit on Snowflake (SoS)*

After submission in Figure 9-14, you see the updated changes.

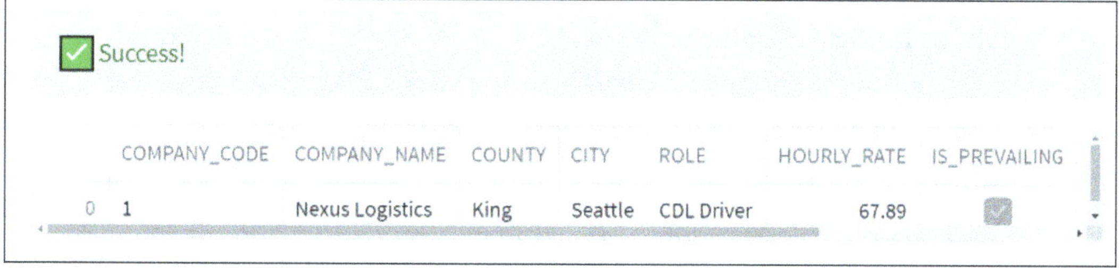

***Figure 9-15.** Streamlit on Snowflake (SoS)*

This Streamlit application directly addresses the pain points identified in the problem statement by streamlining the previously error-prone and time-consuming process of updating union rates. The benefits are significant.

- **Accuracy and efficiency**: Automating data entry eliminates the risk of human error introduced during manual input, safeguarding the integrity of financial calculations and bids.
- **Real-time decision-making**: Immediate updates within the data lakehouse allow sales and analysts to factor in current wage information, driving more accurate and responsive strategies.
- **Agility**: The company gains the flexibility to adapt quickly to changing union rates, optimizing profitability and enhancing its competitive edge.
- **Resource optimization**: Eliminating manual data manipulation frees up valuable time for analysis and strategic initiatives.

This POC demonstrates how Streamlit, deployed within Snowflake, provides a rapid and effective way to solve business problems. Its intuitive interface and seamless integration with Snowflake allow for the swift development of impactful data applications. Of course, this is a POC, and for the team to capitalize on this solution, they must consider the following when bringing this product to production.

- **Advanced validation**: Implement more sophisticated rules for input fields (realistic rate ranges, date checks, etc.)

- **CSV upload**: Provide a CSV upload option catering to users familiar with that workflow, while maintaining data quality checks. Native Streamlit offers a feature called `st.file_uploader`. Unfortunately, this is currently not supported in Streamlit in Snowflake. Check out Snowflake documentation for more information on unsupported features.
- **Reporting and analytics dashboard**: Develop a dedicated dashboard within Streamlit to visualize rate trends and cost impacts, and support strategic insights.
- **Role-based access controls**: Implement security mechanisms to ensure appropriate access to sensitive rate data.

The distribution company has successfully transformed a hidden bottleneck into a competitive advantage. By leveraging a straightforward yet impactful Streamlit application in Snowflake, they have enhanced accuracy, agility, and the intelligent use of their data assets. This highlights the potential for rapid, value-driven solutions using modern data technologies like Snowflake.

Additional Snowflake Documentation

`https://docs.snowflake.com/en/developer-guide/streamlit/limitations`

`https://docs.snowflake.com/en/developer-guide/streamlit/about-streamlit`

Recipe 9-4. Snowflake for Applications

In today's data-driven economy, purpose-built applications are essential to transform raw information into actionable insights, optimize processes, and deliver exceptional customer experiences. Snowflake Data Cloud provides a robust foundation on which to build, deploy, and integrate these innovative applications alongside analytics.

Snowflake's unique scalable architecture positions itself well for powering applications. By leveraging Native Apps, Container Services, Unistore, Data Sharing, Marketplace, and Snowflake's already exceptional database, an ecosystem is created and allows business applications to thrive. The possibilities are endless, but some common examples include applications handling customer interactions, e-commerce transactions, financial record updates, generative AI, document processing, and more. Snowflake's ability to handle high volumes of reads and writes concurrently, with consistent speed and responsiveness, is key for ensuring smooth application experiences. Moreover, Snowflake scales seamlessly without manual intervention as usage and data grow. This liberates businesses from constant maintenance and removes worries about application performance degrading as the business succeeds and grows. Critically, Snowflake's support for a mixture of structured, semi-structured, and unstructured data aligns with real-world applications. They often generate a mixture of traditional transactional records and less rigidly formatted data (like user behavior logs). All this data in one place eliminates the need for complex synchronization pipelines. Applications instantly benefit from the most up-to-date analytics data and insights, enhancing their decision-making capabilities.

Snowflake provides a powerful foundation for a wide array of innovative business applications. From the realm of customer experience, applications fueled by Snowflake can serve up real-time personalized recommendations based on the freshest customer behavior data. This translates into increased engagement, targeted offers, and heightened customer satisfaction. On the operational front, Snowflake can support intelligent tools that oversee manufacturing, forecast equipment maintenance needs, and drive data-driven automation across vital workflows. Snowflake enables fintech applications to blend live market data with granular transactional insights within the finance sector. This unlocks more accurate risk mitigation strategies, enhanced fraud detection, and the ability to pinpoint high-yield investment opportunities.

The subsequent sections explore how Snowflake fuels a spectrum of data-centric applications. We explore application features, looking at how Snowflake drives modern and innovative capabilities, streamlines how organizations extract maximum value from their data assets, and how your organization can get started.

Problem

Consider a mid-cap manufacturing company that prides itself on its streamlined inventory management and customer satisfaction. In a fast-moving data landscape, their investment in on-premises databases and application infrastructure, which aimed to create a system that balanced cost-effectiveness with the ability to handle fluctuating order volumes, was not up to the needs of the business. Initially, workflows seemed to run smoothly enough. Orders flowed in, stock levels updated, and reports were generated. However, under the surface, cracks began to appear, gradually widening into significant operational difficulties.

The first sign of trouble was subtle - a minor mismatch in the stock availability displayed to customer service representatives versus what was physically in the warehouse. Digging deeper into the transactional database while diligently capturing sales and updates lagged in syncing this data with the analytics systems. This delay, though seemingly small, had widespread consequences. Sales teams and account executives faced embarrassing situations, unintentionally promising out-of-stock items to clients. Meanwhile, sales conversations floundered due to uncertain delivery estimations. The once reliable system had become a source of frustration and mistrust.

Forecasting models, unwittingly fed stale data, churned out inaccurate demand predictions. This led to costly surprises and backorders on high-demand items disappointed clients, while overstocks tied up capital in slow-moving inventory. Profit margins took a hit, and the root cause was invisible to standard reporting. Exacerbating the issue, the company's on-premises infrastructure buckled during peak seasons. Systems slowed to a crawl, paralyzing customer service as demand was highest. IT teams scrambled to maintain performance, but the rigid hardware limited their options.

The manufacturer's once-predictable inventory system had become a liability. What was intended to streamline operations was now actively undermining customer relationships, eroding profitability, and handcuffing their ability to grow. A drastic change is needed—a solution that unifies data, scales effortlessly, and empowers real-time data-driven decision-making.

Solution

Snowflake's architecture and features provide a compelling solution to address these challenges. From our manufacturing companies' perspective, let's look at some of the advanced application features Snowflake offers and identify if the requirements set forth by the IT team can be met.

Snowpark Container Services allows you to bring your existing code and workflows to Snowflake in a containerized format, enabling a wider array of development languages and specialized libraries. It's a key component of Snowflake's vision for a unified platform handling structured data and applications. Here's what it encompasses.

Images: Self-Contained Packages

Think of a Snowpark Container Services image as a blueprint for a self-contained application. It packages your code, all the necessary dependencies, and instructions for how to run it, ensuring portability. These images follow industry standards (OCI), meaning they can be built using familiar tools like Docker. This approach gives you tremendous flexibility in development. From a business perspective, this consistency allows you to reliably deploy the same data processes across different environments. Updates become easier, minimizing errors caused by mismatched settings and empowering teams to work seamlessly on these applications.

Image Registry/Repositories: A Structured Library

Snowflake's image registry acts as a secure library for your container images. It follows OCI standards for compatibility and provides a central location for storing, managing, and tracking different versions of your containerized applications. This organized structure is vital for businesses—teams can easily share trusted images, roll back to previous versions if needed, and maintain a clear track record of which applications are in use. This control and transparency are critical for collaboration, ensuring reliability, and simplifying necessary audits.

Compute Pools: Power on Demand

Snowpark Container Services' compute pools provide the horsepower to run containerized applications. Consider them clusters of virtual machines customized specifically for container-based workloads. You have granular control—selecting instance types (standard CPU, memory-optimized, GPU-based, etc.), configuring scaling limits, and even suspending compute pools when not in use, offering flexibility similar to managing Snowflake virtual warehouses. Snowflake seamlessly handles the automatic scaling of your compute pool based on demand and load balances traffic across multiple service instances. This detail is essential for businesses because it ensures that critical data processes and forecasts always have the resources to perform reliably. Peak seasons won't lead to slowdowns or service outages, safeguarding customer satisfaction and maximizing sales potential.

Specification Files: Defining the Blueprint

A specification file is the architectural blueprint for your containerized service within Snowflake. It's a structured file (in YAML format) that outlines everything Snowflake needs to know: the specific container image to use, its location in the image registry, any environmental settings it needs, and how it should connect to other services or data. Experienced Kubernetes users will find these files conceptually similar to Kubernetes deployment manifests. For businesses, specification files offer several advantages. They make creating and modifying services repeatable and predictable. This "infrastructure as code" concept also reduces errors from manual configuration, boosting reliability and overall efficiency.

Services: The Heart of Execution

In Snowpark Container Services, services are the heart of execution—the applications built from your specification files and brought to life on compute pools. You can control these services (creating, updating, etc.) and monitor their status and logs. Importantly, Snowflake's billing model charges only for active compute pools and image storage, not the services themselves. This means you can design the right solution for your business needs. It directly addresses the issues in the original scenario: scheduled services keep inventory data constantly updated, APIs provide the real-time insights needed for fast decision-making, and jobs offer the flexibility to perform complex forecasting or analyses that were previously difficult to achieve.

Jobs: Focused Bursts of Processing

Snowpark Container Services jobs are designed for containerized tasks with a clear start and finish. Think of them as specialized tools that you can trigger when needed (either manually or through automated systems). Upon execution, the job runs on a compute pool and then shuts down when complete. While similar in concept to Snowflake Tasks, jobs offer a key difference: they aren't limited to standard SQL or the existing Snowpark APIs. This flexibility is crucial for businesses that rely on highly specialized forecasting models, custom algorithms, or libraries not yet directly integrated with Snowflake. Jobs open a broader range of analytical possibilities, potentially leading to more accurate insights.

The technology is great, but how does this stack up against the requirements of our manufacturer?

- **End inventory frustrations with near real-time accuracy.** Container services help the manufacturer break down the silos that cause so much friction. Transactional updates to inventory or incoming orders instantly sync across sales, forecasting, and analytics. This eliminates mismatches between stock availability and what the sales team sees. Forecasts stay accurate because they're always based on a single, up-to-the-minute picture of the business, reducing costly errors.
- **Optimize inventory to maximize profits.** Snowpark means the manufacturer's forecasting models always crunch the freshest Snowflake data. No more delays waiting for updates in ETL/ELT cycles! This accuracy allows precise calculation of stockouts, prevents overstocking that eats into cash flow, and ensures the right products are available to fulfill customer demand.
- **Peak seasons? No problem.** Snowflake's elasticity within container services ensures the manufacturer's systems can handle anything. Customer demand surges won't overwhelm their applications. This safeguards a smooth customer experience even during high-volume periods, maintaining the manufacturer's reputation and allowing it to capitalize on every sales opportunity.
- **Build tools tailored to your needs.** Container services empower the manufacturer's development team to create custom applications directly within Snowflake. These might be streamlined dashboards tailored for the warehouse team or a forecasting interface directly linked to ordering workflows. Data-driven decisions become easier across the company, boosting efficiency and potentially uncovering new ways to improve operations.

We have discussed how Snowflake Container Services can support business applications. Now, let's focus on the Native App Framework and how it extends container services to drive additional return on investment (ROI) by moving to Snowflake.

The Snowflake Native App Framework transforms the way you interact with your Snowflake data. It enables the creation of applications that live directly within the Snowflake environment, giving you unprecedented flexibility. Not only can you build insightful Streamlit dashboards, but you can also incorporate custom business logic using Snowpark APIs, JavaScript, SQL, and Container Services. This empowers you to package and share these applications publicly in the Snowflake Marketplace to generate revenue or privately with select partners. The development experience is smooth, offering integrated testing environments, seamless code management with favorite industry-leading tools, and the ability to roll out updates in a controlled manner. Plus, with built-in logging, you can easily monitor and troubleshoot applications.

Transform Data into Actionable Insights

The Snowflake Native App Framework bridges the gap between data and decision-making. Build custom applications (using tools like Streamlit) that are directly embedded alongside your Snowflake data. This creates a centralized hub for visualizations, forecasts, and interactive tools. Users no longer waste time switching between systems or reconciling information from different platforms. They are empowered with real-time, accurate insights to drive better business outcomes.

Streamline Workflows, Minimize Errors

Embed these Snowflake-powered applications directly into your existing systems, creating a seamless, data-driven experience. This eliminates friction caused by juggling multiple tools and reduces the potential for errors resulting from data being out of sync or misinterpreted. Users can confidently analyze the latest information and take decisive actions, enhancing overall efficiency and accuracy across your organization.

Beyond supporting your business, the framework allows teams to turn data and application expertise into a profit center. If you've developed valuable dashboards, forecasting models, or custom tools with wide appeal, the Snowflake Marketplace offers a global platform to reach potential customers. It opens a new revenue stream, transforming your hard work into a monetization stream.

Targeted Collaboration and Customization

Private listings let you selectively share applications with specific partners or clients. This is ideal for several scenarios.

- **Deepen relationships**: Provide exclusive tools that enhance the value you deliver to key partners, strengthening those business bonds.

- **Tailored solutions**: Craft applications that address the unique needs of individual clients, showcasing your adaptability and problem-solving skills.
- **Controlled rollouts**: Test new applications with a select group or phase in updates without disrupting a wide user base.

Overall, the Snowflake Native App Framework and Snowpark Container Services offer a transformative and compelling solution for the manufacturing company.

Note At the time of this writing, Snowflake Native App Framework isn't available on GCP, and Snowpack Container Services is a feature in public preview for AWS only.

By enabling the creation of applications natively within Snowflake, it directly addresses the core issue of data silos. Dashboards, forecasts, and customer-facing tools always reflect the most up-to-date warehouse information, eliminating delays and errors. This restores trust in the inventory system and leads to more accurate sales promises and decision-making. Additionally, leveraging Streamlit and Snowpark within these applications empowers forecasting models to operate on collocated data. It eliminates inaccurate predictions, which optimizes inventory levels and minimizes costly stockouts or overstocking. Furthermore, the inherent scalability of Snowflake ensures that even during peak seasons, customer-facing applications maintain performance, safeguarding customer satisfaction and preventing missed sales opportunities. The Native App Framework also opens exciting new avenues: streamlined workflows by embedding dashboards into existing tools, potential revenue streams by monetizing governed and secure data through the Snowflake Marketplace and strengthening collaboration with suppliers or clients via private listings.

Additional Snowflake Documentation

`https://docs.snowflake.com/en/developer-guide/native-apps/tutorials/getting-started-tutorial#introduction`

`https://docs.snowflake.com/en/developer-guide/snowpark-container-services/overview`

`https://docs.snowflake.com/en/developer-guide/native-apps/native-apps-about`

CHAPTER 10

Machine Learning

Machine learning has seen a remarkable rise in popularity over the past decade, driven by the exponential growth of data and the increasing computational power available. This surge is attributed to several key factors.

- **Data explosion**: The advent of the internet, social media, IoT devices, and other digital technologies has led to an unprecedented amount of data being generated. This data provides the raw material necessary for training machine learning models.
- **Advancements in algorithms**: Significant progress in machine learning algorithms, especially deep learning, has improved the ability to process complex data and achieve higher accuracy in tasks such as image recognition, natural language processing, and predictive analytics.
- **Computational power**: The development of high-performance computing resources, including GPUs and TPUs, has accelerated the training of large-scale machine learning models, making it feasible to handle and process massive datasets efficiently.
- **Open source ecosystem**: The proliferation of open source tools and libraries like TensorFlow, PyTorch, and Scikit-learn has democratized access to advanced machine learning techniques, allowing a broader range of developers and organizations to experiment and innovate.
- **Industry adoption**: Various industries have recognized the potential of machine learning to drive business value through automation, improved decision-making, and new product offerings. This has led to increased investment and research in the field.

This chapter looks at different options and patterns provided by Snowflake to handle your Python and machine learning workflows.

D. Dayton and J. Eipe, *Snowflake Recipes*, https://doi.org/10.1007/979-8-8688-0938-5_10

Recipe 10-1. Snowpark and Third-Party Packages

Problem

Let's assume that you are part of a data team and already use some Python packages as part of your code base. Now that your data team is shifting to Snowflake, you would like to assess if your Python packages could be used in Snowflake.

Solution

Snowpark provides Python runtime along with the default Snowpark libraries from Snowflake.

To utilize external or third-party libraries there are three options.

- Import via Snowflake stage
- Import using Snowflake's new custom package feature
- Import from Anaconda Snowflake channel

Import via Snowflake Stage

Snowflake stages can be used to import packages and Python scripts. You can bring in any Python code that follows the guidelines defined by Snowflake.

- Snowflake-enforced security restrictions
    - You cannot fork a process, but threading is allowed.
    - Access to a file system is not allowed.
    - You can write to the /tmp directory. Each query gets its own memory-backed file system storing its /tmp data. Different queries running parallel won't conflict on the file and cause an overwrite, but if a single query calls the same UDF more than once, then the data is overwritten.
    - You cannot access an external network, but you can access resources through the Snowflake external access integration feature.

- All UDFs and modules must be platform-independent and must not contain native extensions.
- Python UDFs are not sharable.
- Database objects that contain Python UDFs are not sharable.
- Database replication is supported for in-line Python UDFs. However, replication is blocked if a Python UDF has a dependency on a file in a stage.
- Snowflake uses the Python zipimport module to import Python code from stages. As a result, any zipimport limitations are also present with UDFs.

In the code repository, the Python notebook, Chapter-10/snowpark_external_lib_demo_1.ipynb, is an example that demonstrates how to import your Python script via a stage.

Import Using Snowflake's New Custom Package Feature

If the Python library is listed and available in the PyPI repository, it could be imported with just a simple step. PyPI, short for Python Package Index, is the official third-party package repository for the Python programming language.

This can be done in multiple ways, but all use the session.custom_package_usage_config parameter to make it work.

The custom_package_usage_config session parameter provided by Snowflake can force Snowflake to use the pip Python package manager to install the mentioned libraries.

Here is an example use case.

```
session.custom_package_usage_config = {"enabled": True, "force_push": True}
session.add_packages("amazon.ion==0.12.0","snowflake-snowpark-python")
```

Other ways to specify packages are via requirements.txt files.

```
session.custom_package_usage_config = {"enabled": True}
session.add_requirements("./requirements_file.txt")
```

The following describes some of the limitations.

- Python packages (not present on Snowflake's Anaconda channel) are pip-installed locally and imported for use via a temporary remote stage directory. To allow this, pip needs to be present in your environment and the current local user would need Permission to write to a temporary directory is required.
- If you need packages that rely on OS native code, the packages *must* originate from the Anaconda Snowflake channel. If your package relies on dependencies that use native code, Snowpark makes a best-effort attempt to switch to versions present in Anaconda. This might result in versioning incompatibility issues.
- If the package you want relies on OS native code and is not present on the Anaconda channel, it may not work.

In the code repository, the Python notebook, Chapter-10/snowpark_external_lib_demo_2.ipynb is an example of this.

Import from the Anaconda Snowflake Channel

Snowflake has partnered with Anaconda to provide the growing Python community of data scientists, data engineers, and developers with effortless access to open source Python packages to build secure and scalable data pipelines and machine learning workflows.

If you are new to Anaconda, it is a package and environment management system for open source source. It provides a special package manager called Conda package manager. The Conda manager handles dependency conflicts, version control, vulnerability monitoring, and environment management.

Snowflake Snowpark offers a native Anaconda integration that provides built-in access to Anaconda's ecosystem of open source Python libraries, and there is no additional cost to use this feature.

To utilize this feature, all you need to do is enable it by using your orgadmin role and then making sure the package you are looking for is available in the Anaconda repository.

Figure 10-1 is a screenshot of Python worksheets in Snowflake. Note that you can select packages and their versions from the drop-down.

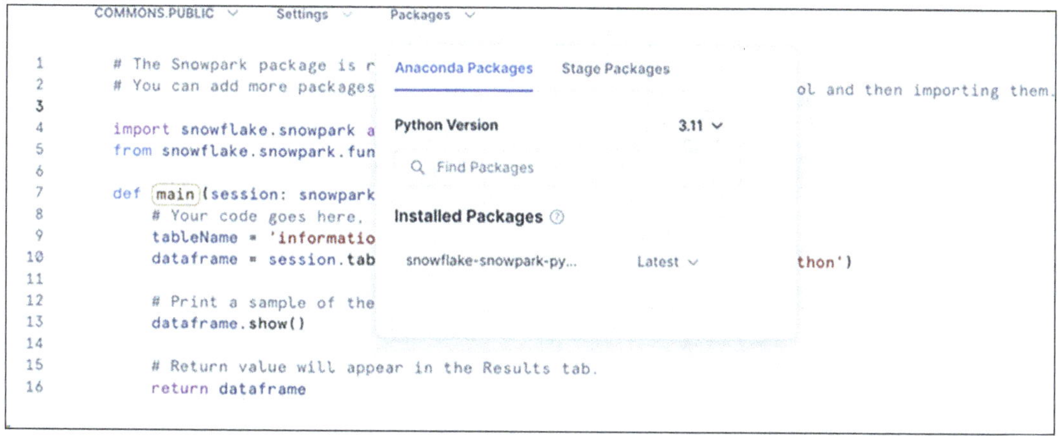

Figure 10-1. *Snowflake Python Worksheet*

Additional Snowflake Documentation

https://docs.snowflake.com/en/developer-guide/snowpark/python/index

https://github.com/Snowflake-Labs/sfguide-intro-to-machine-learning-with-snowflake-ml-for-python

https://community.snowflake.com/s/article/how-to-use-other-python-packages-in-snowpark

---

Recipe 10-2. Machine Learning

This recipe looks at how to run machine learning (ML) code in Snowflake and the different approaches available.

The focus of this section is *not to* delve into the elements of *exploratory data analysis*/ML algorithms or model performance. The focus is on how to get your ML code executed in Snowflake using a simple data set that is familiar to every data scientist and data engineer.

Problem

As a data engineer, you need to run Python ML code in Snowflake from your previous project that used a local machine and an in-house spark cluster.

What are the different ways you can run Python ML code within Snowflake?

Solution

Let's use one of the many housing prices data sets available in Kaggle.

Let's first try to download the data and load it in a Snowflake table to examine and finally create a model to run predictions.

Prerequisites

1. Download the data file from `www.kaggle.com/competitions/house-prices-advanced-regression-techniques/data`.
2. Download the Kaggle data set and upload it in Snowflake.
   a. You can use the Snowflake Snowsight (web UI) to upload a dataset and create a table in a single step.
   b. In this solution, the table is created with the name HOUSE_PRICES_RAW_DATA for the train data set and HOUSE_PRICES_TEST_DATA for the test data set.
3. Create a Kaggle or Google Collab account, or you may set up Jupyter to run locally on your computer.

Approach 1: Data in Snowflake but All Other Compute External to Snowflake

This is the easiest approach for data scientists because all the code can stay in their notebooks, and only the data extraction changes.

The following explains the steps involved.

1. Create a local or an online Jupyter (Kaggle or Google Collab) environment if you haven't already.

2. Install the snowflake-snowpark-python library (you could do that in your Python worksheets).

3. Create a session object and load the data into a pandas DataFrame.

4. Pick up the features.

5. Check for missing data.

6. Perform one-hot encoding for categorical data.

7. Build models.

The code is available in the code repository in the Chapter-10/house-prices-prediction.ipynb file.

A challenge with the approach is that you cannot scale your ML computation using Snowflake's computational capabilities. However, this is the quickest and most convenient approach with the least code changes.

Approach 2: The Data and Model in Snowflake but Compute External to Snowflake

This does not need any demonstration. The idea is that the model can be exported as a file into a snowflake stage to be used for execution later.

Approach 3: The Data, Model, and Code Runs in Snowflake

This approach reads data from Snowflake and runs the code that prepares and creates the model in Snowflake.

Since all execution happens in Snowflake, this can be done directly via Snowflake Snowsight (web UI) worksheets or by running your code locally to register it as a UDF or procedure, which can later be invoked by calling the UDF/procedure from anywhere where the Snowflake session is available.

The following are the steps involved.

1. Create a local or an online Jupyter (Kaggle or Google Collab) environment if you haven't already.

2. Install the snowflake-snowpark-python library (you could do that right from within your worksheets).

3. Create a session object and a Snowflake DataFrame to point to your source data.
4. Add Python packages to the session.
5. Define your input variables like your training table name, source columns that need to be selected and the target column.
6. Create a Snowflake stage to store the model.
7. Create and register Python's train_model stored procedure.
    a. Perform one-hot encoding for categorical data.
    b. Create model.
    c. Upload the model into the previously created Snowflake stage.
    d. Return a dictionary of model information, including absolute mean percentage error.
8. Invoke the train_model stored procedure.
9. Unit test the model and the train_model procedure using a stored procedure.
10. Utilize the model from a UDF.
11. Utilize the model from a vectorized UDF.
    a. Vectorized UDFs use the UDF Batch API, which enables defining Python functions that receive batches of input rows (a.k.a. chunked rows) as pandas DataFrames and return batches of results as pandas arrays or series.
    b. The column in the Snowpark DataFrame is vectorized as a pandas series inside the UDF.

The code is in Chapter-10/house-prices-prediction-snowpark.ipynb, which has details and comments for each code block.

Approach 4: Data, Model, and Code Run in Snowflake and Utilize Snowflake ML Features

Snowflake ML Modeling

This approach is similar to the previous approach but with the added benefit that when you use Snowflake ML API features, it utilizes the Snowflake's multi-cluster (and not just run in a single node in one of your Snowflake warehouses).

This helps to reap the maximum benefit of running in Snowflake Data Cloud.

The steps involved are the same as approach 3 except for adding snowflake-ml-python and snowflake-snowpark-python.

Additionally, you replace all the references to pandas and sklearn with Snowflake API so that everything is run natively in Snowflake.

The code is in Chapter-10/house-prices-prediction-snowparkml.ipynb, which has details and comments for each code block.

Snowflake Model Registry

You could use the Snowflake ML Registry feature to store the models instead of using a stage. The ML Registry acts as a wrapper around this file management via stages, providing versioning and cataloging capabilities.

```
from snowflake.ml.registry import Registry

# assumes you have created a seperate schema and database for storing
models. This is only as a best practice.
reg = Registry(session=session, database_name="ML", schema_name="REGISTRY")
model_version = reg.log_model(model_RFR,
                    model_name="houseprice_estimator",
                    version_name="v1",
                    comment="house prices prediction model",
                    metrics=model_info,
                    sample_input_data=XY_train.drop(target_col).limit(10))
if model_version.model_name:
    logger.info("successfully uploaded the model into snowflake registry")
```

The Snowflake Model Registry lets you securely manage models and their metadata in Snowflake, regardless of origin. The model registry stores ML models as first-class schema-level objects in Snowflake so they can easily be found and used by others in your organization.

The Snowflake Model Registry supports the following types of models.

- Snowpark ML Modeling
- scikit-learn
- XGBoost
- LightGBM
- CatBoost
- PyTorch
- TensorFlow
- MLFlow PyFunc
- Sentence Transformer
- Hugging Face pipeline
- Other types of models via the snowflake.ml.model.CustomModel class

---

Additional Snowflake Documentation

`https://docs.snowflake.com/en/developer-guide/snowflake-ml/snowpark-ml`

`https://docs.snowflake.com/en/developer-guide/snowflake-ml/model-registry/overview`

`https://docs.snowflake.com/en/user-guide/ecosystem-analytics`

Recipe 10-3. Snowpark Container Services

Snowflake's Snowpark Container Services is a fully managed container offering designed to facilitate the deployment, management, and scaling of containerized applications within the Snowflake ecosystem.

This service enables users to run containerized workloads directly within Snowflake, ensuring that data doesn't need to be moved out of the Snowflake environment for processing.

Snowflake uses a modified version of the Open Container Initiative (OCI) runtime execution environment optimized for Snowflake. The OCI was a Linux Foundation project established in June 2015 by Docker and other leaders in the container industry. The idea is that given an OCI image, any container runtime (Docker, Kubernetes, or Snowpark Container Services) that implements the **OCI Runtime Specification** can unbundle the image and run its contents in an isolated environment.

Snowpark Container Services is fully integrated with Snowflake. For example, your application can easily perform these tasks.

- Connect to Snowflake and run SQL in a Snowflake virtual warehouse
- Access data files in a Snowflake stage
- Process data sent from SQL queries

Snowpark Container Services is also integrated with third-party tools. It lets you use third-party clients (such as Docker) to easily upload your application images to Snowflake.

Problem

Your data team has developed a Python API used across many data applications. You are now tasked with using the same in Snowflake without rewriting or changing the API code.

Solution

Snowpark Container Services is a great feature if you already have a containerized application (or one that can be containerized) and want to use Snowflake's computing power. It is also a great option if you are in a situation to use custom code that isn't equipped for Snowpark code for the following reasons.

- You need a long-running service.
- You need to use specific Python libraries not available in Snowflake-provided Conda packages nor could it be bundled using a Snowflake stage.
- You need to use a specific language runtime now natively supported by Snowflake.

To create and run containerized applications in Snowflake, you should create three specific Snowflake objects.

Image Repository

An image repository is used for storing your images. You could use any OCI-compliant client, such as Docker CLI or SnowSQL to access your image in the repository.

Each image registry in a Snowflake account has a unique hostname, which allows OCI clients (such as Docker CLI) to access an image registry using REST API calls.

The general syntax for an image registry hostname is <orgname>-<acctname>.registry.snowflakecomputing.com.

Note that a registry is a service that serves the OCIv2 API, and a repository is a storage unit that you create within the service. A repository is a named location in your account where you store images. You can create one or more repositories in your Snowflake account. For example, DEV, TEST, and PROD repositories can store images during development, testing, and production.

The following is a general syntax for a Snowflake repository URL.

<registry-hostname>/<db_name>/<schema_name>/<repository_name>

Compute Pool

A compute pool is an account-level construct, unlike an image. You cannot have multiple compute pools with the same name in your account.

The minimum information required to create a compute pool includes the following.

- The machine type to provision for the compute pool nodes
- The minimum nodes to launch the compute pool with
- The maximum number of nodes the compute pool can scale to

A compute pool can be in any of the following states.

- **IDLE**: The compute pool has the desired number of nodes, but no services are scheduled. In this state, autoscaling can shrink the compute pool to the minimum size due to a lack of activity.
- **ACTIVE**: The compute pool has at least one service running or scheduled to run on it. The pool can grow (up to the maximum nodes) or shrink (down to the minimum nodes) in response to load or user actions.
- **SUSPENDED**: The pool currently contains no running compute nodes. When you suspend a compute pool, Snowflake suspends all services except the job services. The job services run until they reach a terminal state (DONE or FAILED), after which the compute pool nodes are released.
- **STARTING, STOPPING, RESIZING**: These are transient states when you create or resume a compute pool.

Service

A service represents Snowflake running your containerized application on a compute pool.

There are two types of services.

- **Long-running services**: This is like a web service that does not end automatically. After you create a service, Snowflake manages the running service. For example, if a service container stops, for whatever reason, Snowflake restarts that container so the service runs uninterrupted.
- **Job services**: A job service terminates when your code exits, similar to a stored procedure. When all containers exit, the job service is done.

You provide a name for your service and a service specification file (in YAML format) after you have uploaded your image and created the compute pool.

Let's consider a simple scenario to explore Snowpark Container Services.

Consider a containerized Python app that generates a random sample of personal data using the Python mimesis library.

Step 1: Create the API.

Create a simple API. The code for the this sample is available in the code repository - Chapter-10/snowpark_container_app.

Figure 10-2 shows the command line screen after the python application has been started on your local machine.

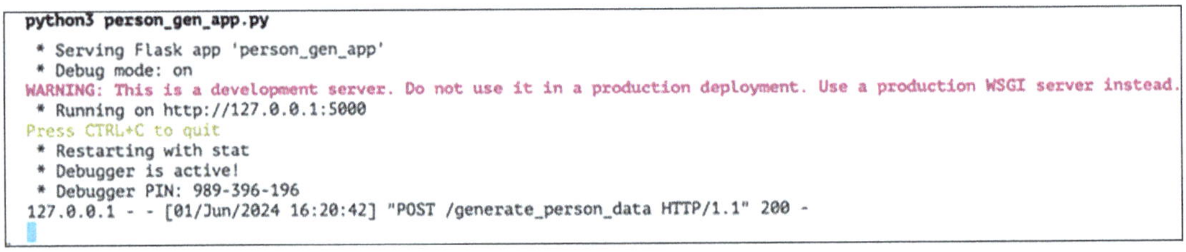

```
python3 person_gen_app.py
 * Serving Flask app 'person_gen_app'
 * Debug mode: on
WARNING: This is a development server. Do not use it in a production deployment. Use a production WSGI server instead.
 * Running on http://127.0.0.1:5000
Press CTRL+C to quit
 * Restarting with stat
 * Debugger is active!
 * Debugger PIN: 989-396-196
127.0.0.1 - - [01/Jun/2024 16:20:42] "POST /generate_person_data HTTP/1.1" 200 -
```

Figure 10-2. *Starting Flask application*

After the server is started you could use curl to test the endpoint.

```
curl -X POST -H "Content-Type: application/json" -d '{"email_domain":
"gmail","gender": "M", "is_employed": "true"}' http://127.0.0.1:5000/
generate_person_data
```

You should see a response similar to the following.

```
{
  "data": {
    "email": "hide1858@gmail.com",
    "name": "Emilio Wheeler",
    "occupation": "Tennis Coach",
    "university": "Bridgewater State University"
  }
}
```

Step 2: Create a Snowflake role with the necessary grants.

Let's create separate roles to create and manage the container service.

```
CREATE ROLE CONTAINER_USER_ROLE;
GRANT ALL on DATABASE COMMONS to ROLE CONTAINER_USER_ROLE;
GRANT ALL on SCHEMA COMMONS.UTILS to ROLE CONTAINER_USER_ROLE;
GRANT ALL on WAREHOUSE XSMALL_WH to ROLE CONTAINER_USER_ROLE;

GRANT CREATE COMPUTE POOL ON ACCOUNT TO ROLE CONTAINER_USER_ROLE;
GRANT CREATE INTEGRATION ON ACCOUNT TO ROLE CONTAINER_USER_ROLE;
SHOW COMPUTE POOLS;

GRANT MONITOR USAGE ON ACCOUNT TO  ROLE  CONTAINER_USER_ROLE;
GRANT BIND SERVICE ENDPOINT ON ACCOUNT TO ROLE CONTAINER_USER_ROLE;
GRANT IMPORTED PRIVILEGES ON DATABASE snowflake TO ROLE CONTAINER_
USER_ROLE;
```

Step 3: Switch to the newly created role.

Let's switch to the newly created role and use the database and schema created previously.

```
grant role CONTAINER_USER_ROLE to role ACCOUNTADMIN;

use role CONTAINER_USER_ROLE;
use database COMMONS;
use schema UTILS;
```

Step 4: Create security integration, access integration and network rule.

```
CREATE SECURITY INTEGRATION IF NOT EXISTS snowservices_ingress_oauth
  TYPE=oauth
  OAUTH_CLIENT=snowservices_ingress
  ENABLED=true;

CREATE OR REPLACE NETWORK RULE ALLOW_ALL_RULE
  TYPE = 'HOST_PORT'
  MODE = 'EGRESS'
  VALUE_LIST= ('0.0.0.0:443', '0.0.0.0:80');
```

```
CREATE EXTERNAL ACCESS INTEGRATION ALLOW_ALL_EAI
  ALLOWED_NETWORK_RULES = (ALLOW_ALL_RULE)
  ENABLED = true;
```

Step 5: Create a compute pool.

```
CREATE COMPUTE POOL IF NOT EXISTS CONTAINER_HOL_POOL
MIN_NODES = 1
MAX_NODES = 1
INSTANCE_FAMILY = CPU_X64_XS;

DESCRIBE COMPUTE POOL CONTAINER_HOL_POOL;
```

Running the compute pool incurs a cost; feel free to suspend it when not in use.

```
ALTER COMPUTE POOL CONTAINER_HOL_POOL SUSPEND;
ALTER COMPUTE POOL CONTAINER_HOL_POOL RESUME;
```

Step 6: Create an image repository.

The image registry service serves the OCIv2 API for storing OCI-compliant container images. Image registries are schema-level objects, and each image registry in a Snowflake account has a unique hostname, which allows OCI clients (such as Docker CLI) to access an image registry using REST API calls.

The general syntax for an image registry hostname is <orgname>-<acctname>. registry.snowflakecomputing.com.

```
CREATE IMAGE REPOSITORY IMAGE_REPO;

SHOW IMAGE REPOSITORIES IN SCHEMA COMMONS.UTILS;
--look for the string in "repository_url" bwbyxua-wn19806.registry.
snowflakecomputing.com/commons/utils/image_repo
```

Make a note of the "repository_url" and save it for later.

Step 7: Create a stage for the service file.

```
CREATE STAGE IF NOT EXISTS specs_person_gen_api
ENCRYPTION = (TYPE='SNOWFLAKE_SSE');
```

You must create a stage that acts as your volumes (shared, persisted file store) from your containers.

Step 8: Create a file named Dockerfile and add the code to build the image (refer to source code).

Start your local Docker service and run the Docker command to build an image from the source directory.

```
docker build -t johneipe/person-gen-api:latest .
```

Test the image by running it and using curl to call the endpoint.

```
docker run -p 8090:8090 johneipe/person-gen-api
```

```
curl -X POST -H "Content-Type: application/json" -d '{"email_domain":
"gmail","gender": "M", "is_employed": "true"}' http://127.0.0.1:8090/
generate_person_data
```

Step 9: Push the image to the Snowflake repository.

We rebuilt the image to make sure it works fine on Snowflake.

```
docker build --platform=linux/amd64 -t johneipe/person-gen-api:latest .
```

Verify the image has been built successfully and tagged.

```
docker image list
```

```
REPOSITORY                                TAG            IMAGE ID
johneipe/person-gen-api                   latest         e04b71586125
```

Next, let's use our Snowflake repository as the remote image repository from Docker.

```
docker login bwbyxua-wn19806.registry.snowflakecomputing.com -u jeipe
```

Next, let's create a new tag of the image that points at our image repository in our Snowflake account, and then push said tagged image.

```
docker tag johneipe/person-gen-api:latest bwbyxua-wn19806.registry.
snowflakecomputing.com/commons/utils/image_repo/person-gen-api:dev
docker image list
```

```
REPOSITORY                                                  TAG              IMAGE ID
bwbyxua-wn19806.registry.snowflakecomputing.com/commons/utils/image_repo/
person-gen-api    dev              e04b71586125
johneipe/person-gen-api
docker push bwbyxua-wn19806.registry.snowflakecomputing.com/commons/utils/
image_repo/person-gen-api:dev
```

Once the Docker push command completes, you can verify that the image exists in your Snowflake Image Repository by running the following SQL.

```
CALL SYSTEM$REGISTRY_LIST_IMAGES('/commons/utils/image_repo');
```

SYSTEM$REGISTRY_LIST_IMAGES
{"images":["person-gen-api"]}

Figure 10-3. *Query Result*

Step 10: Configure and push the service spec file.

Services in Snowpark Container Services are defined using YAML files.

These YAML files configure all the various parameters needed to run the containers within your Snowflake account.

Here is a simple structure of the spec file.

```
spec:
  containers:
    - name: <container name>
      image: <image name>
  endpoints:
    - name: <name>
      port: <TCP port-num>
      public: <true / false>
      protocol : < TCP / HTTP / HTTPS >
```

For this demonstration, the following names for endpoints and containers were used.

```
spec:
  containers:
    - name: person-mockapp
      image: bwbyxua-wn19806.registry.snowflakecomputing.com/commons/utils/
      image_repo/person-gen-api:dev
  endpoints:
    - name: person-gen-api
      port: 8090
      protocol: HTTP
      public: true
```

Once the file is created, you create a stage and use the Snowflake CLI or web UI to upload the file into a stage.

Step 11: Create a service.

Once we have successfully pushed our image and spec YAML, we have all the components in Snowflake to create the service.

To create the service, you need to specify a service name, a compute pool the service will run on, and the spec file that defines the service. Make sure your compute pool is running if it was suspended before you need to resume it.

```
create service PERSON_MOCK_API
in compute pool CONTAINER_HOL_POOL
from @specs_person_gen_api
specification_file='person_gen_api.yaml'
external_access_integrations = (ALLOW_ALL_EAI);
```

Use the GET_SERVICE_STATUS function to test the service.

```
CALL SYSTEM$GET_SERVICE_STATUS('PERSON_MOCK_API');
/**
[{"status":"READY","message":"Running","containerName":"person-mockapp","
instanceId":"0","serviceName":"PERSON_MOCK_API","image":"bwbyxua-wn19806.
registry.snowflakecomputing.com/commons/utils/image_repo/person-gen-api:
dev","restartCount":0,"startTime":"2024-06-02T03:31:27Z"}]
**/
```

Use the GET_SERVICE_LOGS function to view the logs from the service.

```
CALL SYSTEM$GET_SERVICE_LOGS('PERSON_MOCK_API', '0', 'person-mockapp',100);
```

All services running on a compute pool should be suspended before suspending or destroying the compute pool.

```
alter service PERSON_MOCK_API suspend;
alter service PERSON_MOCK_API resume;
```

Step 12: Test the service using remote functions.

```
CREATE OR REPLACE FUNCTION generate_mock_person (gender varchar, email_
domain varchar, is_employed varchar)
RETURNS array
SERVICE=PERSON_MOCK_API      //Snowpark Container Service name
ENDPOINT='person-gen-api'    //The endpoint within the container
AS '/generate_person_data';           //The API endpoint
```

```
CREATE OR REPLACE FUNCTION generate_random_mock_person ()
RETURNS array
SERVICE=PERSON_MOCK_API      //Snowpark Container Service name
ENDPOINT='person-gen-api'    //The endpoint within the container
AS '/test_generate_person_data';
```

```
select generate_random_mock_person();
select generate_mock_person('M', 'gmail', 'true');
```

Let's test the function against rows of data in a table.

```
CREATE OR REPLACE TABLE PERSON_DATA (
    GENDER VARCHAR,
    EMAIL_DOMAIN VARCHAR,
    EMPLOYED BOOLEAN,
    DETAILS ARRAY
);
INSERT INTO PERSON_DATA (GENDER, EMAIL_DOMAIN, EMPLOYED, DETAILS)
    VALUES
        ('M', 'gmail', true, NULL),
        ('M', 'gmail', false, NULL),
```

```
    ('F', 'outlook', true, NULL),
    ('M', 'outlook', false, NULL),
    ('F', 'gmail', true, NULL);

UPDATE PERSON_DATA
SET DETAILS = generate_mock_person(GENDER, EMAIL_DOMAIN, EMPLOYED);

select * from PERSON_DATA;
```

The following are additional considerations.

- MAX_BATCH_ROWS specifies the maximum number of rows in each batch sent to the proxy service. The purpose of this parameter is to limit batch sizes for remote services with memory constraints or other limitations. It specifies a maximum size, not a recommended size. If you do not specify MAX_BATCH_ROWS, Snowflake estimates and uses the optimal batch size.
- Snowflake services only support POST calls. This is by design, and you might need to tweak your functions if it was originally designed as a GET method.

Additional Snowflake Documentation

https://docs.snowflake.com/en/developer-guide/snowpark-container-services/overview

https://www.docker.com/resources/what-container/

Recipe 10-4. Snowflake Cortex

Snowflake Cortex gives you instant access to industry-leading large language models (LLMs) trained by researchers at companies like Mistral, Reka, Meta, and Google, including Snowflake Arctic, an open enterprise-grade model developed by Snowflake.

Since Snowflake fully hosts and manages these LLMs, using them requires no setup. Your data stays within Snowflake, giving you the performance, scalability, and governance you expect.

Snowflake documentation defines Cortex as a suite of AI features that use LLMs to understand unstructured data, answer freeform questions, and provide intelligent assistance.

Cortex opens a variety of features.

- Snowflake Cortex LLM functions
- Universal Search
- Cortex Search
- Cortex Analyst
- Snowflake Copilot
- Document AI
- Cortex Fine-tuning

Let's briefly look at some of the features with the help of an example.

Problem

Consider you are part of a data team working for a multinational fast-food restaurant. Your data team is dealing with semi-structured data of customer reviews from various stores, and you need to extract sentiments for each review to find which store performs better.

Prerequisites

You could use any publicly available data set, such as restaurant reviews.

We used the Yelp dataset, which is publicly available at `www.yelp.com/dataset`.

The Yelp dataset contains many files representing domains like users, businesses, and reviews. The demo focuses on the business and reviews data to utilize it as customer review information.

The schema of the individual JSON data sets is described at Yelp Dataset JSON (`www.yelp.com/dataset/documentation/main`).

Step 1: Download the dataset.

The Yelp dataset is available from yelp.com and sources like Kaggle.

Download and store these two files.

- yelp_academic_dataset_business.json: 113.4 MB
- yelp_academic_dataset_review.json: 5.0 GB

Step 2: Split large files into multiple smaller files.

The yelp_academic_dataset_business.json is 113 MB with 150,346 records, and yelp_academic_dataset_review.json is 5 GB with 6,990,280 records.

The yelp_academic_dataset_review.json file is beyond the 250 MB limit supported by Snowflake local stages. That means if the file size exceeds 250 MB, it cannot be directly uploaded into Snowflake from your local device.

There is a simple Python program (`https://python.plainenglish.io/split-big-json-file-into-small-splits-4e4b1b90e304`) to split JSON files. The code is listed as follows for convenience.

```
import os
import json
#you need to replace the path here with yours
with open(os.path.join('/Users/johneipe/Downloads/archive/', 'yelp_
academic_dataset_review.json'), 'r',
        encoding='utf-8') as f1:
   ll = [json.loads(line.strip()) for line in f1.readlines()]

   #this is the total length size of the json file
   print(len(ll))

   size_of_the_split=25000
   total = len(ll) // size_of_the_split

   #in here you will get the Number of splits
   print(total+1)

   for i in range(total+1):
```

```
    json.dump(ll[i * size_of_the_split:(i + 1) * size_of_the_split], open(
        "/Users/johneipe/Downloads/archive/reviews_split/" + str(i+1) +
        ".json", 'w', encoding='utf8'), ensure_ascii=False, indent=True)
```

Step 3: Load data into Snowflake.

You could use the command line, SnowSQL (CLI client), or the Snowflake web interface to load the data and create tables.

Remember to use the "Strip outer array" option for the JSON file format to parse and load the reviews file.

We loaded both datasets into separate tables in a schema called RAW.YELP.

Figure 10-4 shows the Snowsight screen that displays the tables under the RAW database.

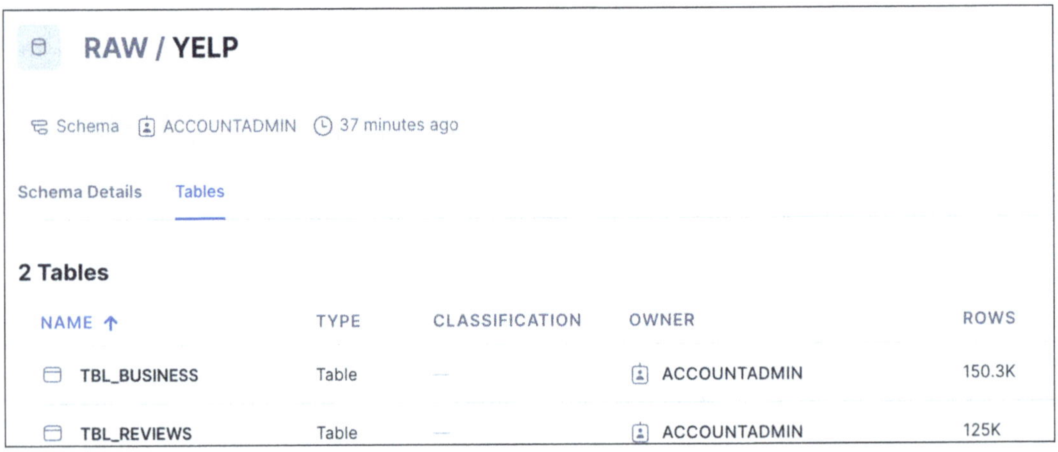

RAW / YELP

Schema ACCOUNTADMIN 37 minutes ago

Schema Details Tables

2 Tables

NAME ↑	TYPE	CLASSIFICATION	OWNER	ROWS
TBL_BUSINESS	Table	—	ACCOUNTADMIN	150.3K
TBL_REVIEWS	Table	—	ACCOUNTADMIN	125K

***Figure 10-4.** RAW tables*

We did not load all the review data, so you see only 125,000 rows instead of 6990280.

The following is the schema of the tables created.

```
create table RAW.YELP,TBL_BUSINESS (
 address VARCHAR
 , attributes VARCHAR
 , business_id VARCHAR
 , categories VARCHAR
 , city VARCHAR
 , hours VARCHAR
 , is_open NUMBER(38, 0)
```

```
  , latitude NUMBER(38, 10)
  , longitude NUMBER(38, 10)
  , name VARCHAR
  , postal_code VARCHAR
  , review_count NUMBER(38, 0)
  , stars NUMBER(38, 1)
  , state VARCHAR
);

create table RAW.YELP.TBL_REVIEWS (
  business_id VARCHAR
  , cool NUMBER(38, 0)
  , review_date TIMESTAMP_NTZ
  , funny NUMBER(38, 0)
  , review_id VARCHAR
  , stars NUMBER(38, 1)
  , text VARCHAR
  , useful NUMBER(38, 0)
  , user_id VARCHAR
);
```

Solution

Snowflake provides a good list of built-in Cortex functions.

These are provided as SQL functions and available in Python (https://docs.snowflake.com/en/user-guide/snowflake-cortex/llm-functions#label-cortex-llm-model-python).

The following summarizes the available functions.

- **COMPLETE**: Given a prompt, returns a response that completes the prompt. This function accepts a single prompt or a conversation with multiple prompts and responses.
- **EMBED_TEXT_768** and **EMBED_TEXT_1024**: Given a piece of text, returns a vector embedding of 768 or 1024 dimensions representing that text.
- **EXTRACT_ANSWER**: Given a question and unstructured data, return the answer if it can be found in the data.

- **SENTIMENT**: Returns a sentiment score, from -1 to 1, representing the text's detected positive or negative sentiment.
- **SUMMARIZE**: Returns a summary of the given text.
- **TRANSLATE**: Translates given text from any supported language to any other.

In this case, we are looking for the SENTIMENT function.

Once the data is loaded in your Snowflake table, you must call the Cortex SENTIMENT function on your data.

```
SELECT SNOWFLAKE.CORTEX.SENTIMENT(text), text FROM RAW.YELP.TBL_REVIEWS
LIMIT 10;
```

Problem

Consider you are part of a data team of a multinational fast-food restaurant.

Your operations team is asking to provide a feature to add search functionality on the review data with options to filter by certain attributes like usefulness.

The search functionality should also be consumable via API. The returned results should include the rating and the business/restaurant name.

Solution

Following the same Yelp data set loaded into Snowflake as part of this recipe, we have two tables available to help set up a Cortex Search service.

But why use Cortex Search rather than a fuzzy search feature like the JAROWINKLER_SIMILARITY function?

Cortex Search enables low-latency, good-quality text search over your Snowflake data with zero overhead of creating and managing the end-to-end process.

- The service automatically indexes and incrementally embeds your data, meaning it only processes changed rows from the underlying data source.
- It combines the strengths of vector search (for retrieving semantically similar documents), keyword search (for retrieving lexically similar documents), and semantic reranking (for ranking the most relevant documents in the result set) into a single search interface.

This approach yields higher quality search results across a variety of Retrieval-Augmented Generation (RAG) search workloads than a vector search or a keyword search alone.

Figure 10-5 shows the control flow for Cortex Search at a high-level. Note that Cortex Search powers a broad array of search experiences for Snowflake users including RAG applications leveraging LLMs.

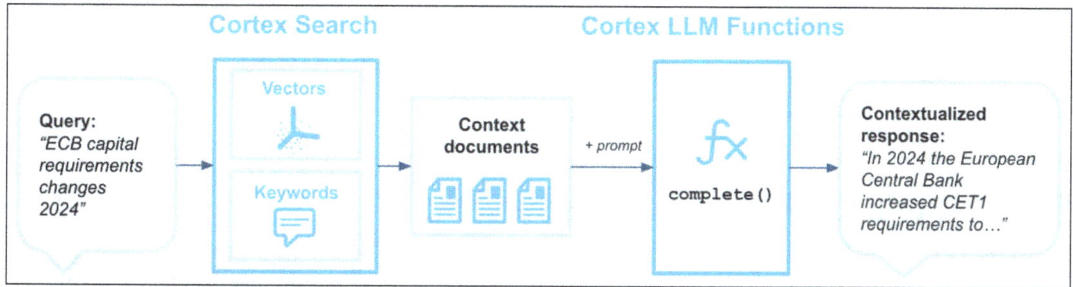

Figure 10-5. *Source: Snowflake Inc*

There are two primary use cases for Cortex Search.

- **RAG engine for LLM chatbots** uses Cortex Search as a RAG engine for chat applications with your text data.
- **Text search** uses Cortex Search as a back end for text search functionality within your application.

Our focus is on text search, and it is very easy to set it up in Snowflake once you have the data available to be queried.

Step 1: Build a base query for the search.

Since we want to retrieve the review information and a few other attributes like stars, usefulness, and details about the restaurant/business, we shall join the review and business table onto business_id.

```
select
 r.text as text
 , r.stars as stars
 , r.useful as useful
 , b.name as name
 , b.city as city
```

```
  , b.state as state
from raw.yelp.tbl_reviews r
inner join raw.yelp.tbl_business b
on b.business_id = r.business_id;
```

Step 2: Enable change tracking.

You must set CHANGE_TRACKING to TRUE for the tables the search service references if the role that creates the search service does not have OWNERSHIP of the source table.

```
ALTER TABLE <table> SET CHANGE_TRACKING = TRUE;
```

Step 3: Build the Cortex Search service.

All the operational complexity of building the search service is abstracted into a single SQL statement for service creation.

Figure 10-6 shows the syntax and structure of the create Cortex Search service.

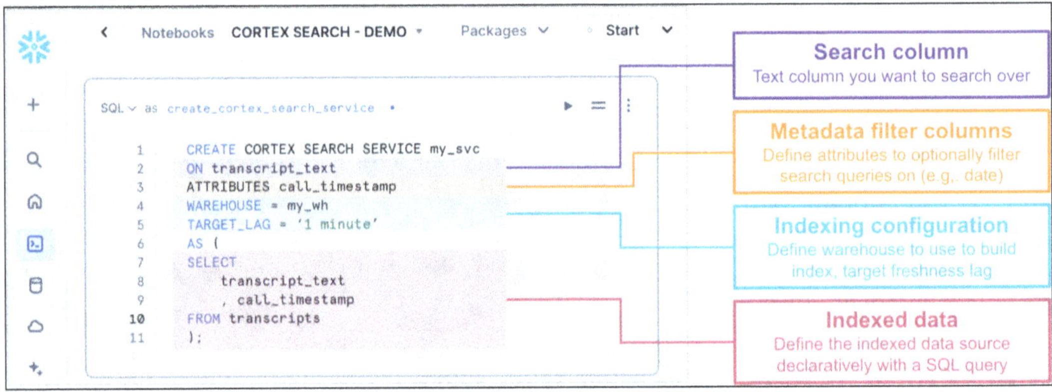

Figure 10-6. *Create Cortex Search Service source: Snowflake Inc Blog*

Once the service is created, it's easy to query it from your application via REST or Python APIs. This includes both applications hosted in Snowflake (e.g., Streamlit in Snowflake) or applications hosted in an external environment.

```
create or replace CORTEX SEARCH SERVICE yelp_review_search
ON text
ATTRIBUTES stars, useful, state
WAREHOUSE = compute_wh
```

```
TARGET_LAG = '1 hour'
AS
  select
   r.text as text
   , r.stars as stars
   , r.useful as useful
   , b.name as name
   , b.city as city
   , b.state as state
  from raw.yelp.tbl_reviews r
  inner join raw.yelp.tbl_business b
  on b.business_id = r.business_id;
```

Step 4: Test the Cortex Search service.

The service can now be quickly tested using the helper function provided by Snowflake: SEARCH_PREVIEW. This function returns the response of the specified Cortex Search service when relevant parameters are passed to it.

```
SELECT PARSE_JSON(
 SNOWFLAKE.CORTEX.SEARCH_PREVIEW(
     'raw.yelp.yelp_review_search',
     '{
       "query": "good chicken wings",
       "columns":[
           "text",
           "stars",
           "useful",
           "name"
       ],
       "filter": {
           "@or": [
               { "@eq": { "stars": "4.0" } },
               { "@eq": { "stars": "5.0" } }
           ]
       },
```

```
        "limit":10
      }'
  )
) as results;
```

This returns the results in less than a second.


```
[] RESULTS

{
  "request_id": "e848876c-3be3-4bbe-9dba-3b0918ad50aa",
  "results": [
    {
      "name": "Crown Fried Chicken",
      "stars": "5.0",
      "text": "The chicken wings here are soooooooo good. They have the real thing! I ordered hot wings
and regular wings and they were both amazing. Flavorful and crunchy. The guy that works there is also very
kind. I wish I lived in the area so that I can get those wings all the time! The prices are moderate.
Definitely worth the stop. This place deserves better ratings.",
      "useful": "1"
    },
    {
      "name": "Ciconte's - Swedesboro",
      "stars": "4.0",
      "text": "Best chicken wings in town.  Zippy chicken sandwich is a winner and the steaks are also
good.  Love this place.",
      "useful": "0"
    },
```


Figure 10-7. *Query Results*

Step 5: Consume the service via Snowpark API or REST API.

You could use the Snowflake Snowpark API as shown below. This could be run from within your Snowsight python worksheet or from your local machine.

```
from snowflake.core import Root
from snowflake.snowpark import Session

CONNECTION_PARAMETERS = {"..."}

session = Session.builder.config(CONNECTION_PARAMETERS).create()
root = Root(session)

yelp_review_search = (root
  .databases["RAW"]
  .schemas["YELP"]
  .cortex_search_services["yelp_review_search"]
)
```

```
resp = yelp_review_search.search(
 query="good chicken wings",
 columns=[
          "text",
          "stars",
          "useful",
          "name"
      ],
 filter={
          "@or": [
              { "@eq": { "stars": "4.0" } },
              { "@eq": { "stars": "5.0" } }
          ]
      },
 limit=10
)
print(resp.to_json())
```

Snowflake also provides a REST endpoint for HTTP access, as of this writing in public preview.

The REST endpoint generated for a Cortex Search service uses the following structure.

`https://<account_url>/api/v2/databases/<db_name>/schemas/<schema_name>/cortex-search-services/<service_name>:<query>`

It is described as follows.

- <account_url> is your Snowflake account URL. (See "Finding the organization and account name for an account" for instructions on finding your account URL.)
- <db_name> is the database in which the service resides.
- <schema_name> is the schema in which the service resides.
- <service_name> is the name of the service.
- <query> is the method to invoke on the service. In this case, the query method.

Additional Snowflake Documentation

https://docs.snowflake.com/en/user-guide/snowflake-cortex/llm-functions

https://docs.snowflake.com/en/user-guide/snowflake-cortex/cortex-analyst

https://docs.snowflake.com/en/user-guide/snowflake-cortex/cortex-search/cortex-search-overview

https://docs.snowflake.com/user-guide/snowflake-cortex/cortex-search/query-cortex-search-service

https://www.snowflake.com/en/blog/cortex-search-ai-hybrid-search/

Index

A

D. Dayton and J. Eipe, *Snowflake Recipes*, https://doi.org/10.1007/979-8-8688-0938-5

E

F

G

H

I

J

K

L

M

N

O

P

Q

R

S

T

U

V, W

X

Y, Z

GPSR Compliance
The European Union's (EU) General Product Safety Regulation (GPSR) is a set of rules that requires consumer products to be safe and our obligations to ensure this.

If you have any concerns about our products, you can contact us on

ProductSafety@springernature.com

In case Publisher is established outside the EU, the EU authorized representative is:

Springer Nature Customer Service Center GmbH
Europaplatz 3
69115 Heidelberg, Germany

www.ingramcontent.com/pod-product-compliance
Lightning Source LLC
LaVergne TN
LVHW080310260326
834688LV00038B/1035
* 9 7 9 8 8 6 8 8 0 9 3 7 8 *